MW01629139

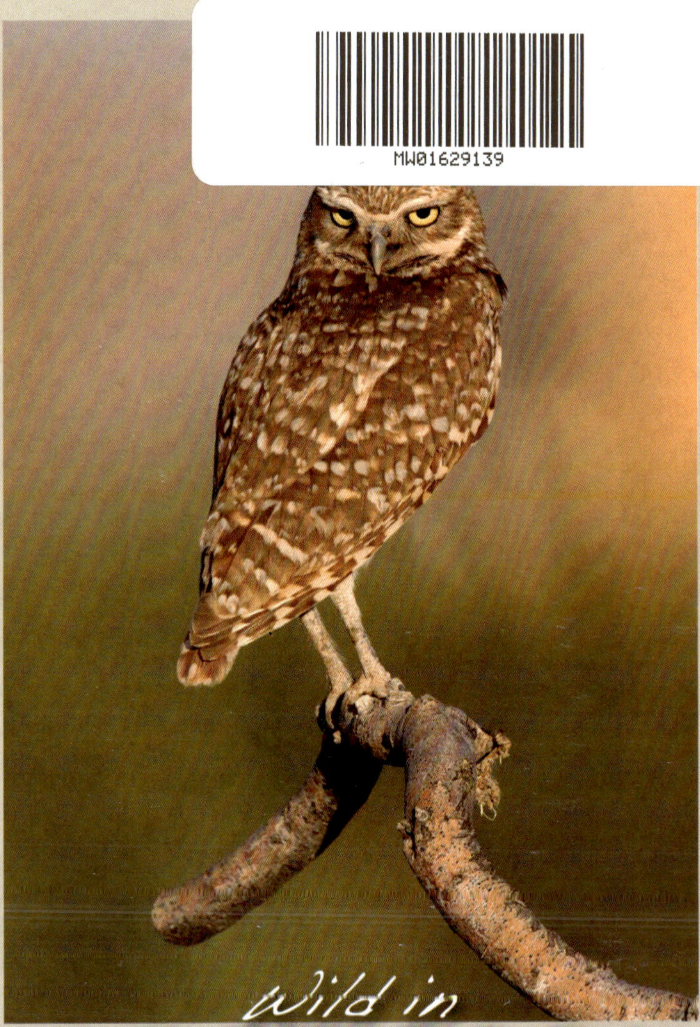

Wild in ARIZONA

Photographing Arizona's Wildlife

A Guide to When, Where, & How

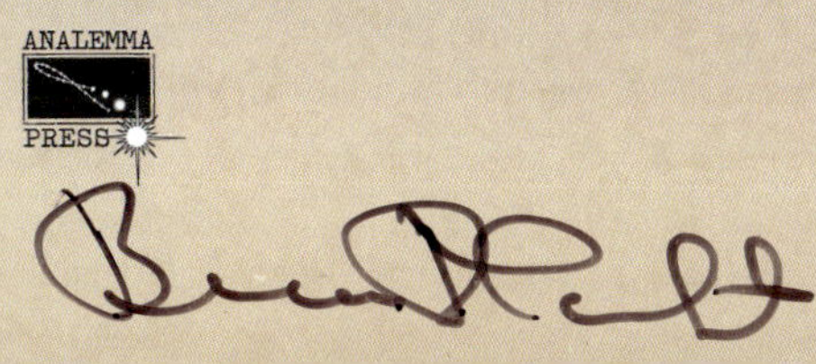

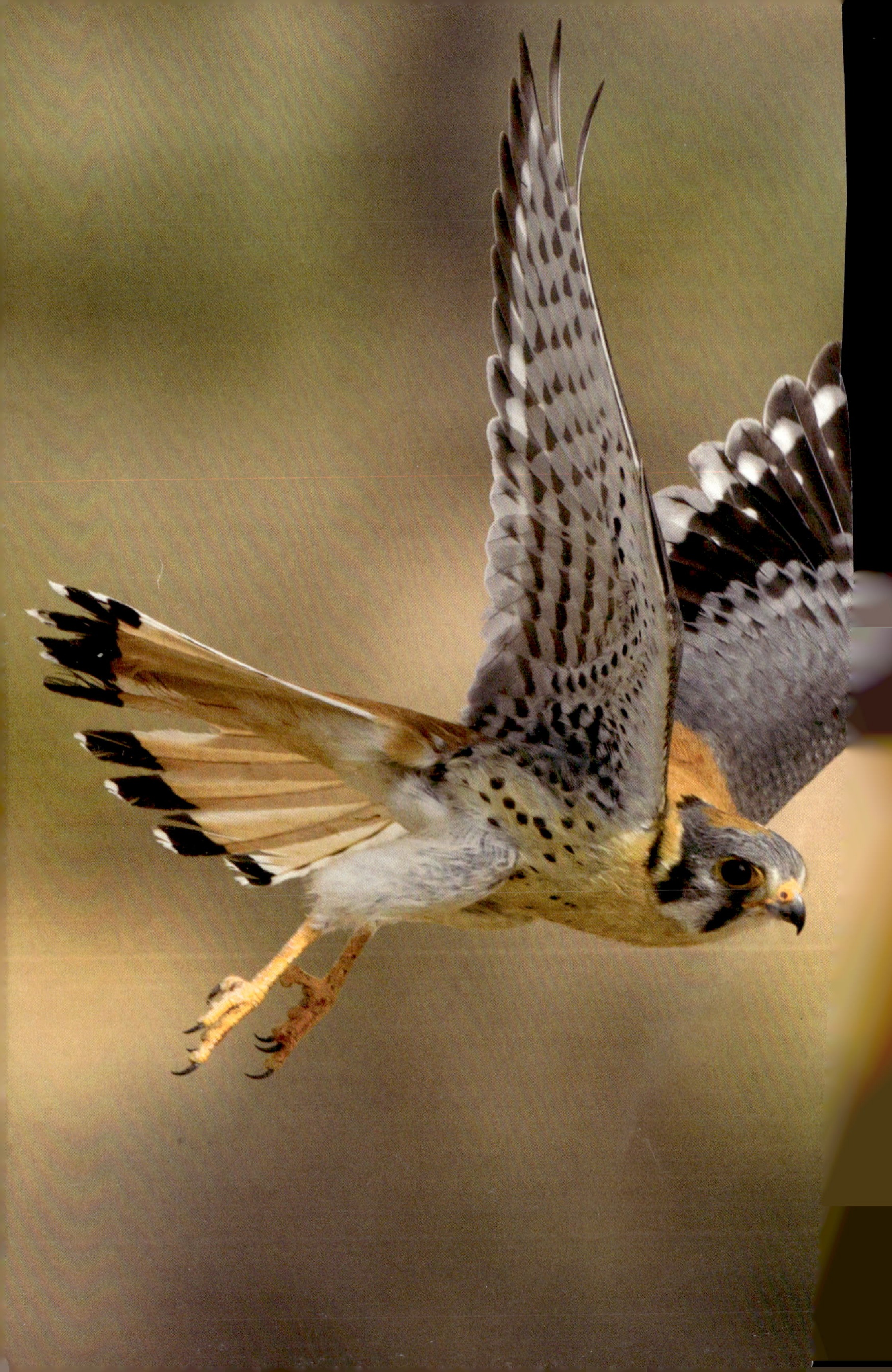

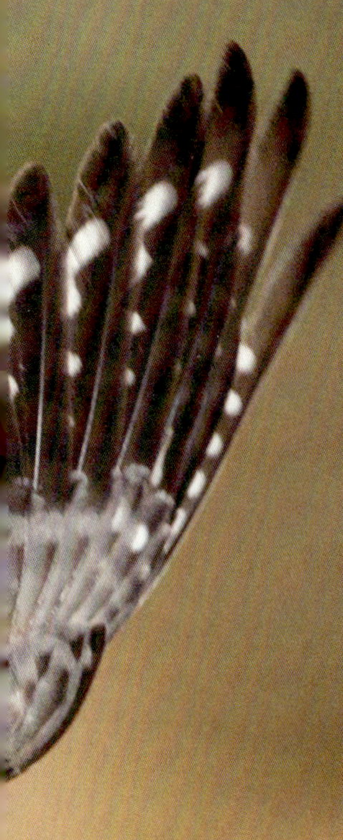

Wild in ARIZONA

Photographing Arizona's Wildlife
A Guide to When, Where, & How

Bruce D. Taubert

Foreword by Colleen Miniuk-Sperry

To my wife, Anne, and daughters, Julie and Katie, for supporting my wildlife adventures and putting up with my leaving early, returning late, bringing all sorts of plants and animals into our home to keep and photograph, and loving our life in Arizona as much as I do.

To my mother, Virginia, and father, Kenneth, for giving me the space to explore the wilds of Wisconsin and always being there to lend a supporting hand as I crossed the many hills and valleys to becoming an adult. Special thanks to my mother for sharing her own photographic interests with me.

To Randy Babb, Bud Bristow, and Barry Mansell for their photographic mentoring, support, and continuous sharing.

–Bruce D. Taubert

Wild in Arizona: Photographing Arizona's Wildlife
A Guide to When, Where, & How

Copyright © 2016 Bruce D. Taubert / Analemma Press, L. L. C., Chandler, Arizona.
Printed in the United States of America.

All rights reserved. No part of this publication may be reproduced in any form or by any means, electronic, mechanical, photocopying, recording or otherwise, without the prior written permission of the publisher.

Text and Photography: Bruce D. Taubert
Editors: Erik Berg and Colleen Miniuk-Sperry
Layout and design: Paul Gill/Gill Photo Graphic

ISBN: 978-0-9833804-6-7
Library of Congress Control Number: 2016937071

Disclaimer: Enjoying the outdoors is not without risk. Because no guidebook can adequately disclose all potential hazards or risks involved, all participants must assume responsibility for their own safety and actions while visiting the locations and participating in the activities suggested in this book. The author and publisher disclaim any liability for injury, suffering, or other damage caused by traveling to and from a location, visiting a recommended area, or performing any other activity described in this book.

The author and publisher have conscientiously tried to ensure this guidebook contains the most accurate and up-to-date information at publication time. However, since names, roads, places, conditions, etc. may change over time, please spend time prior to your outing adequately researching and preparing for your trip.

Also, many of the designations used by manufacturers and sellers to distinguish their products are claimed as trademarks. Where those designations appear in this book, and the author and publisher were aware of a trademark claim, the designations have been printed with initial capital letters or in all capitals.

COVER PHOTO: A ringtail feeds on an agave flower. Canon 50D, 100-400mm at 105mm, ISO 200, f/11 @ 1/250 sec., off-camera flash.

TITLE PAGE PHOTO: An observant burrowing owl sits atop a snag, looking for food or an enemy. Canon 50D, 500 mm, 1.4x teleconverter, ISO 200, f/9 @ 1/500 sec.

OVERLEAF PHOTO: Male American kestrel hunts for lizards to feed its young. Canon 1DMIV, 500mm, ISO 1250, f/9 @ 1/4000 sec., off-camera flash @ -2/3 FEC.

Table of Contents

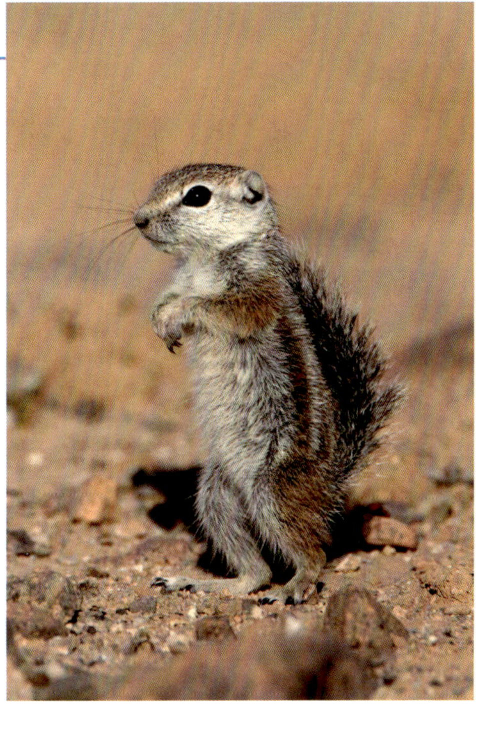

Baby antelope ground squirrel surveys the landscape for predators. Canon 50D, 500mm, 1.4x teleconverter, ISO 400, f/8 @ 1/2000 sec.

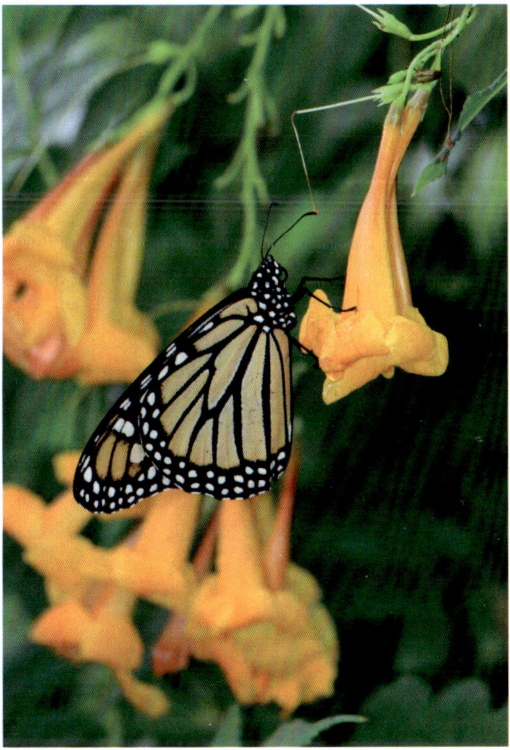

Monarch butterfly feeds from a tecoma flower at Desert Botanical Garden. Canon Digital Rebel XTi, 400mm, ISO 400, f/8 @ 1/20 sec., flash @ -2/3 FEC.

Locations Listing

Northern Arizona — Page 38 (N)

Western Arizona — Page 56 (W)

Central Arizona — Page 76 (C)

Eastern Arizona — Page 130 (E)

Southern Arizona — Page 160 (S)

Wild in Arizona: Wildlife Locations

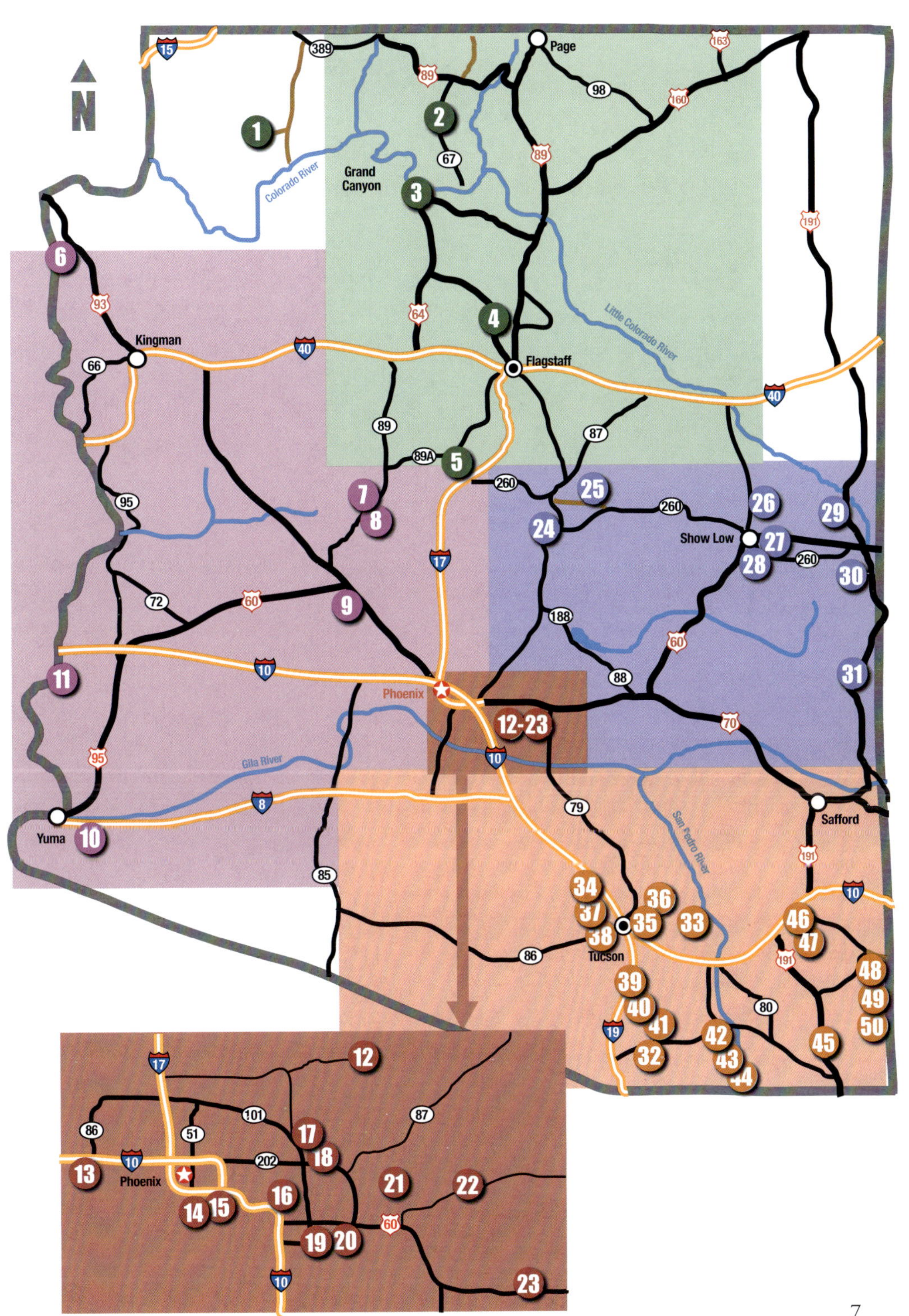

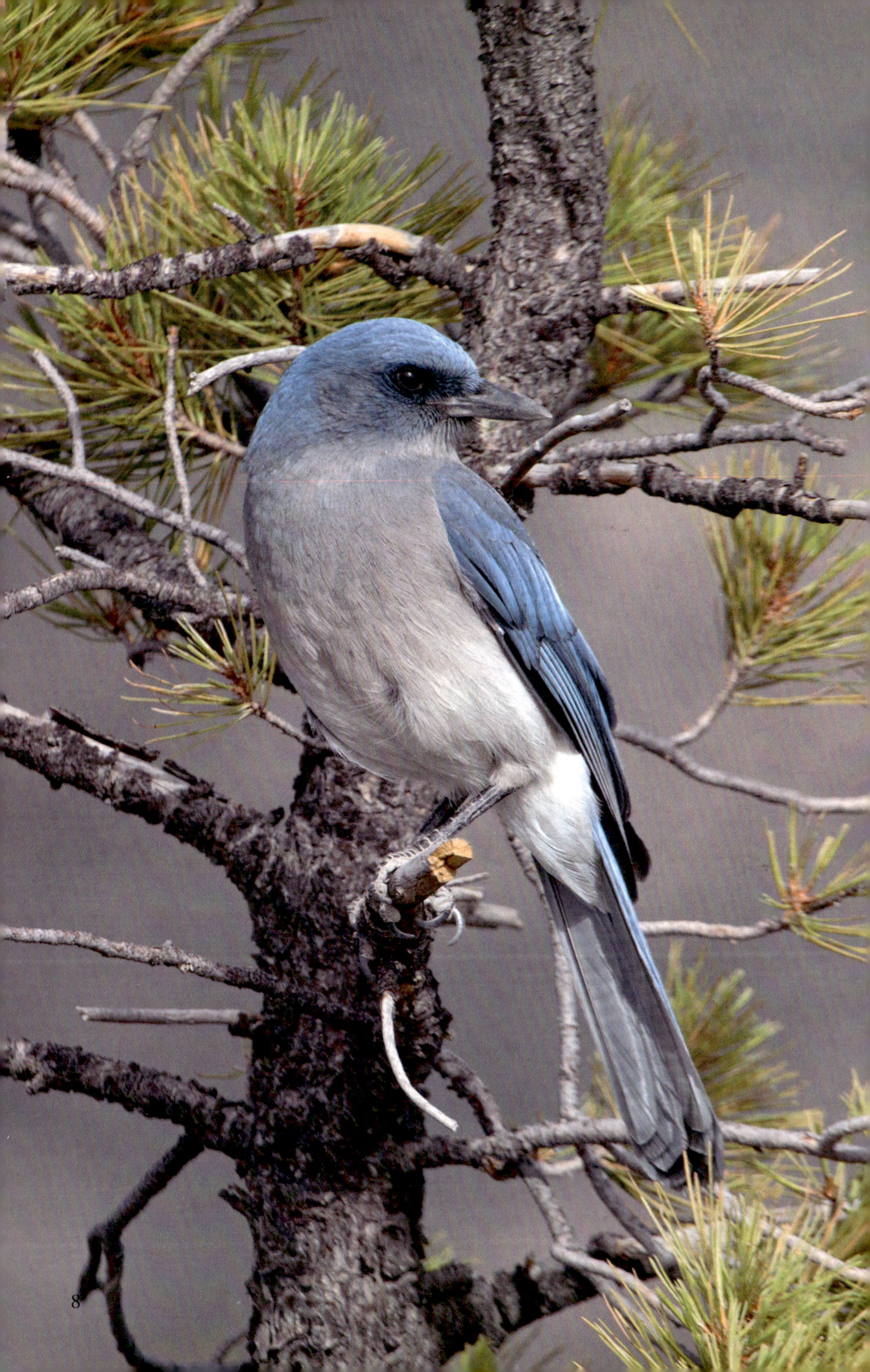

Photography Tips

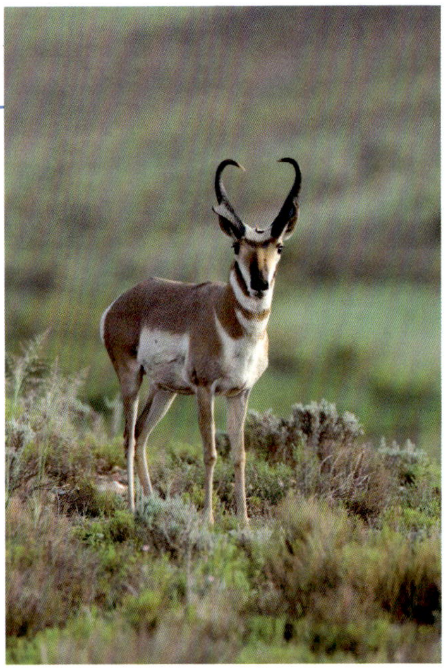

Pronghorn, the fastest mammal in Arizona, keeps an eye out for danger. Canon Digital Rebel XTi, 500mm, 1.4x teleconverter, ISO 200, f/5.6 @ 1/500 sec.

"Making the Photo" Stories

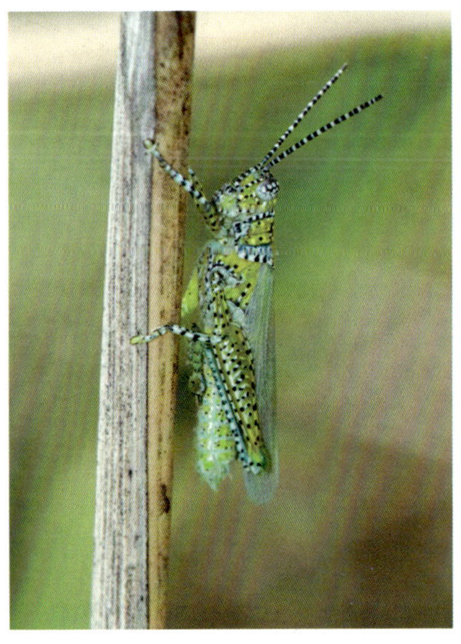

Panther-spotted grasshopper waits for the temperatures to rise to begin its day. Canon Digital Rebel XSi, 100-400mm at 190mm, 500D macro, ISO 200, f/10 @ 1/200 sec., on-camera flash.

OPPOSITE: Mexican jay sits high in a pinyon pine in the Chiricahua National Monument. Canon 50D, 100-400mm at 400mm, ISO 320, f/9 @ 1/640 sec.

Acknowledgements

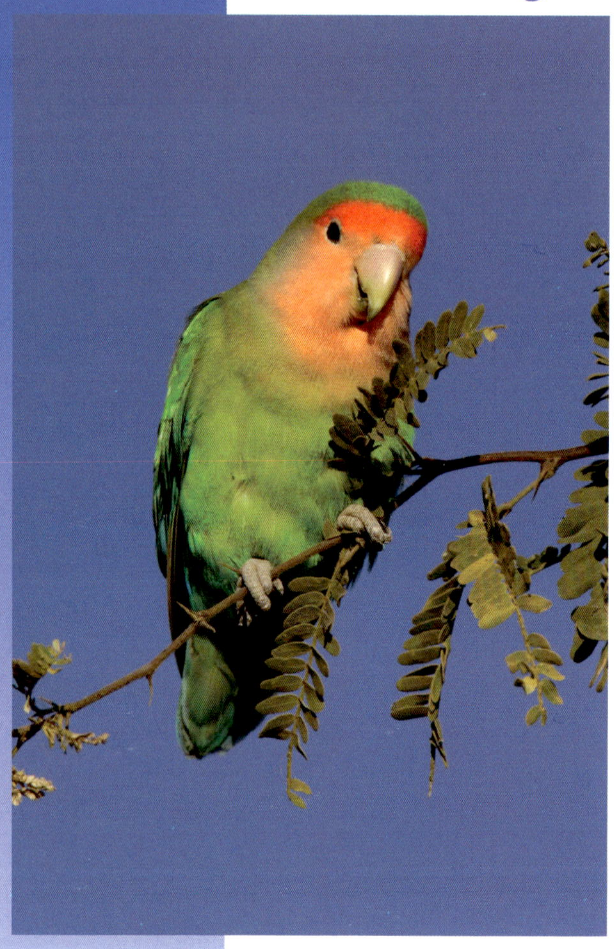

The beautiful peach-faced lovebird may feel more at home in its native land in Australia, but escaped pet birds now thrive throughout the Phoenix metropolitan area. The Riparian Preserve at Water Ranch (page 114) offers the best chance to photograph these small parrots. Canon 20D, 500mm, 1.4X teleconverter, ISO 200, f/8 @ 1/800 sec.

Thanks to Colleen Miniuk-Sperry and Paul Gill for allowing me to share the incredible framework they have developed in the "Wild in Arizona" publications. Colleen and Paul unselfishly guided me through the process of writing the words and selecting the images for this book while, at the same time, "scrubbing my text" and carrying the heavy workload of design, layout, advertisement, publication, and the rest of those many tasks that I was untrained to handle. There is absolutely no doubt that without their massive amount of assistance I could not have completed this book.

Thanks to Erik Berg and Colleen for their word by word, location by location, edit of my work. Their knowledge of Arizona's wildlands is unmatched, and they were able to keep me smiling through a normally trying process. Randy Babb, with assistance from Paul, took on the very difficult task of designing the wildlife symbols.

I am especially grateful to Elizabeth Ruggerio-York for taking the time and effort to review my initial drafts with not only a critical eye for inconsistencies but also with her extensive knowledge of photography. Thanks also to Lori A. Johnson, Linda Shroufe, Amanda Spencer, and Reed Westberg for their keen editorial assistance on the final draft.

The entire Analemma Press team greatly appreciates the financial support and the opportunity to partner with outstanding sponsors of this book: Arizona Highways Photography Workshops, Boyce Thompson Arboretum, Cave Creek Ranch, Hoodman Corporation, PhotoTrap, Santa Rita Lodge, Tempe Camera, and Wimberley. We are equally thankful to those who generously supported our Indiegogo fundraising campaign. Thank you to Colleen's father, Bob Miniuk, for helping to secure sponsor partnerships and for his invaluable sales and marketing expertise.

Finally, a special thanks to Stan Bormann for asking me "why don't you contact Colleen and discuss publishing a book?" Without Stan's coaching, one of my life's dreams of publishing a book on my photographic experiences might never have been accomplished.

Foreword

In the summer of 2010, my friend and fellow nature photographer Paul Gill and I met at the San Tan Brewery in Chandler, Arizona to discuss creating a book together. With excitement swirling in the air, we dreamed about how we could possibly extend our partnership beyond a single title.

"We could produce a whole series about wildflowers, fall colors, wildlife, maybe more," Paul suggested. Then with a theatrical flair, he added, "We could call it 'Wild in Arizona'!"

"Let's do it! Except the only time you or I ever shoot wildlife is if an animal just happens to run into our landscape scenes," I joked with a hearty laugh. "I'm not sure either of us is qualified to publish a book on Arizona's wildlife, but do you know who would knock this type of book out of the park?"

Paul nodded as we answered in unison, "Bruce Taubert."

Although Paul and I were both familiar with Bruce's stunning wildlife photography from browsing the pages of *Arizona Highways* magazine, I had crossed paths with Bruce only once: in 2009 at an Arizona Highways Photography Workshops holiday party (we both led photo workshops with the organization). In a single short conversation, Bruce immediately impressed me with his intelligence, wit, and friendly persona.

At the time, Paul and I had decided to set aside future book titles in order to focus on delivering our original idea—the *Wild in Arizona: Photographing Arizona's Wildflowers* book—in early 2012 (followed by the second edition in early 2015). In May 2013, serendipity struck. Out of the blue, I received a phone call from Bruce while I was stopped at a gas station in Idaho. After we exchanged pleasantries, he said, "I just presented at the Grand Photo Club and Stan Bormann [a mutual friend] suggested I call you about doing a book. What if we did *Wild in Arizona: Photographing Arizona's Wildlife* together?"

Delighted that the stars had aligned, I responded, "Let's do it!"

Since that phone call, Bruce, Paul, and I have collaborated tirelessly to produce this insightful and practical guide hoping to inspire and help readers explore the Grand Canyon State in search of critters big and small. Bruce's unmatched knowledge, unwavering passion, and a sincere commitment to excellence have produced a book that I am profoundly proud to have as a part of our Wild in Arizona series. No matter your camera type, photographic skill level, or prior experience photographing animals, Bruce's helpful tips, tricks, and stories will surely provide sound direction on "when, where, and how" to capture striking images in Arizona.

So if you are ready to get WILD about Arizona's wildlife—as I certainly am now, thanks to this book—pack this guidebook, grab your camera, and enjoy discovering and photographing Arizona's diverse and intriguing birds, mammals, reptiles, insects, and more!

Colleen Miniuk-Sperry
Chandler, AZ
Publisher of the *Wild in Arizona* book series and co-author/photographer of *Wild in Arizona: Photographing Arizona's Wildflowers: A Guide to When, Where, & How*

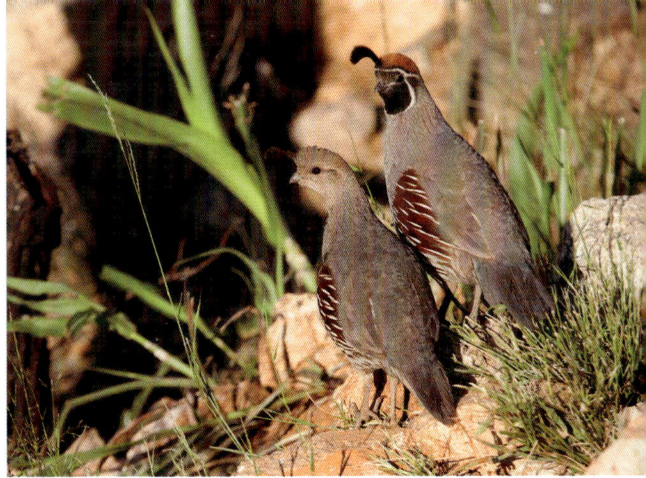

Mom and dad Gambel's quail. Canon 50D, 100-400mm at 350 mm, ISO 400, f/8 @ 1/800 sec.

Introduction

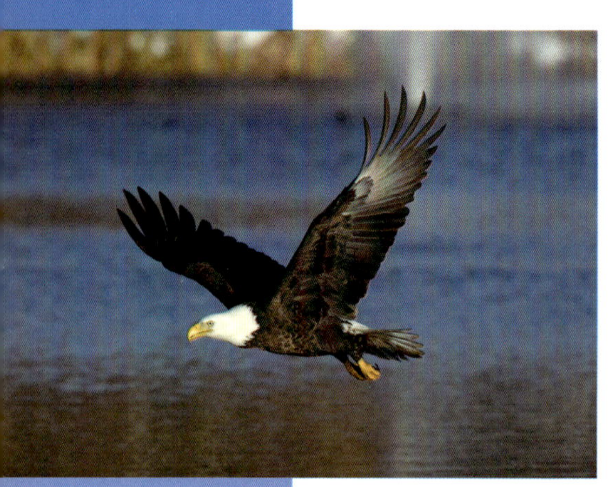

Arizona's population of southwest bald eagles have rebounded well from the edge of extinction. Canon 7DMII, 100-400mm at 400mm, ISO 320, f/7.1 @ 1/2500 sec.

When I take photographs or discuss photography with students, other photographers, or the public, the troubles of our society no longer encumber my mind. Sitting in a blind at dawn watching American kestrels deliver food to their young draws me into the lives of these incredible creatures and allows me the time to be part of the wild world that I have always loved and nurtured. Thanks to photography, I now look at small partitions of vast environments and examine them in a fashion that, without photographs, would have not been possible for me.

Hailing originally from Wisconsin, the jagged horizons and firm ground of Arizona disrupted my linear way of looking at life—the Earth was no longer flat. Cholla cacti taught me hard lessons as I attempted to stroll in the desert in the same way I had strolled in the marshes of northern Wisconsin.

As new experiences shaped new beliefs, my appreciation and love of Arizona's wildlife increased. I began to appreciate Gila monsters (the only venomous lizard in the United States), spadefoot toads that stayed underground for months or years waiting for the monsoonal rains to aid their transportation to the surface, and desert tortoises that lived as long as humans and crawled to mountain tops even four-wheel drive vehicles could not traverse. Mexican gray wolves and black-footed ferrets seemed out of place in this desert environment. I found more hummingbirds and bats than could be photographed almost anywhere in North America. I could examine Mexican species like the elegant trogon, coatimundi, sulphur-bellied flycatcher, Mojave rattlesnake, and the elf owl (the smallest owl in the world) without a passport.

Pronghorn, black bear, Rocky Mountain elk, bighorn sheep, mountain lion, bobcat, bison, and golden eagle interweave with the massive ponderosa pine forests and expansive plains so integral to the uniqueness of the Grand Canyon State. Arizona is as special to me as Namibia, Malawi, Ecuador, or any other exotic location to which I have traveled.

For the past 34 years, I have traveled throughout the state of Arizona experiencing, conserving, and photographing its incredible wildlife. My goal for this book is to transfer my love for both photography and Arizona's wildlife to you. Although I have made photographs of animals in hundreds of Arizona locations, the 50 sites featured in this book are the crème de la crème and all but guarantee to provide you with wonderful experiences and memorable images. By taking images of our wild resources and sharing those images, photographers can impart an appreciation of the state's wildlife and stimulate the public to preserve and protect the beauty that photographers enjoy.

Wildlife photography in Arizona excites, challenges, and rewards, so pack your camera, lots of memory cards, and plenty of batteries, and join me in the great outdoors of the Grand Canyon State.

How to Use This Book

For the most part, wildlife migrations, breeding seasons, feeding areas, and other parts of animals' lives follow a set pattern. Unless severe events such as fire and inclement weather modify their lives, wildlife spend much of their time in the same place conducting the same activities year after year. The purpose of this book is to get you where the animals are at the right time and with the appropriate camera equipment. To assist in planning your visit, each location in this book provides helpful information, including:

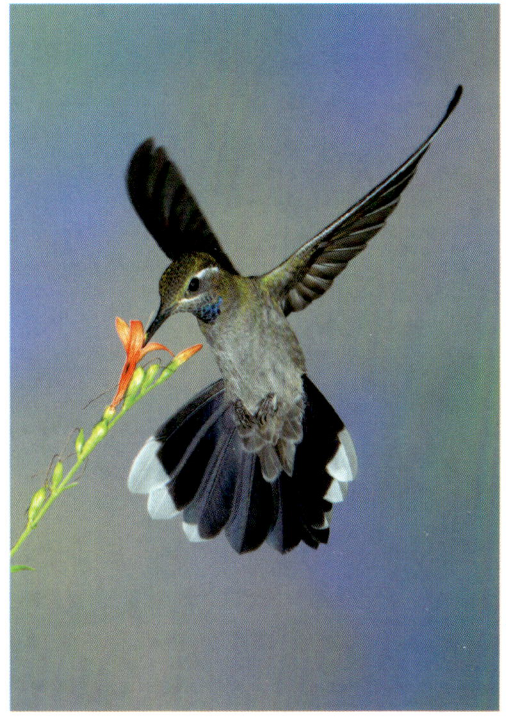

Blue-throated hummingbird feeds on wildflower at Cave Creek Ranch. Canon 1DMIV, 70-200mm at 200mm, 2x teleconverter, ISO 500, f/18 @ 1/200 sec., high-speed hummingbird set-up.

- **Viewing Time:** Though many of the critters appear at other times of the year, the month(s) listed define when to expect the best photographic opportunities for that site. Peak times can vary by a week or two in either direction depending on rainfall patterns, fire, and other environmental conditions.
- **Ideal Time of Day:** Photographers can record images of wildlife at any time of day, but the suggestions provided in this book represent the best opportunities for making effective images in appropriate lighting and considering wildlife behavior at each location.
- **Vehicle:** Most of the sites in this book are accessible using a passenger vehicle. "Any" means that almost any vehicle will get you to the listed location. Some roads require specialized transportation such as high-clearance and/or four-wheel drive vehicles. If the route's surface is bumpy or consists of loose dirt, take at least a two-wheel drive high-clearance (2WD HC) vehicle. In the rare occasion where the roads are extremely difficult (steep, rocky, or muddy), take instead a four-wheel drive high-clearance (4WD HC) vehicle. At two locations, a boat is necessary.
- **Hike:** The sites identified as "Easy" only require that you get out of the car, walk a short distance, and start taking images. If the site requires a long walk on steep, uneven, or difficult terrain while carrying heavy equipment, I refer to it as "Strenuous." Everything in between is considered "Moderate."

A good photographer is ready for any opportunity—and in Arizona you will find no shortage. Each location emphasizes the primary animal(s) you will likely observe (and hopefully, photograph). At a few special sites, the abundance of critters available to photograph prevents listing them all. However, I have attempted to describe as many as possible.

Instructional "Photography Tips" are presented throughout the book to assist the photographer in learning tried-and-true techniques that will assist them in getting the most out of their photographic experience and lead to a greater understanding of wildlife photography.

Wildlife Seasons and Migrations

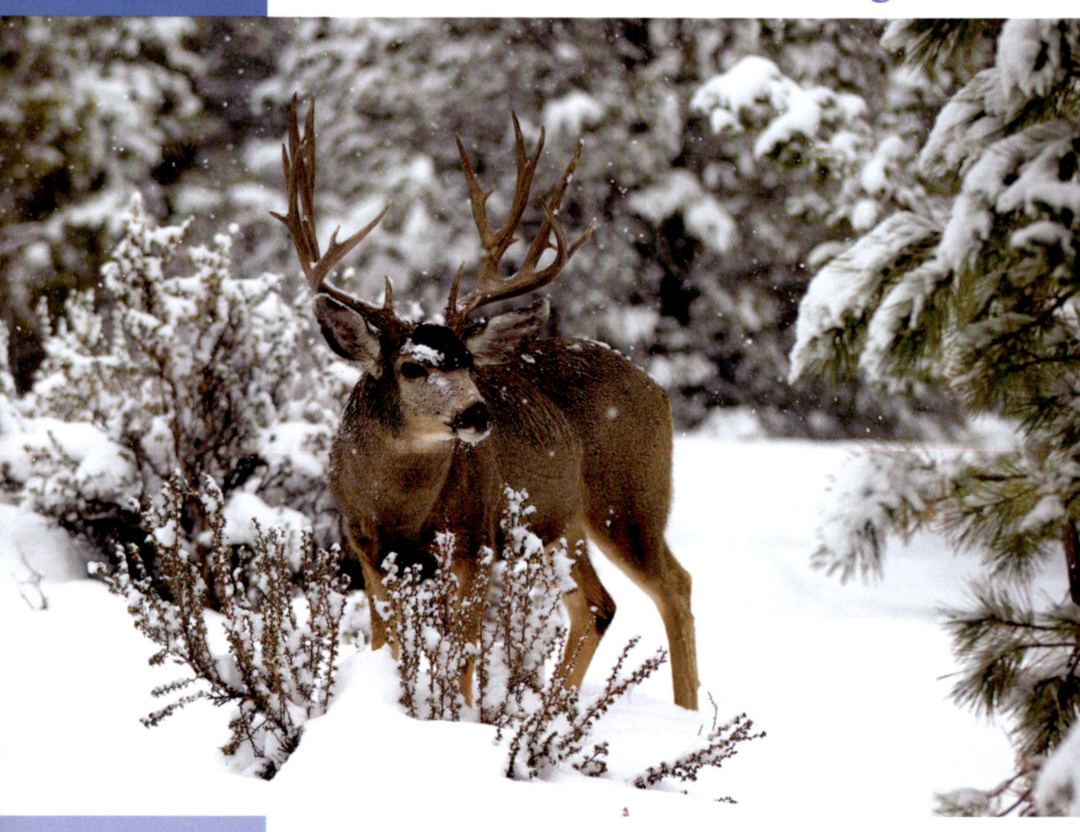

Rocky Mountain mule deer in rare Grand Canyon snowstorm. Canon 1DMIV, 70-200mm at 75mm, 2x teleconverter, ISO 400, f/8 @ 1/1200 sec.

Species from across North America merge in Arizona to provide a potpourri of photographic opportunities. In no other state can one find the likes of Rocky Mountain elk, desert bighorn sheep, trogons, horned lizards, boa constrictors, California condors, fifteen species of hummingbirds, twenty-eight types of bats, and more rattlesnakes than you can shake a stick at—all within a four-hour drive! Along the border with Mexico, southern species such as ocelot, jaguar, elegant trogon, javelina, Coues white-tailed deer, and many more transient creatures cross into Arizona.

Finding wildlife to photograph requires photographers to know not only where to go, but also when. Searches for Rocky Mountain elk in Flagstaff during winter snowfalls may fall short because they have moved to lower elevations to access warmer temperatures and better food sources. Locating broad-tailed hummingbirds anywhere in Arizona from November through March would also prove disappointing since their entire population overwinters in the highlands of Mexico and south to Guatemala. In order to photograph wildlife successfully, one needs some basic knowledge of wildlife movements and the impact of seasonal rainfall during the different months of the year.

Migrations

In Arizona, there are two types of migrations—geographic and elevation.

Geographic migrations are those where animals move across the landscape to find adequate food, water, breeding grounds, and shelter. Of the 554 species of birds found at one time or another in Arizona, about 304 breed here. The 250 species of birds that do not breed in Arizona are migrants, which only pass through the state to reach their summer breeding areas. For example, rufous hummingbirds breed in the northwestern United States, western Canada, and southeast Alaska, and only pass through Arizona as they trek to those locations. Northern bird migrations are most active from March through early June and from September through October as birds make their way back to the wintering grounds. Although bird migration patterns are extremely complicated, most migratory birds found in the Southwest follow the coast of California, the mountains of Arizona, or along the river/riparian pathways.

Classic snowbirds winter in Arizona while breeding in the cool of the northern United States, Canada, and Alaska. These types of migrant birds form a much shorter list, but include such notables as ring-necked ducks, wigeon, scaup, sandhill cranes, and many of the "pond" ducks.

Another set of migratory birds do not spend the winter in Arizona, but rather come here to breed in the warmer months. For instance, red-faced warblers breed in Arizona from March through November, but spend their winters in central and southern Mexico.

Finally, purely native Arizona birds call Arizona home year-round, and do not migrate. Photographers can enjoy photographing cactus wrens, curve-billed thrashers, mockingbirds, red-tailed hawks, Anna's hummingbirds, Gambel's quail, and roadrunners (to name a few) in any season.

A large subset of native Arizona birds does not make geographic migrations, but rather make elevation migrations. Elevation migrations indicate a species moving to lower elevations in winter where food, water, and livable temperatures are more available (or the reverse in the summer). For example, the bridled titmouse breeds at elevations of 7,000 feet (2,134 m), but some of these birds make annual movements during the winter to lower riparian areas where food is more readily available and nighttime temperatures are more moderate.

Many of Arizona's large mammals make rather predictable elevation movements each year. A case in point: weighing up to 700 lbs. (318 kg.), Rocky Mountain elk need the cool days and nights at the highest elevations of Arizona to survive the summer. These higher elevations also see a greater rainfall than other, lower elevation, habitats and therefore, produce more vegetation on which the elk feed.

As winter arrives, life changes for Arizona's elk. Nights become bitterly cold within their ponderosa pine summer homes. Vegetation freezes and hides beneath the snow. Just moving from food

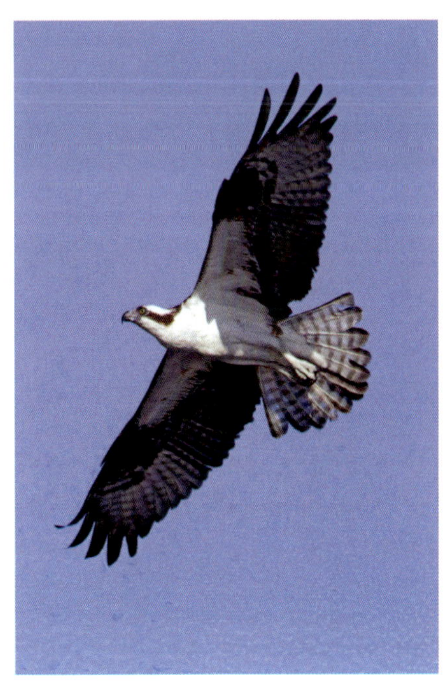

Osprey soars on the lookout for fish. Canon 1DMIV, 70-200mm at 200mm, 2x teleconverter, ISO 640, f/7.1 @ 1/2000 sec.

Wildlife Seasons and Migrations

plot to food plot takes more energy than elk can often spare. Therefore, when winter arrives in full swing, elk herds make mass migrations to lower elevations—sometimes as low as pinion-juniper habitats, where the grass is not covered by snow and temperatures are more tolerable.

Some deer and pronghorn populations make the same general movements as the elk. Extensive studies of the migration patterns of the Rocky Mountain mule deer population in the North Kaibab National Forest show that if they did not move to lower elevations during the winter, the herd would die out.

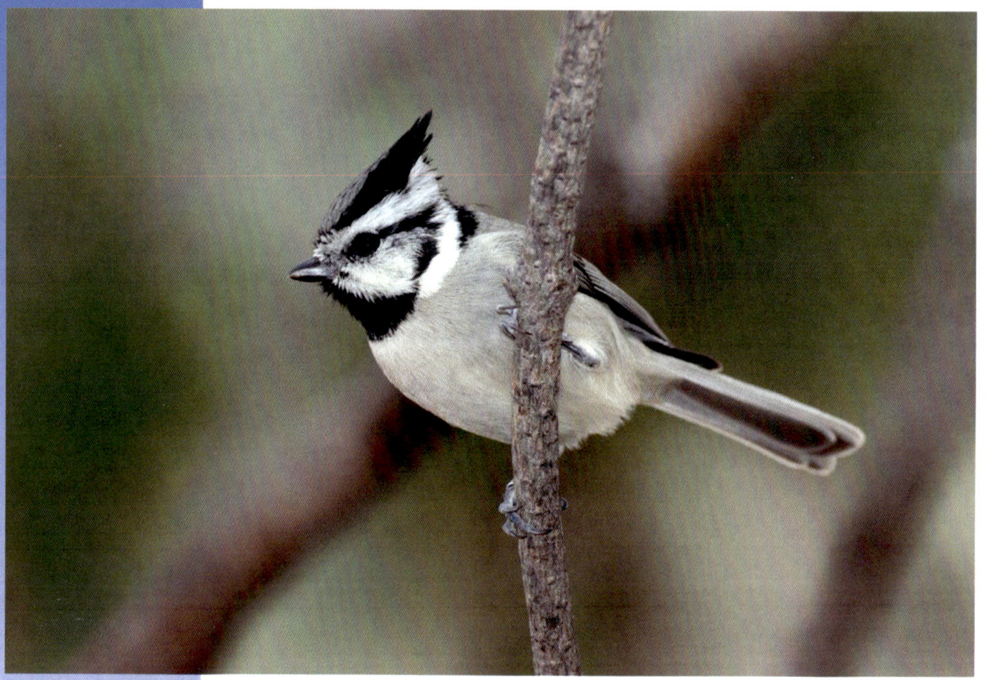

The diminutive bridled titmouse is one of many birds in Arizona that make seasonal migrations to find food, water, shelter, and a place to breed. Canon XTi, 500mm, ISO 400, f/6.3 @ 1/60 sec., on-camera flash @ -1/3 stop FEC.

Seasonal Rainfall

Depending on location, the average annual rainfall in Arizona varies from 3.3 inches to almost 30 inches. Southwestern Arizona tends to have the lowest amounts of rain, while areas at higher elevations in mountainous terrains receive the most. Rain in Arizona falls in two distinct patterns—winter and summer.

Winter rainfall comes to Arizona with large weather fronts originating in California from November through March. Winter rains encourage the development of many of Arizona's plants. The bountiful plant growth and glorious spring wildflower bloom only happens after sufficient winter rain. When abundant grasses and other plants grow, animals lay more eggs or produce more babies than they do after drier winters. Young animals are also better able to hide from predators, giving rise to greater population numbers. Pregnant deer, elk, pronghorn, javelina, and other large mammals feed on more abundant and higher quality food sources and can better carry their young to birth (and feed them afterwards).

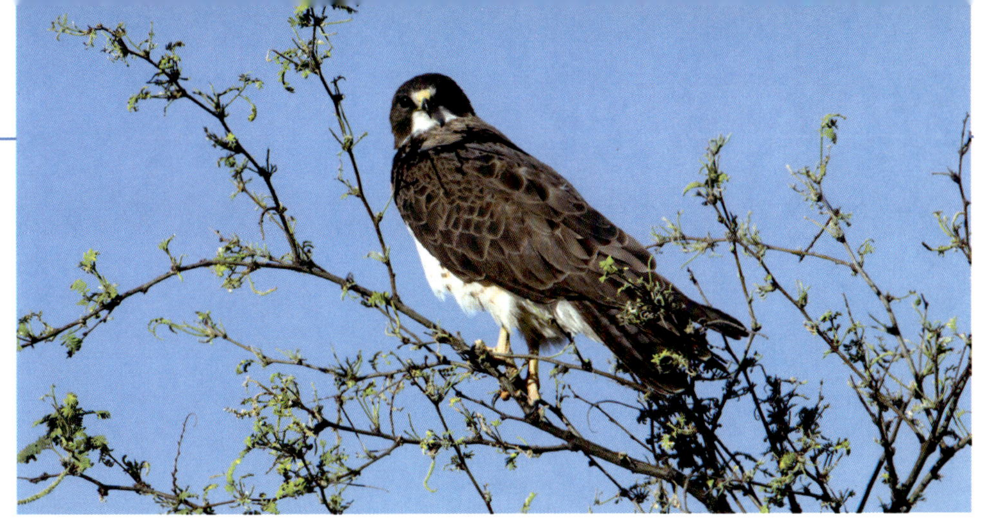

A Swainson's hawk rests atop a mesquite tree after its several-thousand-mile migration from Argentina to Arizona. Canon 7DMII, 500mm, 1.4x teleconverter, ISO 400, f/10 @ 1/1250 sec.

During years when winter rains are low or lacking, Arizona's resident wildlife suffers and populations decline. Resident birds have shorter breeding seasons, produce fewer offspring, and die at a greater rate when winter rains are poor.

The amount of rain during the cool season also affects migratory birds. In times of high winter moisture, the desert fills with flowers and insects necessary to feed many of the migrant birds. Lacking this seasonal moisture, though, a shortage of plants (and the insects that feed on them) leads to migrant birds hurrying through Arizona looking for "greener pastures."

The second peak rainfall period in Arizona occurs from July through September. Referred to as "monsoon season," torrential, but spotty, rains caused by moist air from Mexico and the Gulf of Mexico moving north and meeting the hot dry desert air (in excess of 100 degrees F (38 degrees C)).

Normal or above-average monsoon rains trigger a second growth of plants, which offer a critical food source for the offspring of many mammals. By the time the monsoon rains hit, most mammals have weaned their young, who are then forced to find food on their own. If the summer rains produces small amounts of precipitation, many of the young die or face an approaching winter in poor health.

Arizona Wildlife Activity

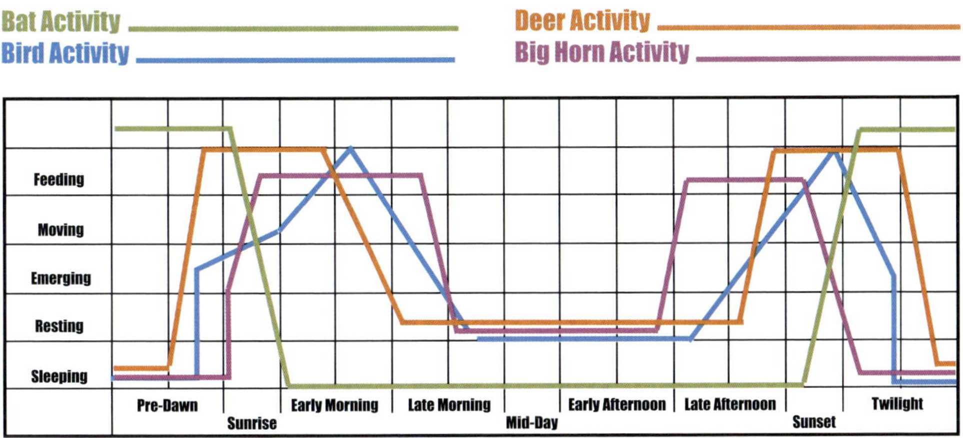

Hide and Seek: Camouflage

In the wildlife kingdom, predators always try to find their next meal, while prey try to avoid becoming that meal. Predators and prey have spent eons developing mechanisms that will, hopefully, keep them alive a little longer—at least long enough to breed and keep their species thriving.

For example, rabbits' fur appears a dull brown color, which allows them to disappear in the shadows. The cunning coyote is similarly colored for concealment and possesses great vision and hearing. If the coyote spots the rabbit, the rabbit becomes lunch. If the rabbit stays well concealed, the coyote goes hungry.

Most wildlife considers people (especially those wanting to get close enough for a great photograph) as threats. They do not understand that we are passing for a mere moment in time. If they would just let us quickly make that award-winning image, we would be out of their lives forever!

In order to get close and record a successful image, photographers need to conceal themselves from many of their subjects. Fortunately for photographers, for thousands of years, hunters have needed to camouflage themselves from their quarry and have developed many (if not all) of the tools photographers need to disappear into the landscape.

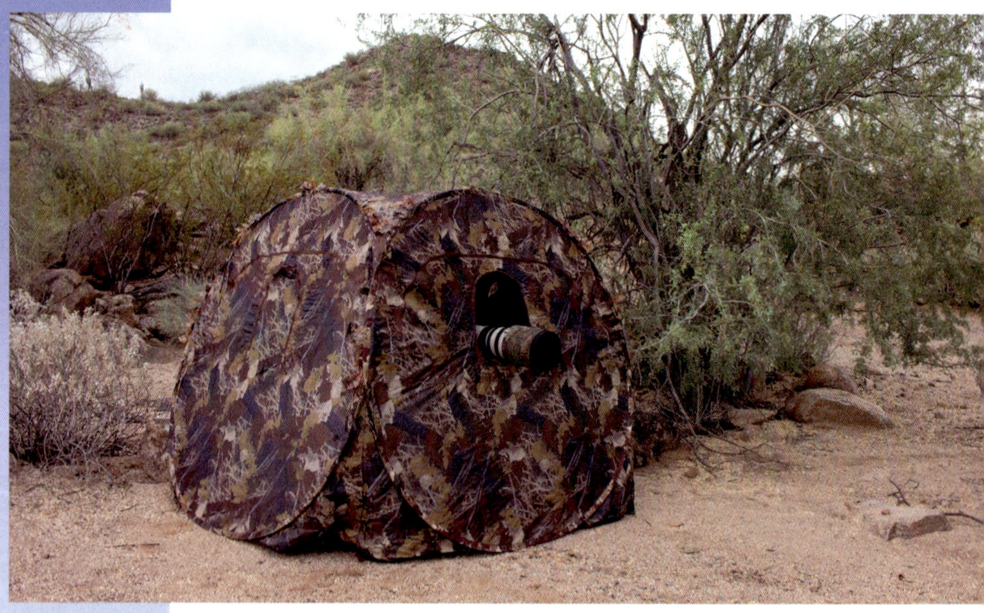

I favor the Ameristep Doghouse hunting blind (above) and the Kwik Camo photographic blind (opposite right) to conceal myself when photographing shy species of wildlife.

The hunting blind serves as the best and most widely available tool to hide photographers. Constructed from sturdy material with complex earth tone colors, these small tent-like structures assemble quickly and "blind" wildlife effectively from our presence.

Although many configurations are available for purchase, the Ameristep Doghouse blind offers the best solution for shutterbugs. This blind is durable, fast and easy to set up, and light enough to carry long distances. Built for two people, it provides photographers sufficient room to change lenses, shift around to get more comfortable, or just stretch. (Sitting

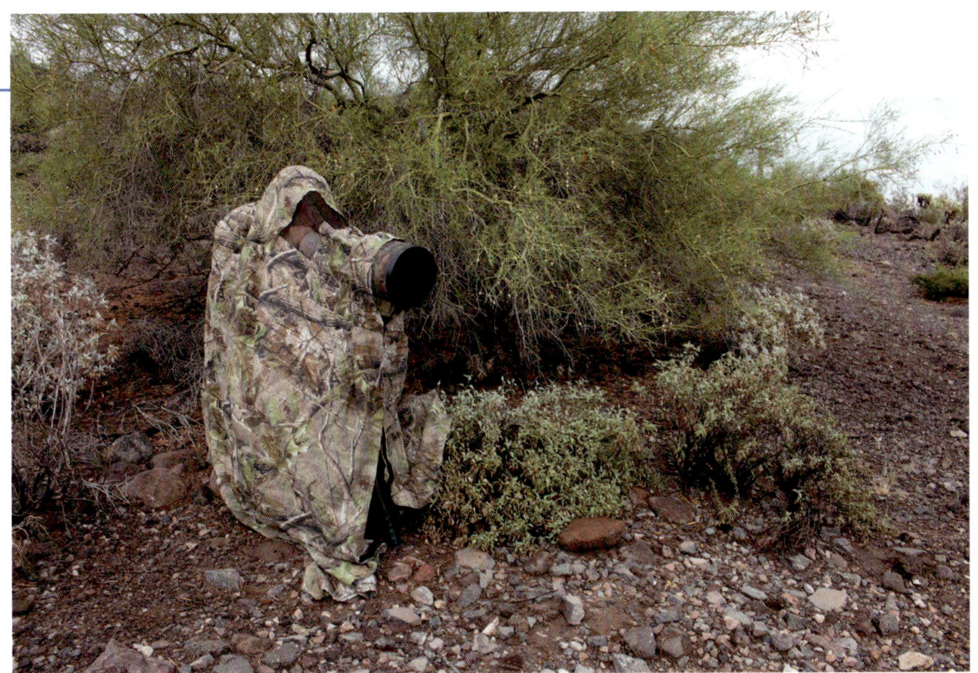

in a blind for hours is hard work!) One-person blinds exist, but offer cramped quarters for a wildlife photographer toting a large pile of gear and tip over easily when the photographer needs to change positions.

For short duration shoots, a bag blind offers the quickest set up and best portability. Constructed from heavy-duty camouflage material, the bag blind (I prefer Kwik Camo) features a hole for your lens and hook-and-loop fasteners to secure the bag around the photographer. If feasible, place your blind in a shadowed location with the sun hitting the back of the blind (and reflecting off the front of the subject). Look for a clean foreground and a background free of distracting branches, grass, or reflecting surfaces to mar your photograph.

In addition, your vehicle makes a great blind. For example, the burrowing owls found at the South Verado Way (see page 82) nest along a canal system next to roads. They are habituated to vehicles zooming past, but are wary of animals (including you) walking in their territory. If you drive up to one of these burrowing owls (or other road-familiar critters elsewhere), slowly pull over and stop the engine. Gradually roll down the window and place your lens on the sill (preferably with a bean bag over your window to stabilize your lens). By remaining quiet in the car, you will allow the burrowing owls (or other animals) to continue their normal activities in short order.

If you are unable to use a blind, but still want to appear inconspicuous, wear dull, natural-colored clothing (e.g., brown, green, black). Find a shaded place and situate yourself in front of some vegetation to hide your silhouette from your subject. Use the shade and/or background foliage to camouflage your presence. Then, sit quietly and still for at least 30 minutes. If no animal activity appears within that time frame, the critters probably left the area.

The ghillie suit can also improve your camouflage arsenal. Originally made for military snipers, these wearable covers feature loose strips of camouflaged material attached to a coat and trousers, breaking up the user's outline by creating jagged non-linear edges. Look for suits made from nylon or some other type of non-shedding material. If you wear a suit constructed from burlap (a popular material for ghillie suits), the suit will shed dust that will stick to your lens, face, and camera sensor. To blend into the land, simply don your suit, walk to your pre-scouted site, and sit (preferably on a chair) in the shade.

Rules and Regulations

Arizona hosts a melting pot of land ownerships. During your photographic adventures, expect to tread on federal, state, private, or Native American soil. Before you travel, review the specific rules and regulations for that area. In addition to rules and regulations restricting access, several entities restrict the commercial use of images taken on their lands and control how to treat the wildlife you might photograph.

The Forest Service, Bureau of Land Management, and Bureau of Reclamation allow a broad range of access to the properties they manage. In most areas, permits are not required for individual photographers to enter these properties or to take photographs (unless it is for commercial purposes). Guiding workshops or conducting a photo shoot that disturbs the activities of other users typically requires a commercial permit. In addition, watch for road closures and off-road limitations.

The State Land Department treats state lands in Arizona as private (even though they are public). To enter state lands for any purpose, the department requires a permit. However, anyone possessing a hunting, fishing, or trapping license can enter state lands for that stated purpose. The State Land Department issues annual recreational permits for activities such as camping, hiking, and photography. As with the federal regulations, obey road closures and off-road travel rules.

Private property is just that, private! If you wish to enter private property to take photos, ask the owner first. If the owner does not respond, do not go on their lands. Unfortunately, some private lands are difficult to distinguish from public lands. Land owners are responsible for posting "No Trespassing" signs on their property, but if you inadvertently end up on private land and the land owner asks you to leave, you must depart immediately—signs or not.

Access to Native American properties in Arizona is governed by the rules and regulations developed by the various tribes and the federal government. Unless the properties have obvious legal access points (such as Monument Valley Navajo Tribal Park), do not enter without explicit permission. Some tribes sell recreational permits, which allow for specific activities including hunting, fishing, hiking, camping, photography, etc. Without such permits, though, any trespass is illegal.

Several entities manage wildlife. For example, the United States Fish and Wildlife Service (USFWS) in coordination with the Arizona Game and Fish Department and landowners

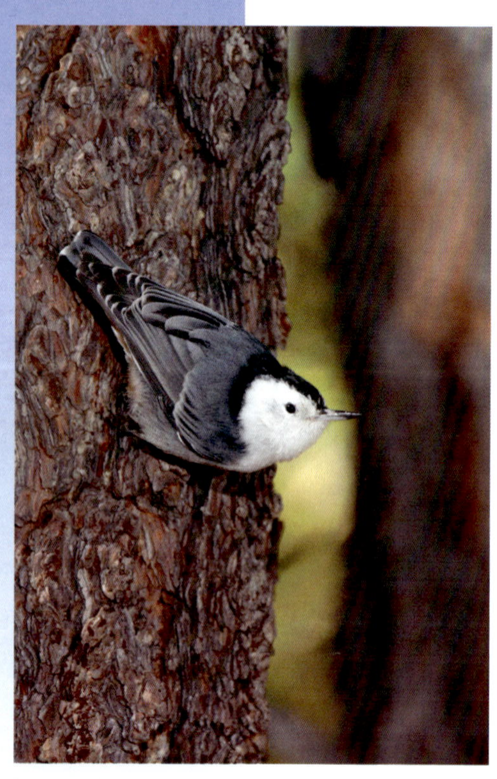

White-breasted nuthatch holds onto the side of a pine tree. Canon 50D, 500mm, 1.4x teleconverter, ISO 250, f/9 @ 1/250 sec., on-camera flash @ -1/3 FEC, Better Beamer.

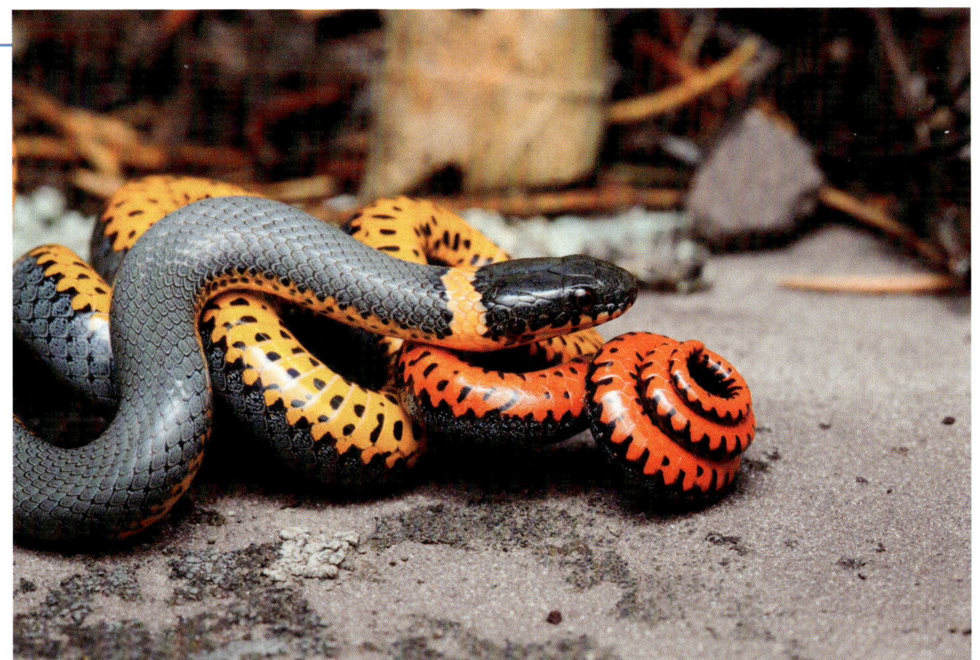

The coiled red tail of a ring-necked snake is meant to scare off potential predators. Canon 5DMII, 100mm macro, ISO 400, f/16 @ 1/200 sec., on-camera flash @ +2/3 FEC.

manage endangered species. Animals falling on the Federal Endangered Species list are afforded special protections and, in general, should not be approached or harassed. Photographers can take photographs from a distance, as long as the normal activities of the endangered species are not affected.

The USFWS also monitors all migratory birds, which cannot be harassed or harmed in any manner, nor their nests disturbed, without appropriate permits. Much confusion surrounds these regulations, so read them for yourselves and ask questions if you have any.

The Arizona Game and Fish Department oversees most other wildlife not considered endangered or migratory (unless they are frogs or game fish in ponds on private lands—the rules are confusing and copious).

Photographers who have purchased a fishing license can handle most amphibians (frogs, toads, salamanders). Similarly, shutterbugs in possession of a valid hunting license may handle most small mammals (rats, mice, voles, etc.) and snakes.

Although the Department of Agriculture regulates the use of insects, they are primarily concerned with managing insects' effects on agricultural crops--not with photographers making images of them. As such, no overarching regulations prevent you from capturing, handling, or photographing insects (except a few species protected by other federal regulations).

To ensure that your actions are legal, read all of the literature provided by landowners and wildlife management agencies. Above all else, call them to ask questions. Do not disturb any wildlife unless you know it is legal to do so and you can prove the legality to law enforcement authorities. It is best to err on the side of caution and hold off taking a photograph or approaching an animal if you are in doubt.

Commercial Wildlife Sites

Few areas are as world famous for birdwatching and wildlife photography as the southeast portion of Arizona. To fill the demand for easy access and comfortable wildlife recreation, several private landowners have built businesses around providing a place to visit and relax while watching and photographing wildlife. Many of these businesses offer rooms for rent (some with private bathrooms and kitchens) for the visitor that wants to be close to nature for extended periods of time.

The owners of the properties referenced in this book are not only businesspeople but also conservationists and wildlife enthusiasts. As such, they have developed some of the best places in the state for wildlife to visit. All of the sites provide animals with a wide range of food, water, and habitat which results in a rich diversity of species to observe.

In this book, I have only included sites where I know the owners personally and have photographed on their properties. Before visiting these private locations yourself, call ahead to let the owners know your intentions. Many of the sites are open to the public for a small fee, but some are only available if you purchase either rooms or services. Each location's website will give you details of the services they offer.

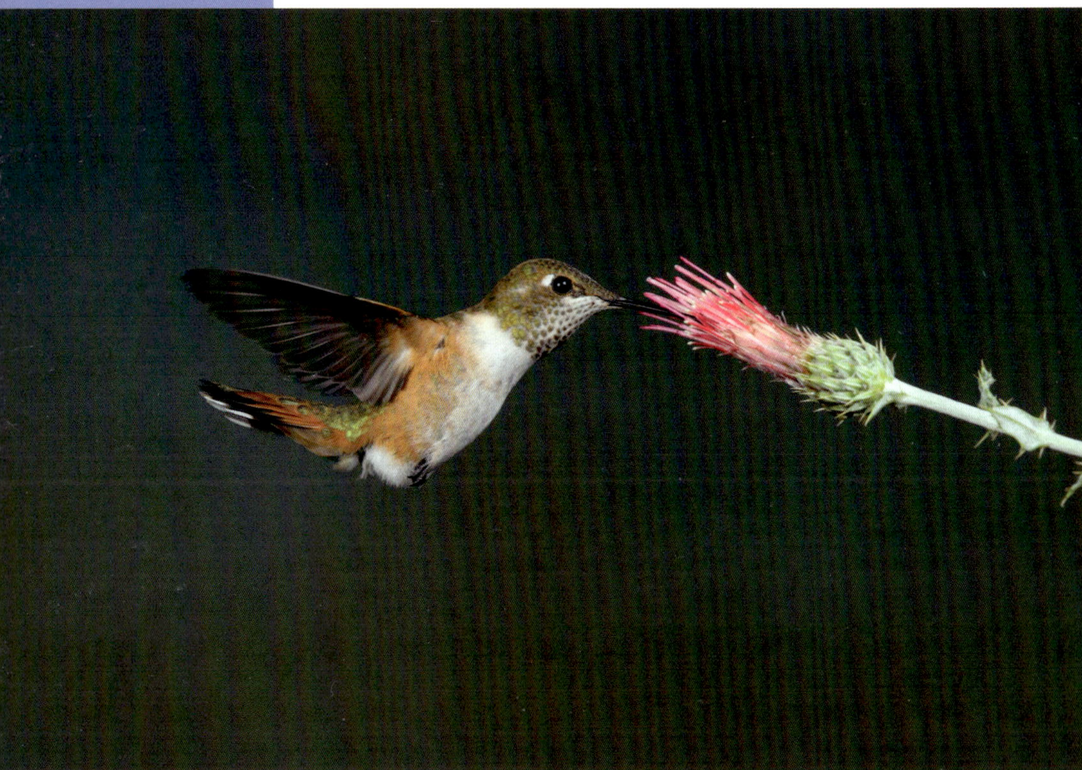

Female rufous hummingbird feeds from thistle flower. Canon 1DMIV, 400mm, ISO 400, f/13 @ 1/250 sec., high-speed hummingbird set-up.

Etiquette

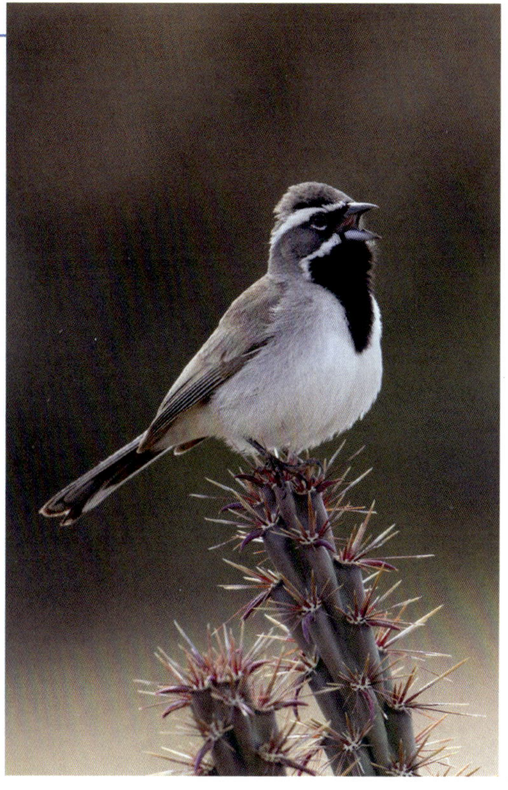

Backlit black-throated sparrow defends his territory atop a staghorn cholla. Canon 7DMII, 100-400mm at 400mm, ISO 400, f/7.1 @ 1/400 sec.

Many of the locations listed in this book feature areas where the wildlife is approachable and photography is unlikely to negatively impact them. In all circumstances, wildlife photography should be an activity for which there are no negative consequences. As photographers, we should leave an area with no trace of our presence and no harm to the wildlife that we love enough to put on our canvas. There are two levels of etiquette in wildlife photography—one for the animals and lands, and one for other land users and photographers.

The "Leave No Trace" principle suggests that the land should not suffer as a result of our activities. Simply put, take out what you took in (with some basic bathroom exceptions). Do not litter. Not only is littering illegal, it also presents the photographic community in a poor light.

When our presence disturbs an animal's behavior, not only have we potentially harmed the individual (and possibly their young), but also our images will not represent natural events. It is illegal (and poor form) to scare birds just to obtain images of them in flight. Likewise, do not take an action that makes a four-legged critter run just so you can take an exciting shot.

Stay away from bird nests, or set up remote cameras and camouflage your presence. If you think you are disturbing a bird at a nest, you probably are and should leave. Contact the local Arizona Game and Fish Department office or the U.S. Fish and Wildlife Service to familiarize yourself with the legal issues associated with being around bird nests.

Always remember that the public has just as much right to be close to an animal as the photographer does. Be considerate of other people's needs and desires, and give them the benefit of the doubt. If they disturb your photographic opportunity, take it as a learning experience and move on to a new location.

Be kind and considerate of your fellow photographers. If another photographer sets up before you get to a particular location, do not interfere with them. Go somewhere else or return another day. Never step in front of another photographer unless you first ask permission to do so. Be quiet when around other photographers. Practice the old adage "treat others the way you would want them to treat you."

No photograph is important enough to justify breaking these etiquette rules. Images from wildlife photographers should teach the world about the animals we love and elicit greater appreciation of these animals by the public and decision makers. Wildlife photographers should aim to gain respect because of what their images do to conserve and preserve wildlife.

Photography Basics

Throughout this book, I have made specific suggestions about how to set up your camera in order to give you the best opportunity to capture a particular image. In this section, I hope to provide the reader with an understanding of how cameras operate, the technical jargon associated with various operations, and how to use them to best effect. Although this book focuses on digital cameras (which collect light with a sensor), most of the technical aspects I describe here also apply to film cameras.

RAW or JPEG

When making a photograph, the camera records an image in one of two formats: RAW or JPEG.

RAW is not an acronym but represents an image file that holds all of the uncompressed data that the sensor collects. RAW files are large and take up considerable space on a memory card. One downside to using a RAW file is that a photographer must first convert the photograph to either a JPEG or TIFF file before the image can be printed, emailed, displayed on the web, or used by other applications. That said, in post-capture processing software, photographers can easily modify RAW files to correct for things like sharpness, picture style, and white balance (where the color in the final photograph might appear more accurate and representative of what you saw when you made the image).

JPEG stands for "Joint Photographic Experts Group." This image file is much smaller than a RAW file as it does not contain all of the original data that the sensor collects. A computer-generated mathematical model determines which data to

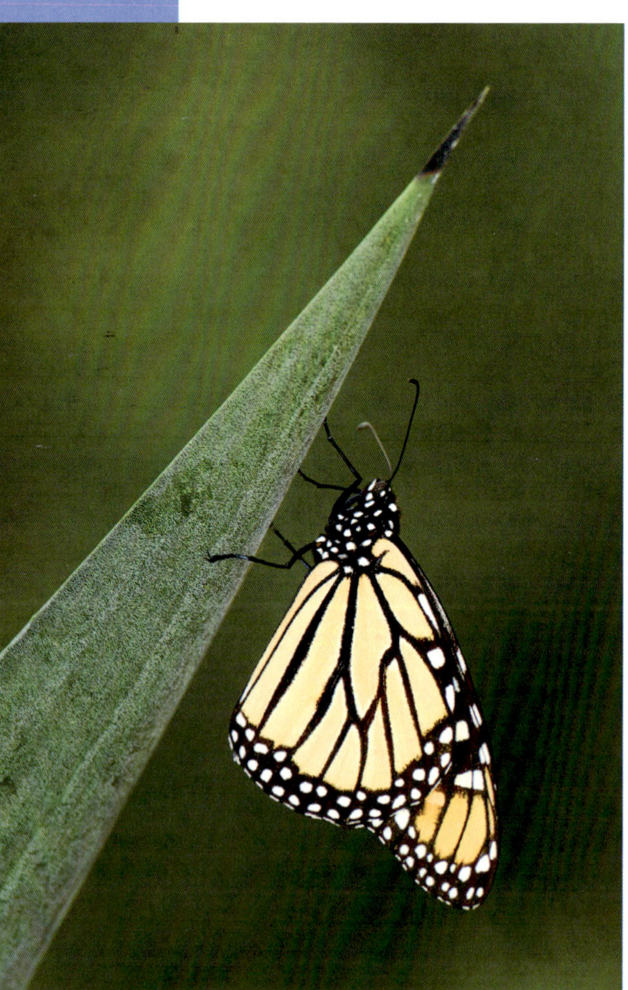

At the Desert Botanical Garden, I positioned the camera so that the agave spine transected much of the frame and the butterfly appeared off-center according to the Rule of Thirds. Canon Digital Rebel XTi, 400mm, ISO 400, f/5.6 @ 1/100 sec., on-camera flash @ -2/3 FEC.

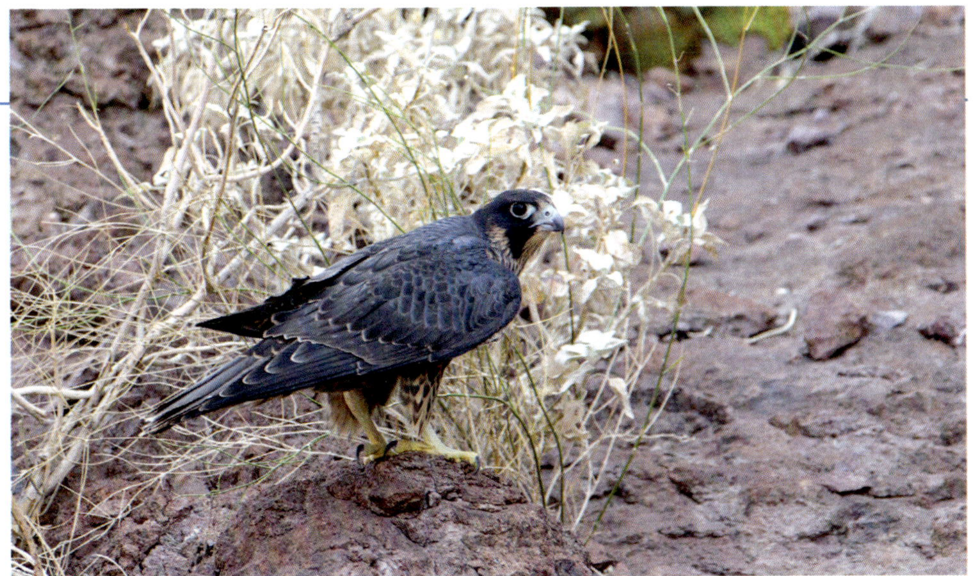

Photographed from a boat, this peregrine falcon rests along the banks of the Colorado River below the Hoover Dam. Canon 1DMIV, 70-200mm at 200mm, 2x teleconverter, ISO 640, f/6.3 @ 1/800 sec.

keep and which data to throw out. In general, when the camera saves an image in JPEG, the model analyzes the image to determine similar colors among pixels. For example, if 1,000 pixels have a similar color of red, a JPEG file will only have one data point for those 1,000 pixels. In reality, each of those pixels probably differ even though the JPEG image records them as being the same. When saving an image as a JPEG, the sharpness, picture style, white balance, and other characteristics are loaded into the file. Therefore, any future modifications (i.e., via post-processing) might degrade the image's quality. Shoot RAW files if you seek to maintain the maximum amount of control over the final disposition of your work.

If you are uncomfortable using RAW files exclusively, shoot in both RAW and JPEG. By shooting both formats, you keep the ability to use RAW in the future while turning the JPEG file into an email file or the like.

Exposure

When making an exposure (taking a picture), the photographer pushes the shutter button, and the lens diaphragm opens to allow light from the subject to be collected by the digital sensor. To control how much light the camera collects during an exposure and obtain the desired results for motion and depth of field, the photographer can adjust three camera settings: aperture, shutter speed, and ISO speed (which photographers often refer to simply as "ISO").

When changing exposures by either doubling or halving the ISO, shutter speed, or aperture size, the photographer adds or subtracts a "stop" of light respectively. Most cameras operate by changing stops in one-half or one-third increments. If you add a stop of light, the digital sensor will receive twice the amount of light. Concurrently, if you take away a stop of light the sensor will only accumulate half the amount of light.

Most Digital Single-Lens Reflex (DSLR) cameras feature different camera modes which enable varying levels of control over the operation of the camera's exposure settings. Photographers can adjust exposure in commonly used modes like program (P), aperture priority (Av), shutter speed priority (Tv—time value—or S), manual (M), and bulb (B). In automatic (A) or when using the scene modes (those modes represented by symbols, such as a flower for macro, a mountain for landscapes, and a stylized runner for sports), the camera does not allow the photographer to make modifications to many of the exposure settings and camera features.

(Continued)

Photography Basics

Aperture

When setting the aperture, the photographer determines how wide to open the lens diaphragm and thus how much light to collect on the sensor. Apertures can range from f/1.0 to f/96 ("f" stands for f-stop and describes the specific aperture setting). The f-stop value is determined by dividing the length of the lens in millimeters by the opening of the diaphragm in millimeters. For instance, if a 100mm lens is set at f/5.6 then the size of the diaphragm opening will be approximately 17.85 millimeters.

Each lens possesses varying minimum and maximum aperture settings. For instance, a lens with aperture settings from f/1.8 to f/22 will have a maximum aperture of f/1.8 (the lens opening as wide as possible) and a minimum of f/22 (the diaphragm opening as small as possible). A small aperture lets in less light than a wide one. When changing from one f-stop to another, the amount of light that hits the sensor either doubles or halves because the surface area of the lens opening either doubles or halves.

To summarize aperture settings and their impact on an image:

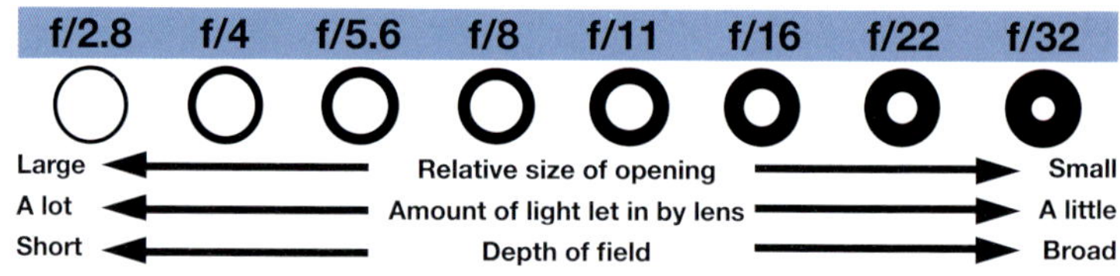

Shutter Speed

Shutter speed refers to the length of time, in seconds, that the camera lens is open and allows light to hit the sensor. Depending on the camera

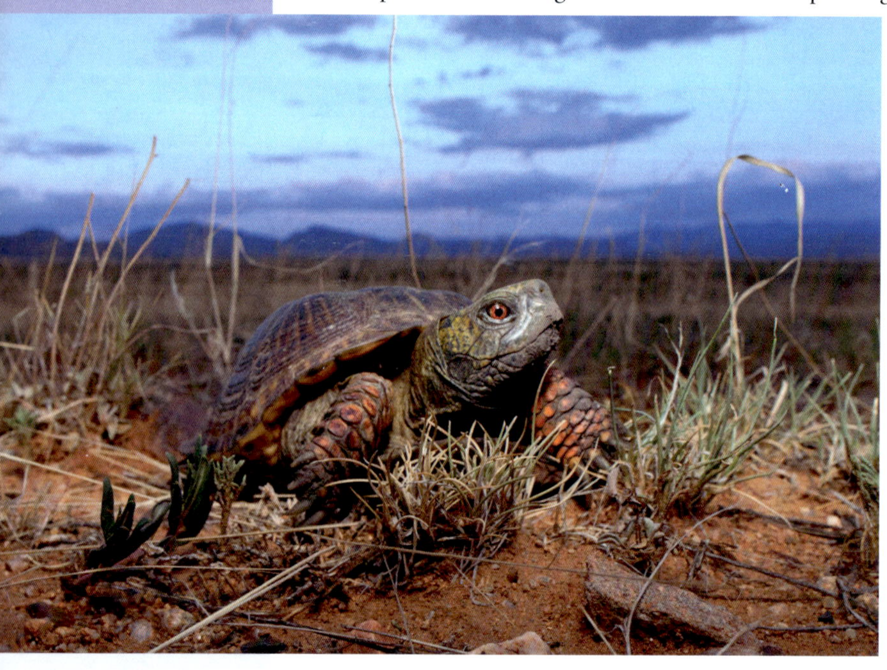

Desert box turtle in the grasslands of southeast Arizona. Canon 5DMII, TS-E 24mm, ISO 800, f/13 @ 1/30 sec., on-camera flash @ -1 1/3 FEC, flash diffuser.

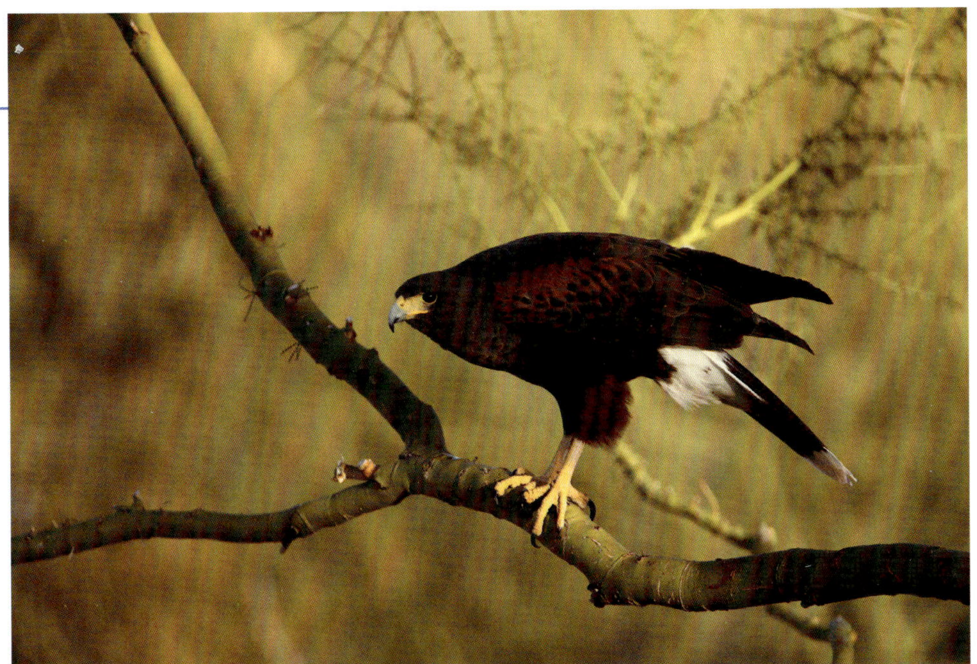

The Harris's hawk is adapted to the high temperatures in the Sonoran Desert and can hunt on days that exceed 100 degrees F (38 degrees C). Canon 5DMII, 500mm, 1.4x teleconverter, ISO 200, f/5.6 @ 1/250 sec.

make and model, shutter speeds generally range from 1/8000 second to 30 seconds.

When shutter speeds are represented as fractions of a second, they are in graduated steps that either allow twice or half the amount of light to shine on the sensor. Similar to aperture, when changing from one shutter speed stop to the next, the amount of light that hits the sensor either doubles or halves.

The faster the shutter speed, the better chance of freezing action and rendering a sharp image. For example, to stop the action of a moving bird, use shutter speeds faster than 1/1000 second. When working with critters that are less active, instead try shutter speeds in the neighborhood of 1/250 second, but only when your camera is securely set on a tripod. To blur an animal's movement, shoot at speeds less than 1/100 second. No matter how you stabilize the camera (tripod, beanbags, etc.) it is difficult to obtain an image that is very sharp if the shutter speed is slower than one-half of a second.

Here are some examples of various shutter speeds and how they affect the way motion is depicted in an image (note that the numbers without the tick mark are fractions of a second and those with a tick mark are in seconds):

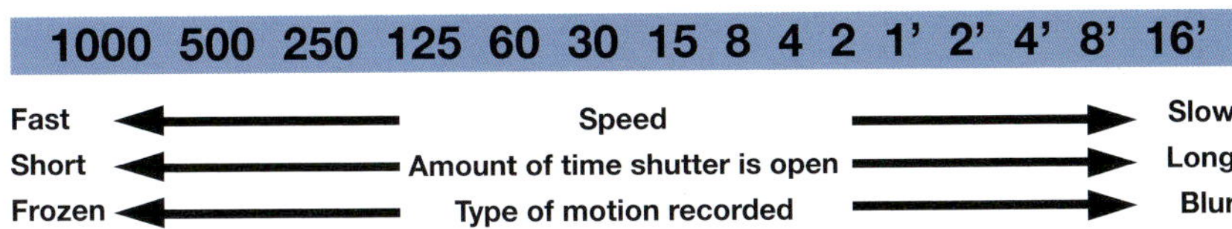

ISO/Film Speed

ISO stands for "International Standards Organization" and is a worldwide standard measure for how sensitive film or digital sensors are to light. On digital cameras, the ISO settings vary, and depending on camera make and model, photographers can set ISO speeds from 25 to over 100,000. Lower ISO numerical settings are less sensitive to light than higher settings and therefore require a longer exposure to capture a correctly exposed image.

(Continued)

Photography Basics

Increasing ISO sensitivity results in an increase in "noise," which renders a photograph with a grainy look and oftentimes creates random red, green, and blue spots throughout the frame.

Similar to aperture and shutter speeds, when changing the ISO from one stop to the next, the sensitivity of the sensor is either doubled or cut in half.

ISO/film speed can be summarized as follows:

50	100	200	400	800	1600	3200	6400

Low ←	**Sensor's sensitivity to light**	→ **High**
Slow ←	**Speed in recording image**	→ **Fast**
Longer ←	**Exposure times**	→ **Shorter**
Lower ←	**Noticeable grain/noise**	→ **Higher**
Greater ←	**Ability to enlarge without noise**	→ **Lesser**

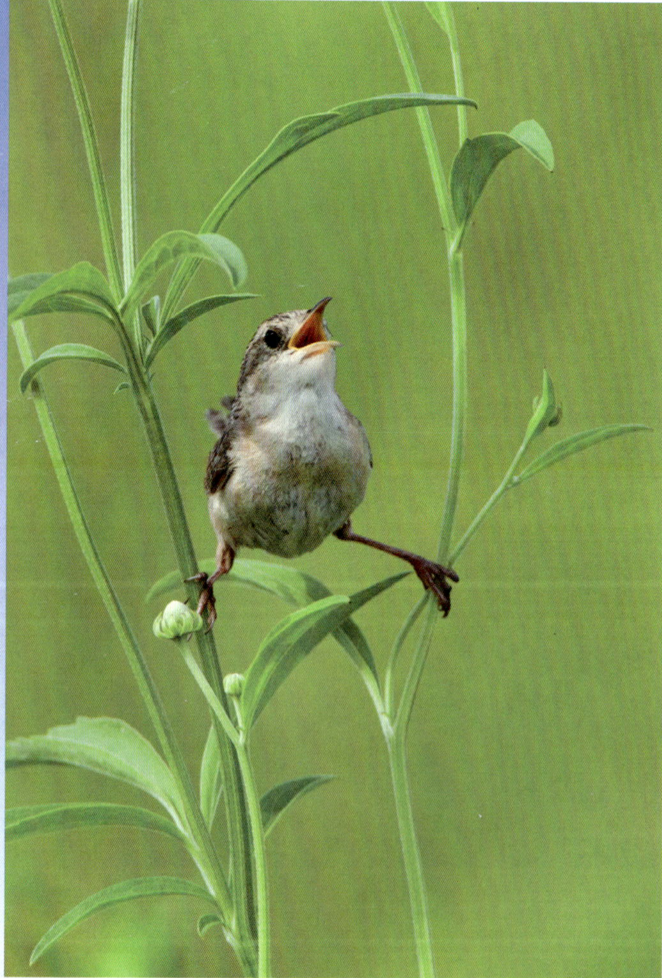

Marsh wren calls to defend its territory. Canon Digital Rebel XSi, 500mm, 1.4x teleconverter, ISO 200, f/7.1 @ 1/400 sec.

Putting It All Together

When the photographer creates a photograph, the camera uses ISO, shutter speed, and aperture to determine the amount of light the digital sensor records. If the photographer changes any of these settings, then the camera modifies the amount of light recorded on the sensor. Any combination of aperture, shutter speed, and ISO speed that allows the camera to record the same amount of light is called an "equivalent exposure."

For example, if you wanted to photograph a flying bird, and your camera is set at f/8.0 at 1/250 second, with ISO 200, the flying bird will likely not appear sharp or frozen in your final image due to a relatively slow shutter speed. By increasing the shutter speed to 1/1000 second or faster, the camera will reduce the amount of light that hits the sensor by two stops. If you snap the image using this shutter speed setting with your original aperture and ISO speed settings, your camera will not collect enough light, and your image may be severely underexposed. To increase the amount of light the sensor sees, open up the aperture or increase your ISO speed (or both).

The Histogram

A histogram is a graphical display of data using bars or lines to represent frequency distributions. Commonly used to show statistics such as height at different ages, longevity, and the value of stocks over time, digital camera makers adopted this approach to assist the photographer in recording a properly exposed photograph.

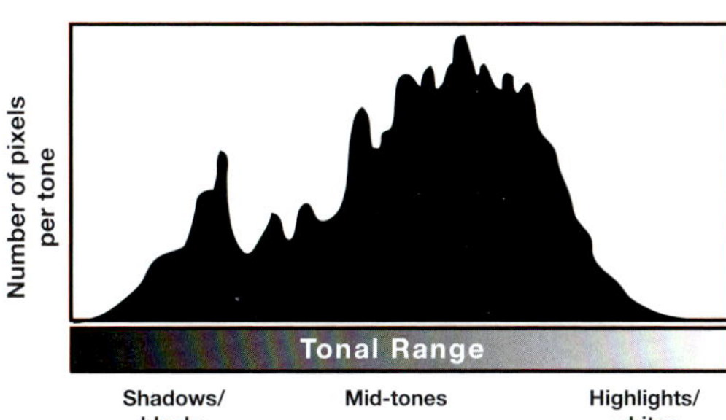

Camera histograms show the relative number of pixels at each point along the brightness spectrum with darker pixels on the left, mid-tone pixels in the center, and lighter pixels on the right. Black pixels appear against the left-most side of the histogram, while pure white pixels appear against the right-most side.

In a daylight scene, the histogram should display a "normal distribution" of tones across the image with a few pixels appearing on the right, a few on the left, and most distributed in the middle.

The distribution of tones in a histogram for an image recorded at night will look very different than those made during the day. Under dark skies, the histogram should show many black pixels on the left. The rest of the histogram will represent those pixels that are not black.

If, after creating the image, the histogram shows too many pixels pegged to the right (called overexposure) or left (called underexposure)—or if the histogram does not appear to represent the true tones in the scene—then you should correct the exposure. Underexposure occurs when the camera has not gathered enough light during an exposure and the image appears too dark. Overexposure occurs when the opposite happens: the camera collected too much light, and subsequently, the image appears lighter than desired.

If your camera has the "Highlight Alert" function available and enabled, the overexposed spots will appear as blinking black and white areas when you review the photograph on your LCD screen, which indicates the camera has not recorded data in those places. Currently, post-processing software cannot recover or recreate this lost information.

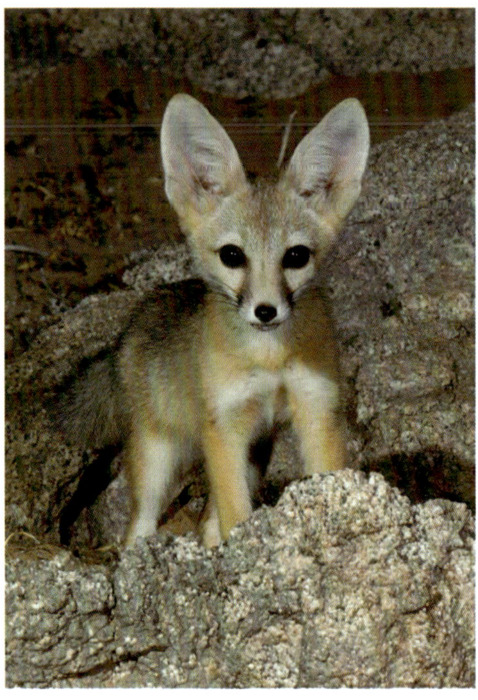

To photograph this young kit fox, I set two remote flashes close to the den. Canon 1DMIV, 70-200mm at 292mm, 2x teleconverter, ISO 400, f/20 @ 1/250 sec., off-camera flash.

(Continued)

Photography Basics

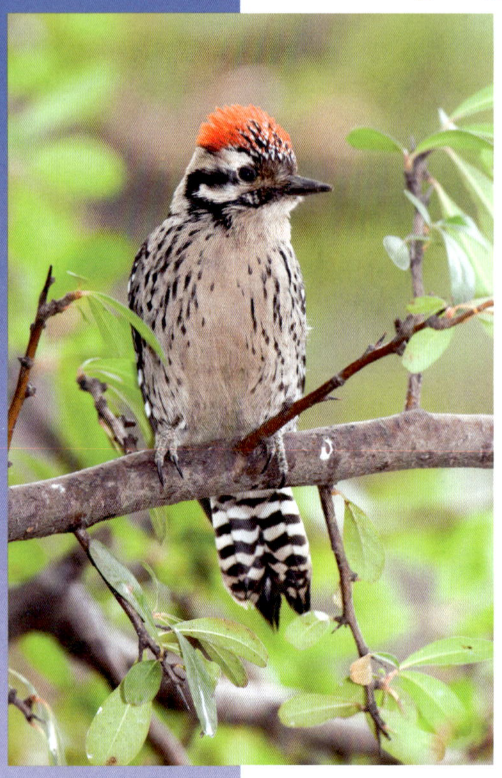

Ladder-backed woodpecker. Canon Digital Rebel XTi, 500mm, ISO 400, f/6.3 @ 1/125 sec., on-camera flash.

In all camera modes except automatic, manual, and the various scene options, the camera's Exposure Value (EV) compensation function gives the photographer the option to darken or lighten the photograph as needed. In manual camera mode, set the shutter speed, aperture, and ISO using the light meter as your guide. After setting the exposure, you must change one of these three settings to add or subtract light. EV compensation does not function in manual camera mode.

For backlit images, a correct histogram may show an abundance of white tones on the right. For instance, when photographing backlit wildlife at sunrise or sunset, the histogram should display some of the tones pegged to the right; otherwise, the subject will be underexposed. Although verboten in most landscape photography, an overexposed backdrop for some wildlife images is acceptable. Despite difficulties in using while photographing moving subjects, one could explore techniques such as HDR or the use of filters to deal with any unwanted over-exposed areas in the image.

White Balance

All sources of light give off different color temperatures. As we look at the world, we do not always see these color casts because our brain automatically corrects for them. When photographing wildlife in the shade, your eyes may observe beautiful reds, oranges, and greens, but the camera may not record them. The white balance setting helps your camera make internal adjustments to render the light in your photograph similar to the way you see it. Although the Auto White Balance (AWB) setting does a reasonable job of correcting the color for most scenes, matching the common white balance settings such as daylight, shade, overcast, flash, fluorescent, and tungsten to the respective actual lighting conditions will yield more accurate and consistent results. For full control over color balance, use manual white balance.

Shooting in RAW format makes it easier to adjust white balance later when post-processing on your computer.

Focus

Most lenses have two ways to focus: manual focus and autofocus. When set in manual focus, the photographer turns the barrel of the lens to obtain focus before they take a photo. In autofocus, the photographer depresses the shutter button halfway, and the camera and lens work together to obtain a sharp focus. Most DSLR cameras use a set of mirrors and motors to obtain focus by measuring when two light beams converge—a very fast process called "phase detection." However, mirrorless cameras use "contrast detection" which moves the lens back and forth to determine where the contrast is highest—a relatively slow process.

Most cameras offer a variety of ways to autofocus, including single-shot, AI Servo (or "continuous focus"), and AI Focus. If the photographer uses single-shot focus, as soon as the camera obtains focus, focusing stops. When the photographer depresses the shutter button, it takes about 55 milliseconds to initiate the shutter and take the image. If the subject moves even a few inches between when the camera obtains focus and when the photographer takes the exposure, the image will be out of focus.

This problem disappears when the photographer sets the focus mode to AI Servo, where the lens continuously changes focus as the subject moves. In AI Servo, the lens changes its focus distance as the subject changes its position.

AI Servo is especially useful in wildlife photography since wild animals are constantly on the move as they fly, run, swim, or jump from one place to the next. You only need to keep a few of the camera's focus points on the critter while depressing the shutter button halfway before blasting away.

When photographing an animal, the most critical part to focus on is the eye. The eye serves as the opening to the soul, and you can determine—to some degree—their state of mind, personality, and beauty as well as add a sense of drama to the image. The piercing eyes of predators add a sinister look to our images while the wide-open eyes of a bugling elk relay a feeling of anxiousness. Without the eye in focus in your photograph, you lose this critical connection between the animal and the viewer of your image.

Drive Modes

Cameras possess several different drive modes, which determine how the camera moves from one frame to the next as you take an image:

1. Single-shot: When you press the shutter button, only one image records.
2. High-speed continuous shooting: As long as you continue to depress the shutter button, the camera will continue to take images, one after the other. This process stops when the photographer releases the shutter button or the camera's recording buffer is full.
3. Low-speed continuous shooting: Similar to high-speed continuous shooting mode, but records fewer frames per second.
4. Self-timer: After depressing the shutter button, the camera will wait a brief period of time (typically 2 to 10 seconds) and then take a single image.
5. Silent shooting mode: Allows for quieter shooting.

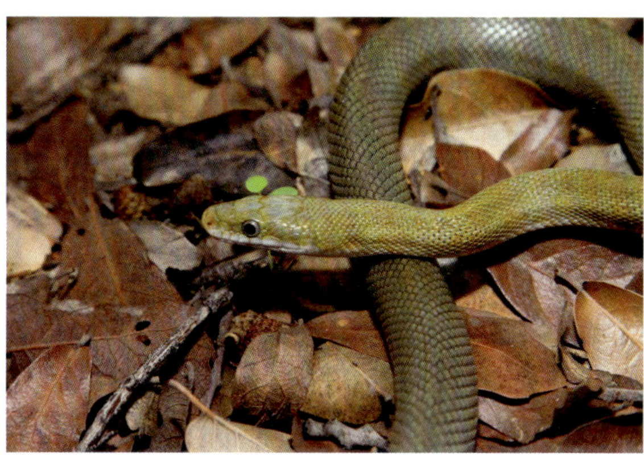

Green rat snake slithers among the leaves in Madera Canyon. Canon 1DMIV, 24-105mm at 105mm, ISO 500, f/10 @ 1/250 sec., on-camera flash.

Preparing for Your Photo Shoot

As much as light is a photographer's palette, our equipment is our canvas and brushes. Photographers can spend hours upon hours investigating the behavior of our subjects, listening to the weather, traveling to a location, and getting to a spot at just the right time of day, but unless you have the necessary equipment to capture the moment, you will become frustrated by the results.

To start, any of the most recent models of Canon, Nikon, or Sony Digital Single-Lens Reflex (DSLR) cameras take wonderful photographs. If you plan to focus on bird or large mammal photography (in contrast to reptiles, amphibians, insects, or small mammals), look for a "cropped frame" camera (also called an Advanced Photo System type-C, or APS-C, camera). The reduced-size sensor in these cameras causes a cropping of the image and an effective magnification of the subject relative to the aspect ratio (also referred to as the crop factor and focus length multiplier, or FLM). With Nikon and Sony APS-C sensor cameras, the crop factor equals 1.5, so the FLM increases by 50%. For Canon, the crop factor amounts to 1.6, so the FLM boosts by 60%. For example, using a cropped frame camera, your 400mm equates to an effective focal length of 600mm with a 1.5 crop ratio (or 640mm with a 1.6 crop ratio). When you want to make your subject as large as possible in the viewfinder and you cannot physically get close to the subject, an increase of 50% or more can help tremendously. Another benefit? APS-C cameras can help you save money on expensive telephoto lenses. A full-frame sensor camera can also take wonderful wildlife photographs, but you may have to purchase more expensive lenses and/or

A diamondback rattlesnake in full strike. Canon 5DMII, 24-105mm at 40mm, ISO 400, f/18 @ 3 sec., three remote-controlled high-speed flashes, Phototrap infrared trigger.

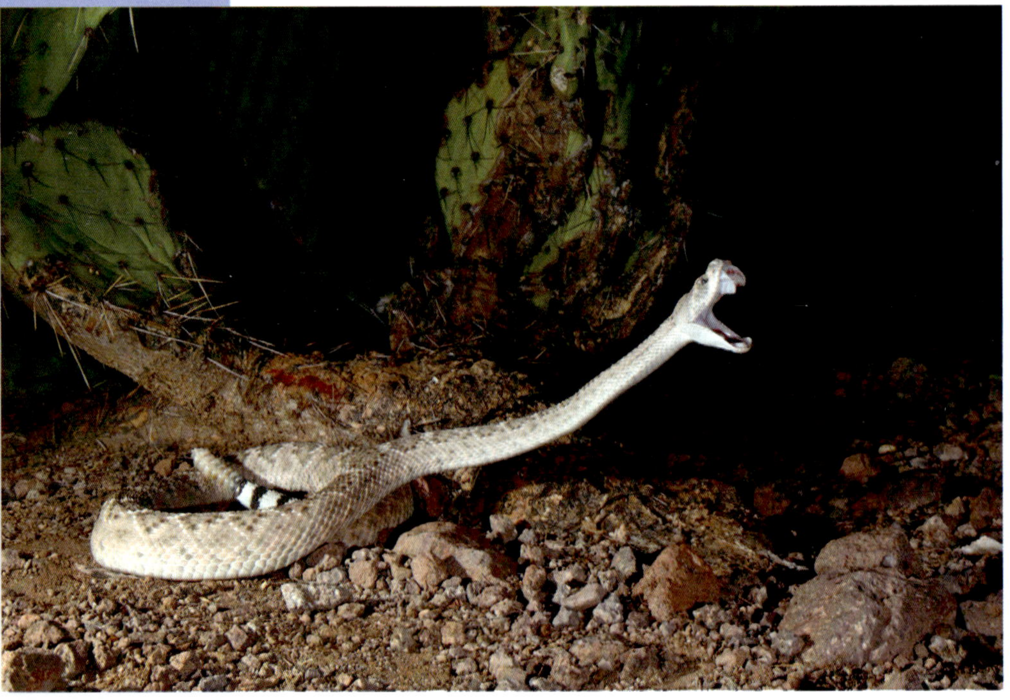

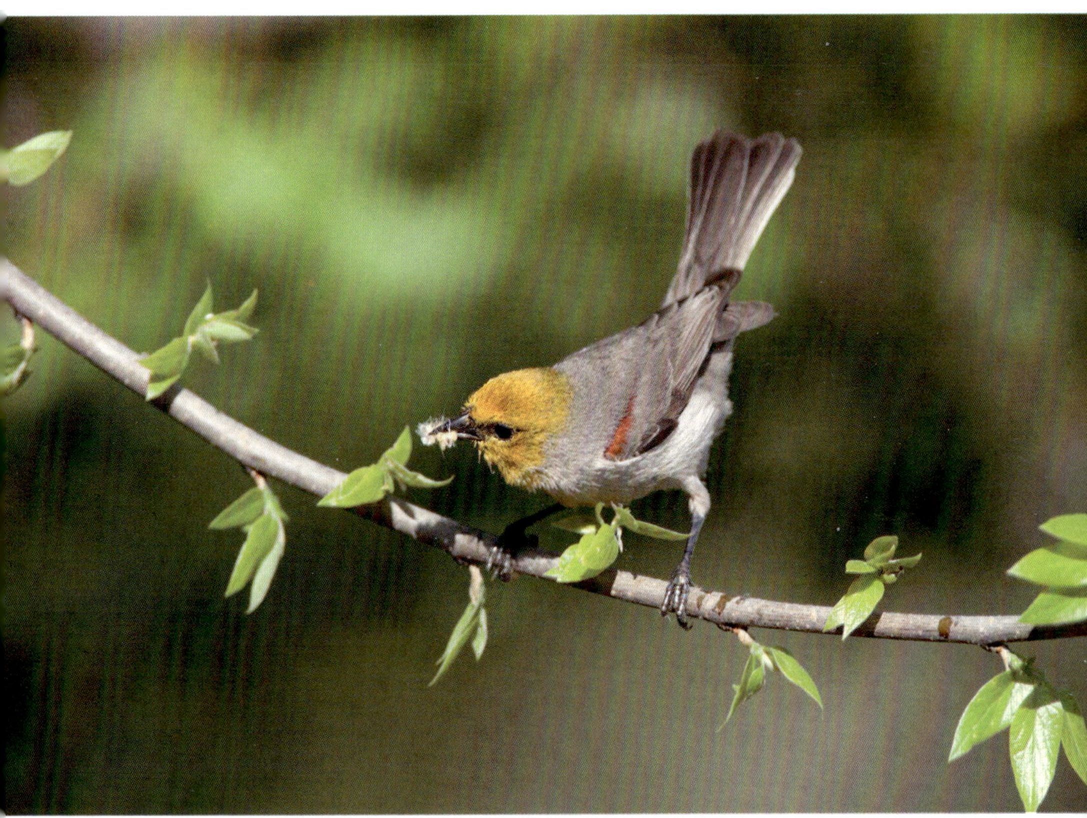

Verdin gathers food for its babies. Canon 50D, 500mm, 1.4x teleconverter, ISO 400, f/7.1 @ 1/1250 sec.

get closer to your subject to obtain the same magnification.

When selecting a camera for wildlife photography, look for the following optional features:

1. Live view
2. AI Servo (or Continuous Focus) capabilities
3. Mirror lockup
4. A buffer rate of at least 12 images when using relatively fast memory cards
5. The ability to use a remote or wireless cable release

Hands down, the most useful lens for wildlife photography is the 400mm lens (which is available from camera manufacturers and various third-party brands in zoom and fixed focal lengths). This long focal length allows the photographer to take photographs of flying birds, fighting bears (without getting too close!), hummingbirds, butterflies, rattlesnakes, bugling elk, and almost every other critter you might find. On a full-frame camera, the 400mm lens translates to a magnification of 8x (similar to most binoculars). To increase this magnification even more, couple your lens with a 1.4x or 2x teleconverter (which will increase the effective focal length of your 400mm lens to 560mm or 800mm, respectively).

To use a 400mm lens for macro photography, add a teleconverter, close-up lens, or extension tubes to allow the lens to provide a closer focal distance from the subject. Or, better yet, take advantage of close-focusing telephoto zoom lenses.

Whether photographing birds, large mammals, reptiles, amphibians, or fish, photographers need to hold their camera steady and be ready for action. Cheap, wobbly tripod legs and difficult-to-use support heads can lead to disaster (i.e., bad images). Look for carbon fiber tripod legs,

Preparing for Your Photo Shoot

General Gear Checklist

Birds and Big Game:

- ❑ Camera
- ❑ Lenses ranging from 70mm to a 500mm
- ❑ 1.4x and 2x teleconverters
- ❑ Tripod (without a center column)
- ❑ Ball head
- ❑ Wimberley gimbal tripod head
- ❑ Flash
- ❑ Better Beamer flash extension
- ❑ Camera rain gear (if rain threatens)
- ❑ Bean bag
- ❑ Memory cards
- ❑ Extra charged batteries
- ❑ Camera backpack
- ❑ Photo blind

Lizards and Small Mammals:

- ❑ Camera
- ❑ Lenses ranging from 24mm to 200mm
- ❑ Macro lens
- ❑ Extension tubes
- ❑ 2x teleconverter
- ❑ Tripod (without center column)
- ❑ Ball head
- ❑ Off-camera flash
- ❑ Flash bracket
- ❑ Flash stands with flash-sized ball heads
- ❑ Bean bag
- ❑ Reflector
- ❑ Diffuser
- ❑ Camera backpack
- ❑ Specialized items, such as snake hooks, tarps to lie on, rubber boots, etc.

Macro:

- ❑ Camera
- ❑ Macro lenses
- ❑ Extension tubes
- ❑ Tripod (without a center column)
- ❑ Ball head
- ❑ GorillaPods
- ❑ Clamps for GorillaPods
- ❑ Off-camera flash
- ❑ Macro twin light flash
- ❑ Macro flash bracket
- ❑ Flash diffusers
- ❑ Focusing rail
- ❑ Hoodman Hoodloupe
- ❑ Artificial backdrops
- ❑ Reflector
- ❑ Diffuser
- ❑ Tent diffuser
- ❑ Various clamps
- ❑ Card table
- ❑ Camera backpack

Night:

- ❑ All of the equipment for lizards and small mammals
- ❑ Headlamps and/or flashlights (preferably with a red-filter)
- ❑ Phototrap infrared camera trigger

Banner-tailed kangaroo rat near Oracle. Canon 5DMII, 70-200mm at 200mm, ISO 400, f/13 @ 1 sec., off-camera flash @ -1 FEC.

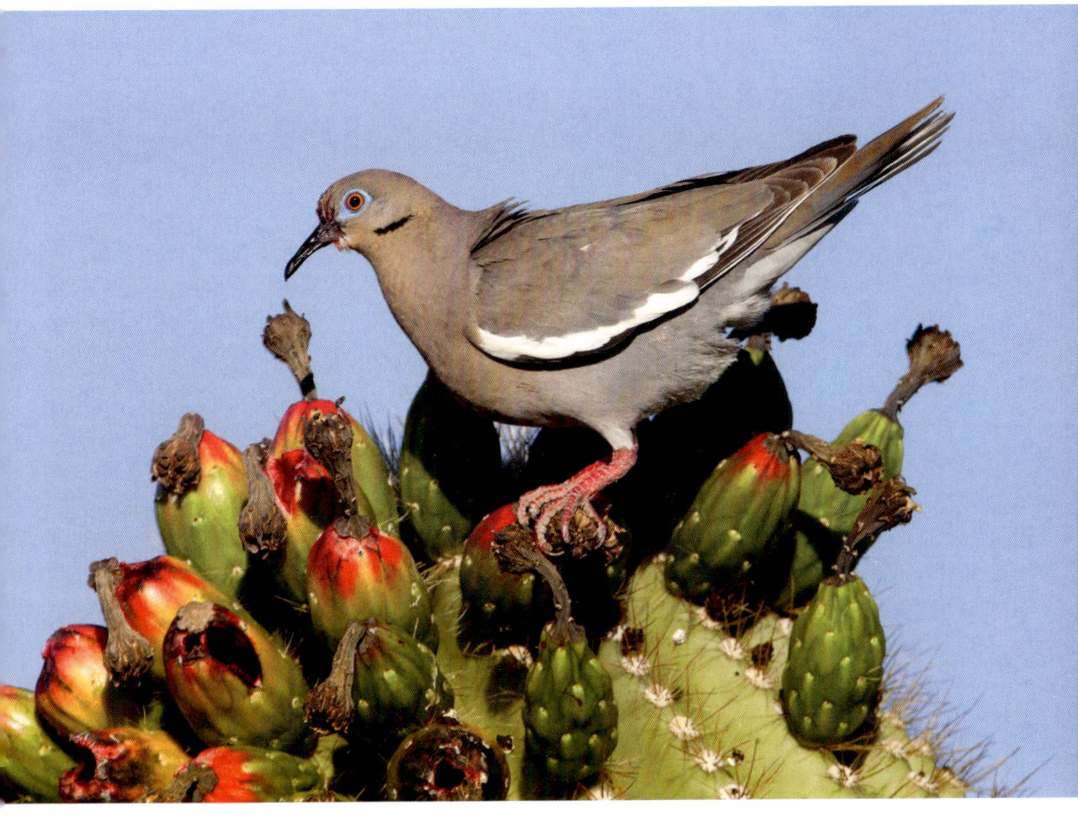

By pairing a long lens with a teleconverter, I obtained a compressed perspective of a white-winged dove feeding on saguaro cactus fruit. Canon 40D, 500mm, 1.4x teleconverter, ISO 200, f/10 @ 1/320 sec.

which are lighter, stronger, and more resistant to harsh environments (especially salt water) than their aluminum counterparts. Purchase legs rated to hold twice the weight you expect to use (for example, if your telephoto lens and camera weigh 12 pounds (5.4 kg), then use carbon fiber tripod legs rated to hold at least 24 pounds (10.9 kg)). In addition, find a set of tripod legs without a center column. Center columns, when raised above the top of the legs, only offer one support point, which makes it more difficult for the tripod to keep the camera steady, especially with long, heavy lenses. If you purchase tripod legs that extend to a height of at least 62 inches (157.5 cm) or greater, a center column is unnecessary.

Only two types of tripod heads support the types of activities a wildlife photographer will encounter—a ball head and a gimbal head (unless you are shooting video, forget the tilt and pan heads). A ball head features separate knobs for setting the tension, loosening the entire ball, and panning. Manufacturers also make gimbal heads that allow your camera and lens to sit in perfect balance and easily move in both horizontal and vertical directions. As with tripod legs, get a ball head rated to support twice the weight you expect to use.

Finally, I have a rule of thumb about tripod heads: if it takes more than a few brain cells and greater than one second to operate, throw it away! The last thing you want to do is use two hands and take precious time getting ready to take a photograph (as the decisive moment with your animal subject simultaneously occurs).

Before you travel to a location, first consider the type of photographs that you might take, and then pack only what is necessary. To determine what gear to bring, consult the Gear Checklist on the left and consider which of the four wildlife photography situations you might encounter.

Caution!

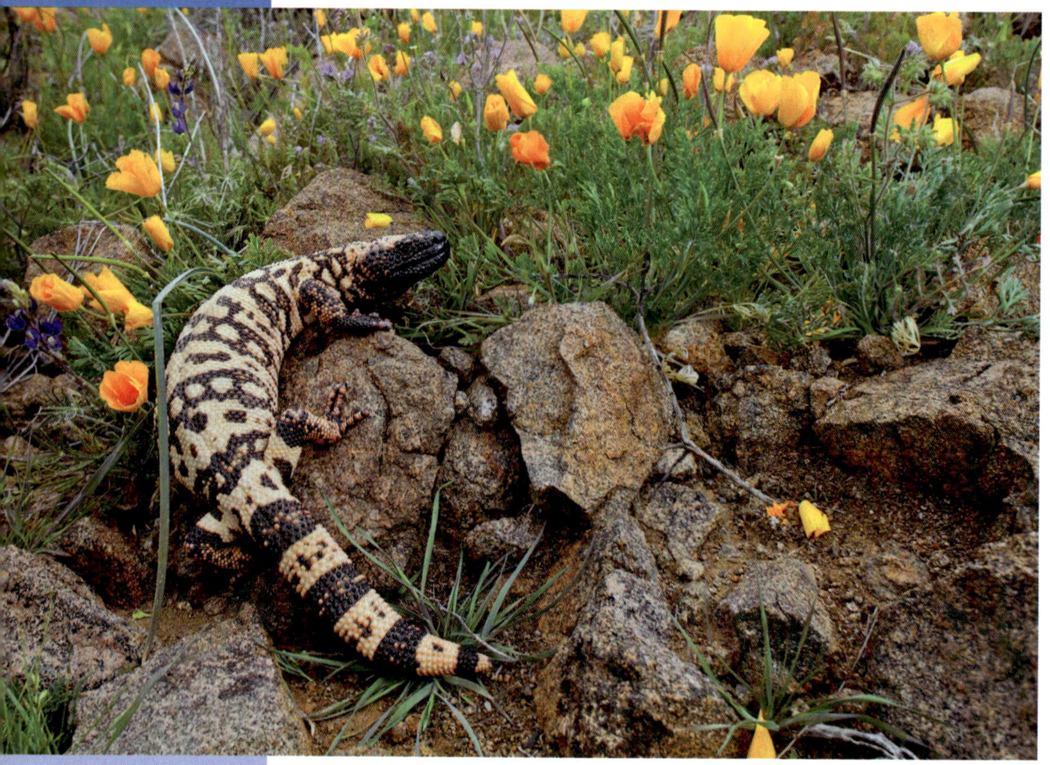

Gila monster searches for quail eggs in a field of poppies. Canon 5DMII, TS-E 24mm, ISO 400, f/11 @ 1/125.

Dangerous Quarry

Some wildlife can be dangerous and might (given an opportunity) attack and injure the unwary photographer. The most dangerous photographic subjects include elk, bison, deer, bear, mountain lion, bobcat, coyote, and rattlesnakes. Angry elk or bison will attack with the intent to do bodily harm, and a mountain lion or bear can be deadly. A close encounter with a coyote or bobcat could leave you with a nasty, or worse yet, rabid, bite. Follow the rule of thumb, "If you think you are too close, then you probably are." Let common sense prevail and use the equipment I suggest in this book. For instance, when photographing elk or bison, I suggest using a telephoto lens and not a wide-angle lens! Keep your distance not only to keep safe, but also to take an image of wildlife behaving naturally and not affected by your proximity.

Many species of wildlife carry pests or parasites on or in them. For example, the lowly flea carries the plague (yes, the black plague) as it sits on the backs of prairie dogs or other unlucky mammals. Deer mice can transmit hantavirus, a specialized, sometimes lethal, virus. Many mammals can carry rabies. More great reasons to keep your distance!

Things that Poke, Sting, and Bite

Arizona is home to a variety of cacti, scorpions, bees, wasps, poison ivy, biting ants, snakes, and other potentially dangerous and annoying things.

The sting of a few species of scorpions is poisonous to the point that the health of infants and the elderly is at risk. Some humans are extremely allergic to bee stings. Do not disturb rattlesnakes as they are unpredictable and grumpy. The venom from a rattlesnake bite, although not often fatal, can cause severe damage to skin and muscle tissue.

Always watch where you are stepping, placing your hands, lying or sitting down, or setting up a blind. Be careful when you pick up discarded equipment or move a piece of wood or vegetation used as a prop.

Watch Your Step

In order to take great images, photographers may have to crawl over boulders, walk across shifting desert sands, sneak around at night in cactus-infested places, wade in knee-deep muddy water, or climb to the top of steep mountains. Just traversing many of these places can be a chore, but add cameras, tripods, blinds, etc., and getting from one place to another can be downright difficult.

Before you go into the field, research the area. If possible, take a friend to enjoy the trip and provide a little bit of a safety net. Always tell someone the locations of your travels and take a cell phone with you—do not leave it in the car.

Weather

Before embarking on your outing, check with local and national weather stations and websites. Inches of snow can fall in a few hours, making many of Arizona's roads impassable. During the winter months, always take an extra blanket, a source of heat, plenty of food and water, and a method of communicating with the authorities or family. If you find yourself caught in one of these winter storms, authorities recommend staying with your vehicle.

Annual monsoon thunderstorms are unpredictable and dangerous. Monsoon thunderheads expel great quantities of rain over a very short time frame. The ground cannot absorb all of the water, so normally dry desert washes can fill with water in a flash. Stay out of washes during these monsoon events, even if it is not raining where you are. If you are downstream from a major monsoon rainstorm, expect the washes to fill with water.

Arizona-Mexico Border

Both illegal immigration and smuggling continue on the land adjacent to the border between Arizona and Mexico. When traveling through southern Arizona, expect to see U. S. Border Patrol vehicles and to pass through a few highway checkpoints. Be careful and check with locals to see if any current issues exist.

Staying Hydrated

Arizona is a hot and dry state, so drink plenty of water. It takes at least two gallons of water to replace what a person loses in the heat of the day during moderately strenuous activity.

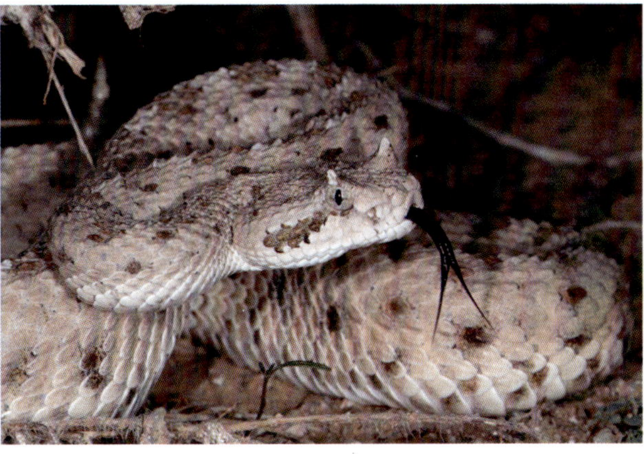

Sidewinder rattlesnake uses his forked tongue to taste the air. Canon 50D, 100mm macro, ISO 200, f/16 @ 1/250, on-camera flash.

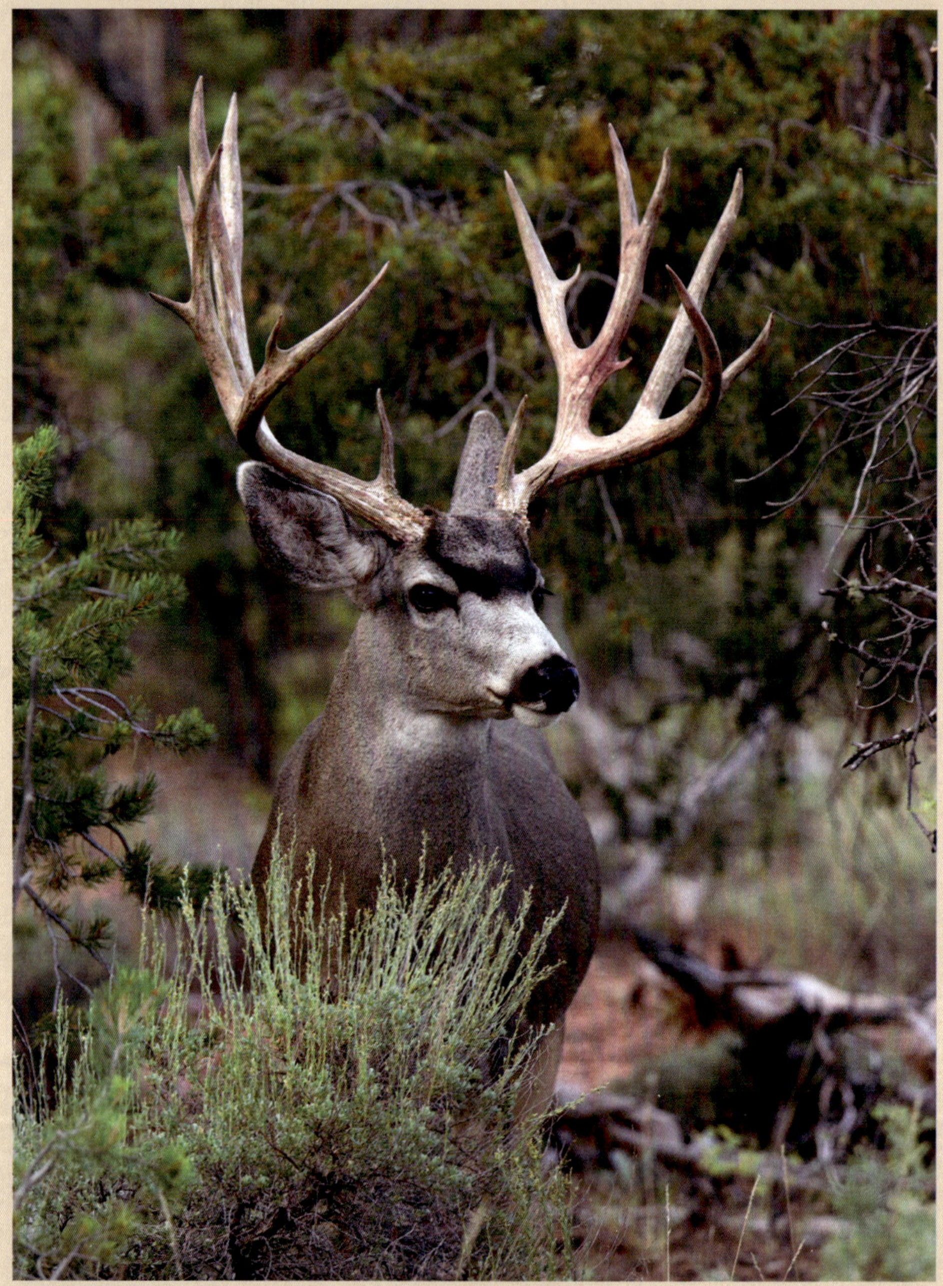

Rocky Mountain mule deer on the South Rim of the Grand Canyon National Park. Canon 1DMIV, 70-200mm at 165mm, 1.4x teleconverter, ISO 1250, f/5 @ 1/400 sec.

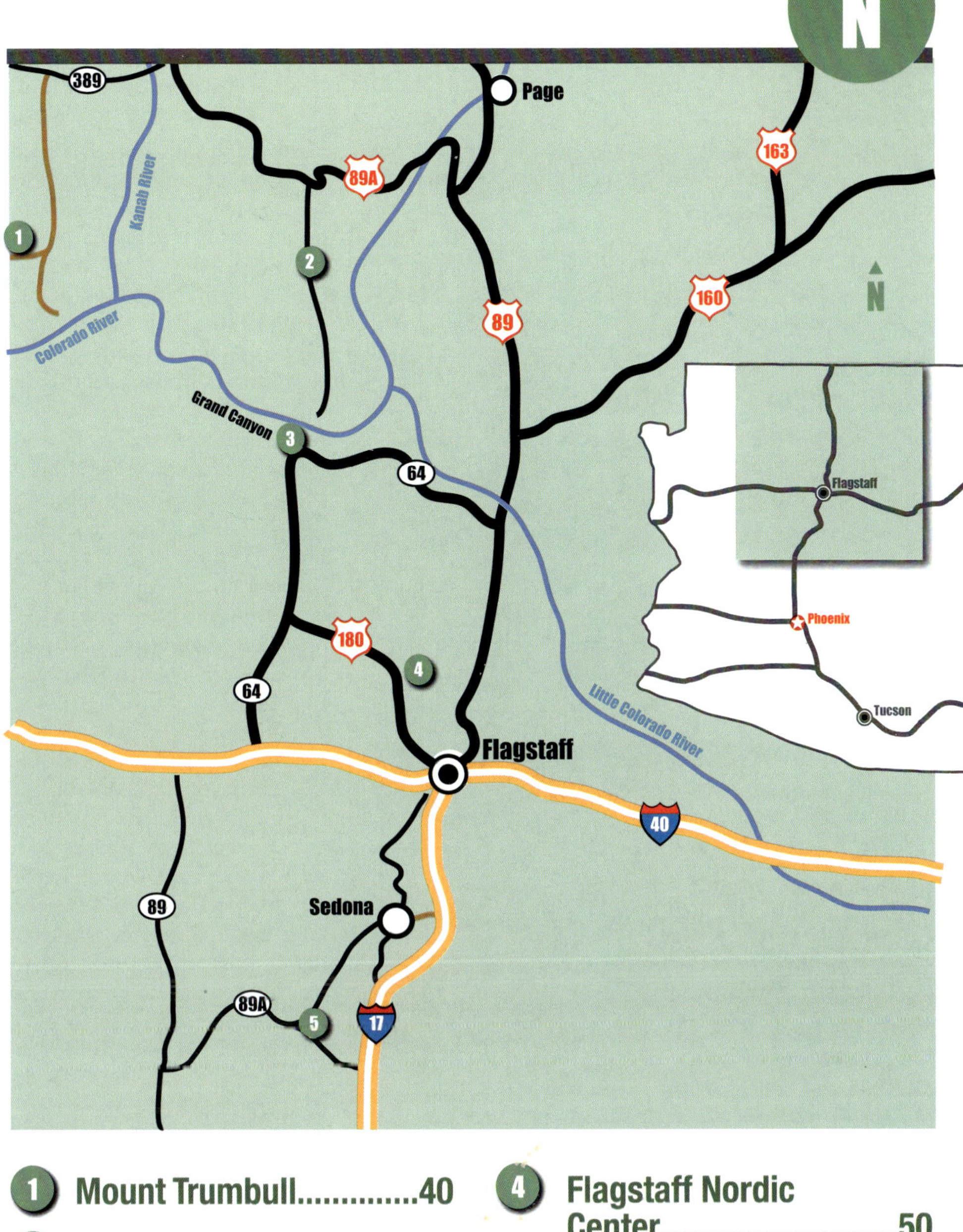
N
389
Page
163
89A
Kanab River
1
2
160
89
Colorado River
Grand Canyon
3
64
Flagstaff
Phoenix
Tucson
180
4
64
Little Colorado River
Flagstaff
40
89
Sedona
89A
5
17

NORTHERN ARIZONA

Mount Trumbull

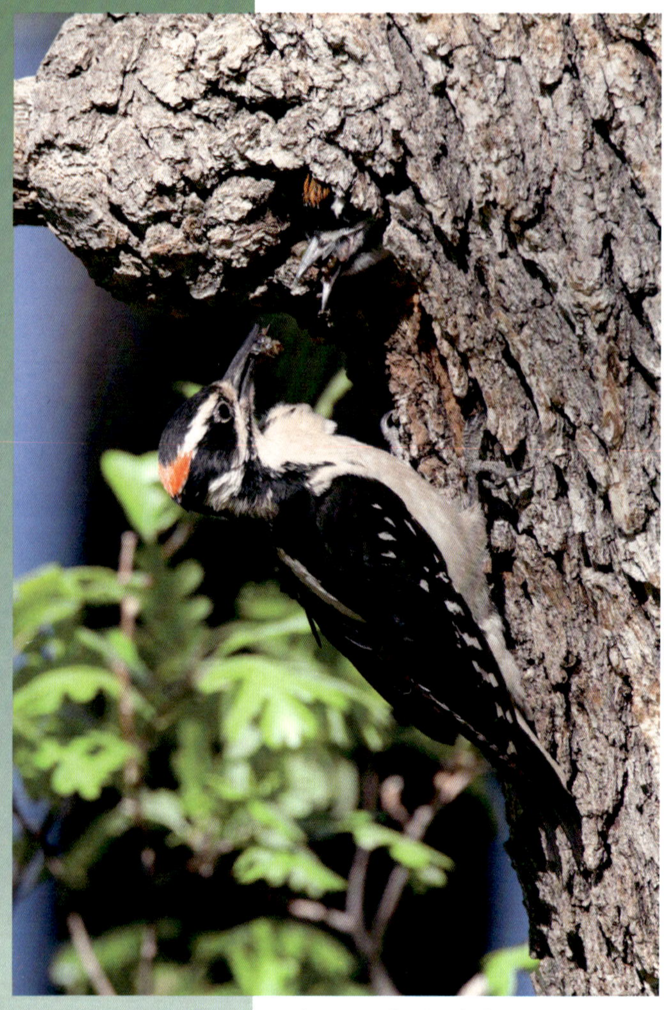

A hairy woodpecker feeds its young in the cavity of an oak tree. Canon 50D, 500mm, 1.4x teleconverter, ISO 400, f/10 @ 1/320 sec.

Tucked far away in the Arizona Strip in the northwestern corner of the state, Mount Trumbull is known as the premier place to find large mule deer, traditional cattle ranching, and huge expanses of ponderosa pine. With habitats ranging from pinyon-juniper woodland to mixed conifer forests, this portion of the state sees a bird diversity found in few other locations and thus also offers one of the best places to photograph cavity-nesting birds in Arizona. During the breeding season, shutterbugs can easily photograph woodpeckers, bluebirds, flickers, sapsuckers, and many others. After arriving, use powerful binoculars to search the oak thickets for bird activity. Find their hang out spots, but do not disturb the birds.

Once you have located a nest, observe which branch the birds land on before they take food to the nest. Set up a blind so that you can photograph them as they go to their perch or as they rest next to the nest. Many of the birds in the oak thickets are cavity nesters and the holes are only a few feet from the ground. Harsh shadows may appear in the dense understory, so photograph a well-lit bird in front of a clean, dark background.

Use continuous autofocusing locked on the bird's eye to ensure sharp focus. Use a **telephoto lens** with a **tripod** to hone in on your subject. Either use a **flash extender system** or place an **off-camera flash** outside the blind to add fill light to the scene. Young birds in the nest will frequently stick their heads out of the hole in anticipation of the adults bringing food—a great time to take a great image.

Because photographing around Mount Trumbull can be a multi-day adventure, set up a comfortable camp near the Nixon Administrative Site. Fill up with gas, ice, food, and water in Colorado City before you begin this adventure.

VIEW TIME
April to July

IDEAL TIME OF DAY
Sunrise to sunset

VEHICLE
4WD high-clearance

HIKE
Easy to moderate

1

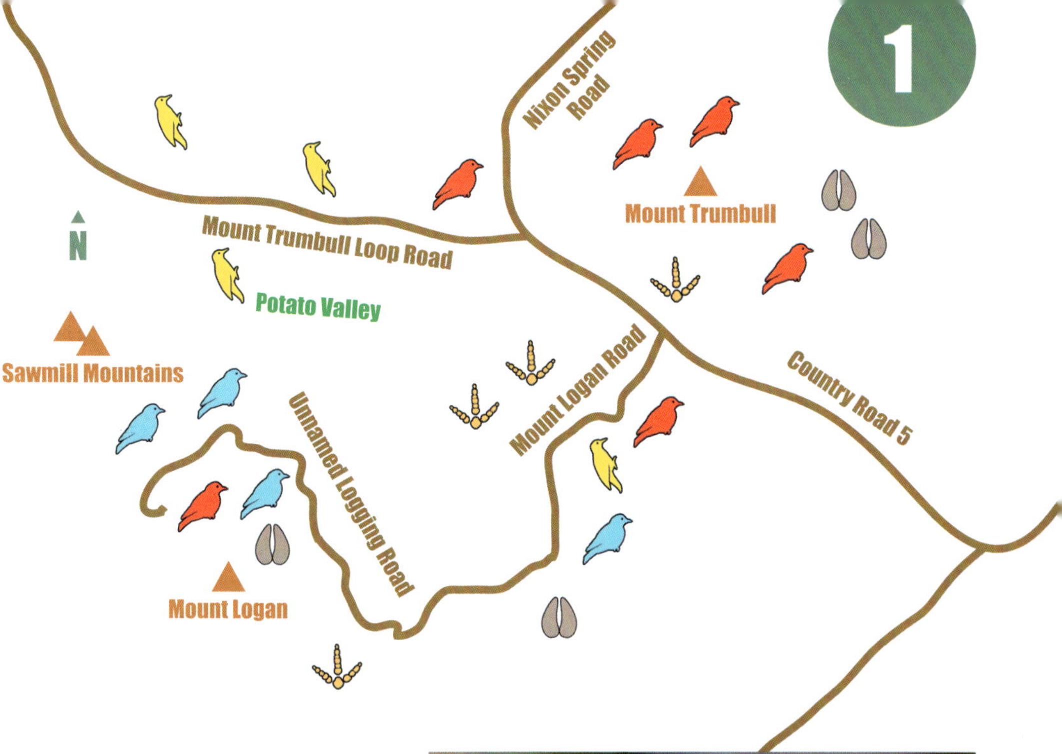

Nixon Spring Road
Mount Trumbull
Mount Trumbull Loop Road
N
Potato Valley
Sawmill Mountains
Mount Logan Road
Country Road 5
Unnamed Logging Road
Mount Logan

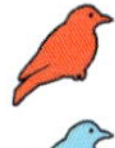 **Songbird**

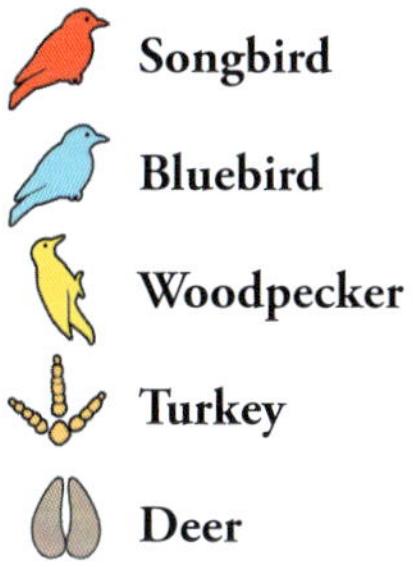

 **Bluebird**

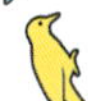 **Woodpecker**

 **Turkey**

 **Deer**

DIRECTIONS:

From Colorado City, follow AZ 389 south for 3.5 miles (5.6 km). Turn right onto County Road (CR) 5. Travel 37.2 miles (59.9 km) to where CR 5 joins Mount Trumbull Loop Road. Continue on CR 5/Mount Trumbull Loop Road for 15 miles (24.1 km) to the Mount Logan Road turnoff. At the turnoff, begin looking for a good camping site.

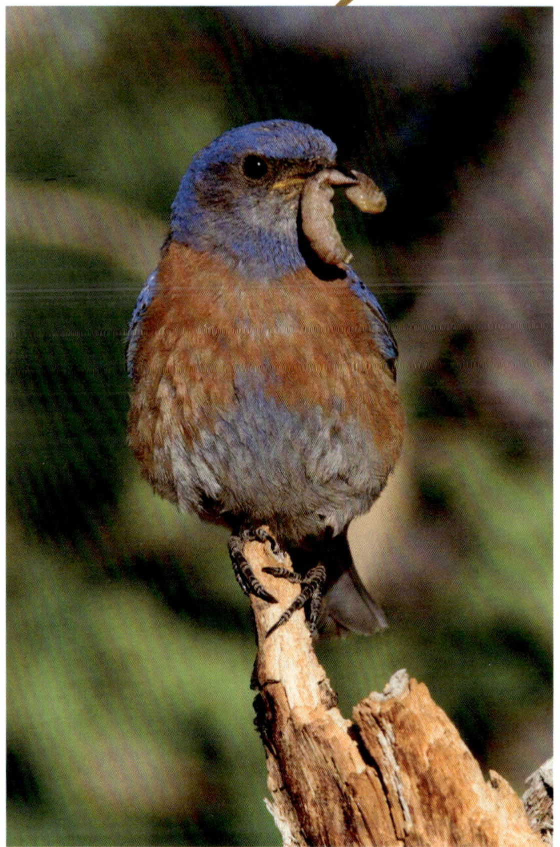

Western bluebird prepares to carry a grub to its hungry young on Mt. Trumbull. Canon 20D, 500mm, ISO 400, f/9 @ 1/640 sec.

Grand Canyon Highway

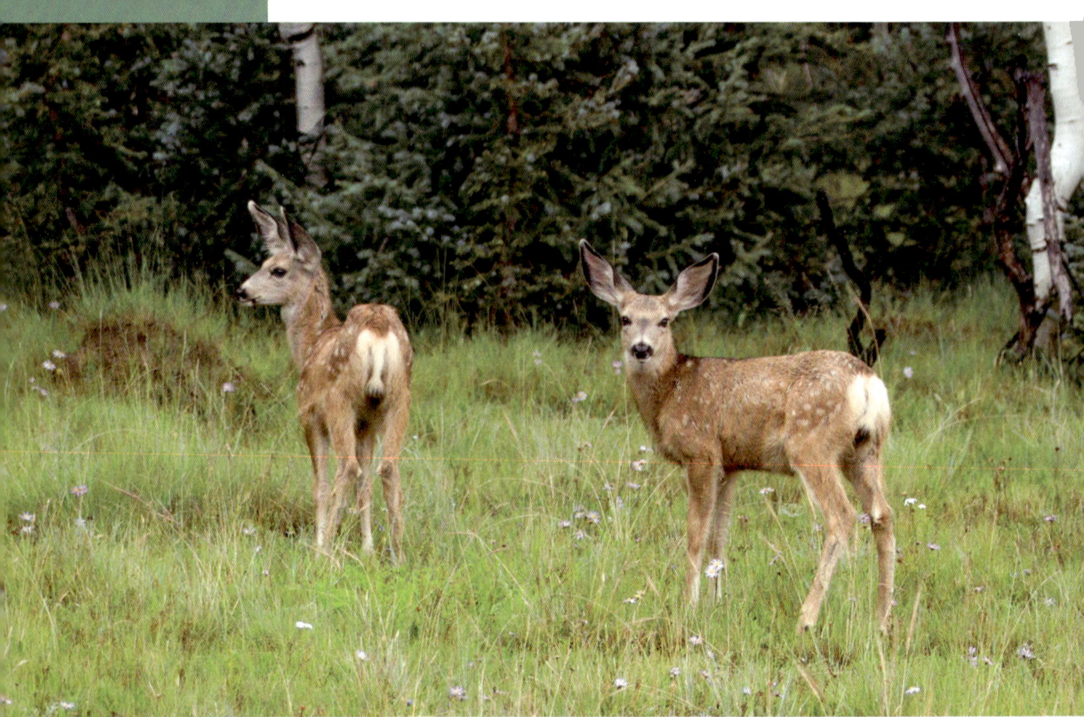

Rocky Mountain mule deer fawns near the roadside along Highway 67. Canon 7D, 70-200mm at 165mm, 2x teleconverter, ISO 400, f/5.6 @ 1/200 sec.

VIEW TIME
May to October

IDEAL TIME OF DAY
Sunrise to sunset

VEHICLE
Any

HIKE
Easy to moderate

Traveling south from Jacob Lake to the North Rim of the Grand Canyon, the 44-mile-long (70.8-km) Grand Canyon Highway (Highway 67) traverses through ponderosa pine, spruce fir, and mountain meadows at nearly 8,000 feet (2,438 m) in elevation. One might believe they were in Montana or Wyoming and not the desert state of Arizona! American bison, a large population of Rocky Mountain mule deer, unique small mammals, and deep winter snows only add to the illusion of a more northern climate.

Along FSR 461 and west of the Kaibab Camper Village (just south of the town of Jacob Lake), a vigorous population of Uinta chipmunks scurries about and feeds off of abandoned food in the campground. Restricted to the Kaibab Plateau, these cute little rodents are relatively easy to approach, but are best photographed from your vehicle. Place your **longest telephoto lens** on a **cropped frame camera** (or utilize a **1.4x** or **2.0x teleconverter** with a **full-frame camera**). Turn image stabilization on while balancing the lens on a bean bag on the car window sill.

Then, shoot in bursts of five or so images in high-speed continuous shooting mode. When you initially press the shutter button, your finger causes some camera movement. As you keep the shutter button depressed, the camera absorbs your motion, resulting in less camera "shake" during the next few images. As such, some photographs in the series will appear sharper than others.

Return to the highway and continue south about 25 miles (40.2 km). Keep a lookout for American bison (buffalo). Hard to miss, these one-to-

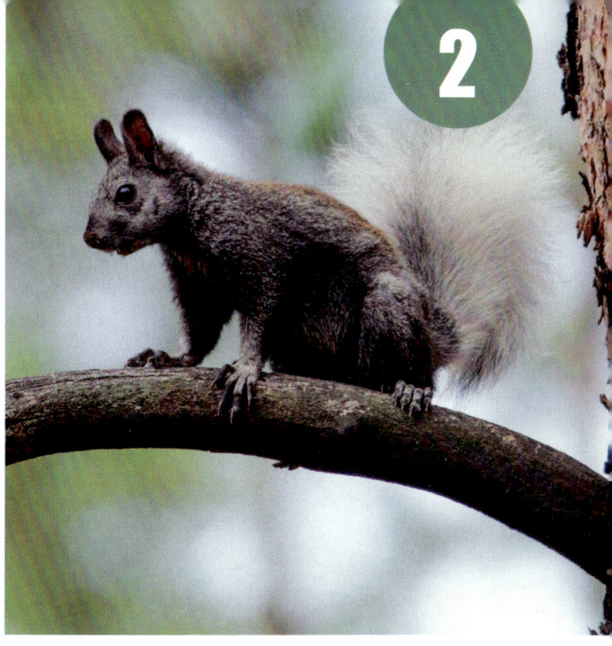

two-thousand-pound (454- to 907-kg) grazers travel in herds and feed on the succulent meadow grasses. Most bison simply keep their heads to the ground eating away at the grass and do not present a very photogenic subject. Look for mothers and calves interacting for the best images.

Although bison do not seem afraid of any predator on the Kaibab Plateau, they are one of the most dangerous animals in North America. Keep your distance, and stay in your vehicle while photographing if they come close to the road. After locating a herd, use a telephoto lens (and a tripod or bean bag for added stabilization).

Found exclusively in Arizona, the Kaibab squirrel perches on a limb of a ponderosa pine tree while showing off its unique black body and white tail. Canon 7D, 70-200mm at 285mm, 2x teleconverter, ISO 800, f/6.3 @ 1/250 sec.

While traveling the campgrounds and gravel roads, keep an eye out for the Kaibab squirrel. Unlike any other tassel-eared squirrel, this unique sub-population showcases a black body and a white tail. Most of the time, Kaibab squirrels on the ground will make a mad dash for the safety of a tree as soon as a vehicle slows down next to them. Once up in the branches, they must feel safe and invincible in the trees because, quite often, they have a habit of sitting on a low branch (and chirping at the photographer)—making them quite easy to see and photograph. To photograph these fast-moving creatures, use a telephoto lens set to a wide aperture, autofocus, and high-speed continuous shooting mode.

Rocky Mountain mule deer fawns step onto the scene starting in July. At daybreak and dusk, fawns and their ever-present mothers often intermingle near the road. Drive the highway, carefully observing the edges of the forest and meadow for their presence. Photograph these precious animals from the vehicle using a telephoto lens supported by a bean bag.

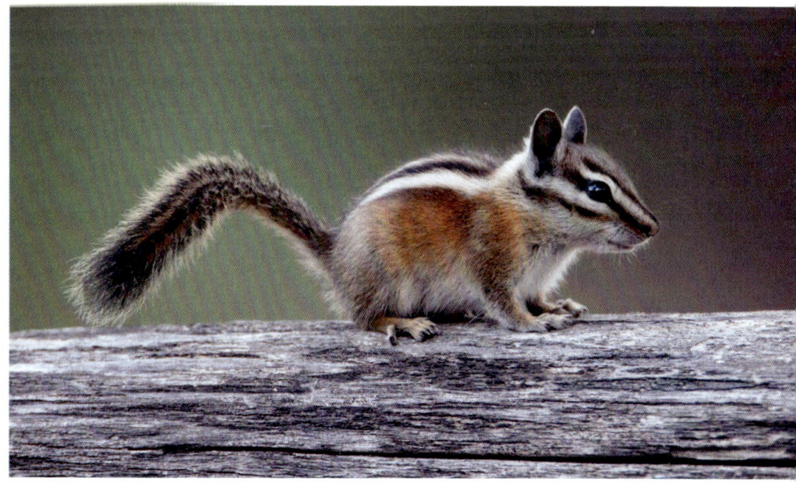

Uinta chipmunk at Kaibab Camper Village. Canon 50D, 500mm, 1.4x teleconverter, ISO 800, f/5.6 @ 1/200 sec.

Grand Canyon Highway

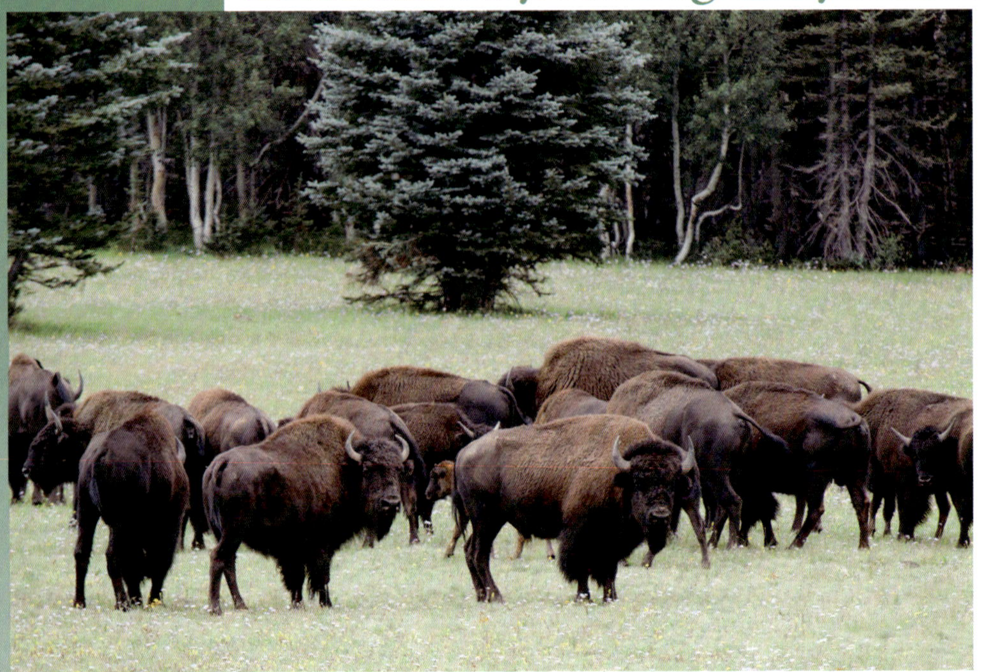

Bison herd grazes along Highway 67. Canon 50D, 100-400mm at 400mm, ISO 400, f/8 @ 1/200 sec.

PHOTO TIP 1

Rules of Composition

Centuries ago, well before the invention of photography, painters and other artists developed "rules" for composition. By following these artistic guidelines to effectively place your subject and supporting elements within your photograph, you can create an image that is more effective and visually moving.

One of the most common compositional aids, the Rule of Thirds, suggests the photographer keeps the primary subject out of the center of the frame. Instead of creating static and symmetrical balance, this rule advises a photographer to place the most important elements of an image off-center; specifically near the intersections of an imaginary tic-tac-toe grid, which divides the frame into nine segments. By placing an important part of the scene (horizon, animal, tree, clouds, etc.) at the intersection, or along the edge of, one of the imaginary horizontal and vertical lines, you create asymmetrical balance within your image. Asymmetrical balance brings the various elements together visually and leads to increase movement of the viewer's eye around the frame.

Although the Rule of Thirds will go a long way in assisting the photographer to create a great image, you should also consider the Rule of Space (or as I like to call it, "room to move"). This rule suggests you place more room in front of a subject than behind it, especially for subjects like wildlife that often appear animated. Even when the animal sits in repose, leave more room in the frame in front of the direction they look, as a viewer's eye will follow the direction of their gaze.

2

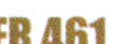

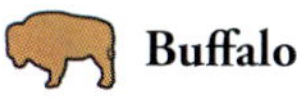

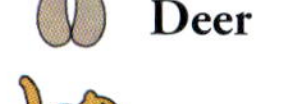

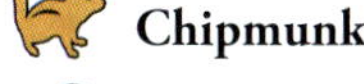

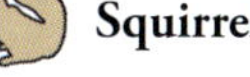

DIRECTIONS:

To find the campground with the Uinta chipmunks, travel 0.3 miles (0.5 km) south on AZ 67 from Jacob Lake and then turn west on Forest Service Road 461 for 0.8 miles (1.3 km) to Kaibab Camper Village. For the best bison and deer locations, travel south from Jacob Lake for 44 miles (70.8 km). By the first of November or following an early snowstorm, AZ 67 closes until the spring.

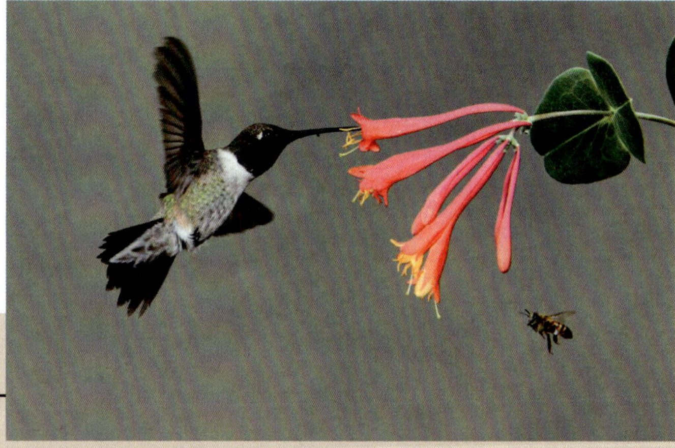

Black-chinned hummingbird approaches a flower for a little bit of pollen at the same time (and focal plane) as a honey bee. What luck! Canon 1DMIV, 100-400mm at 160mm, ISO 500, f/13 @ 1/250 sec., high-speed hummingbird set-up.

In addition to implying forward motion, photographers should give airborne animals a little more space below them than above. For example, when photographing an owl in a tree, the image looks more realistic, and has more visual punch, if the photographer positions the owl in the upper third of the frame (so the bird has "room to fall"). Similarly, as you compose, position most terrestrial wildlife lower in the frame. Case in point, when photographing a pack rat, place the critter towards the bottom of the scene so it has "room to jump."

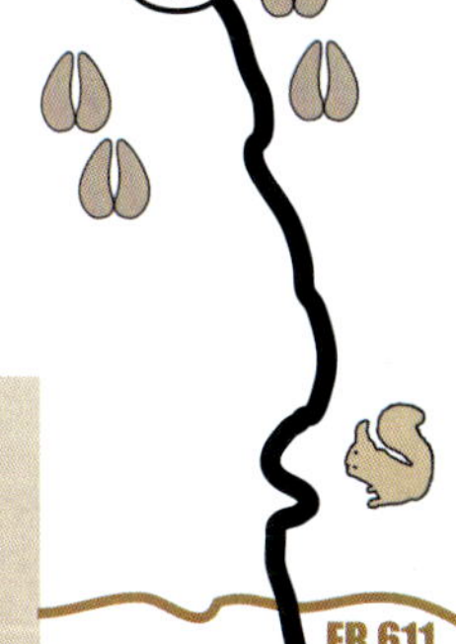

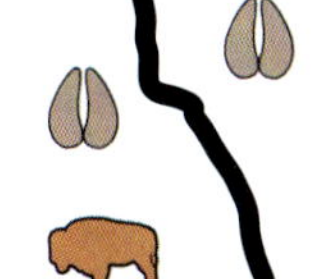

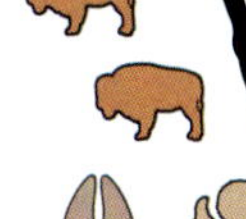

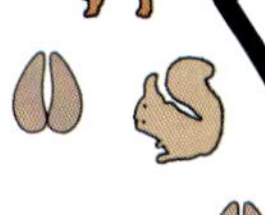

NORTHERN ARIZONA

SOUTH RIM
Grand Canyon National Park

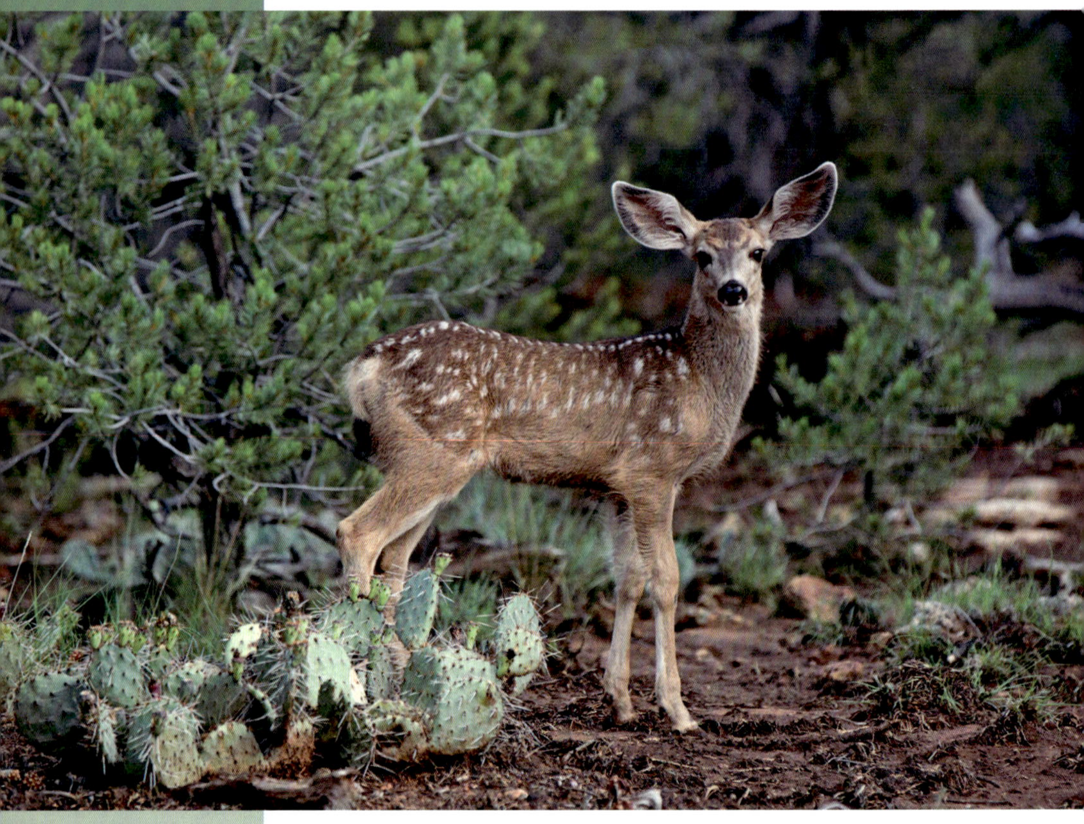

ABOVE: A newly-born Rocky Mountain mule deer fawn at the South Rim of the Grand Canyon. Canon 5DMIII, 70-200mm at 140mm, 2x teleconverter, ISO 800, f/5 @ 1/120 sec.
RIGHT: Rocky Mountain mule deer in a snow storm at Grand Canyon National Park. Canon 1DMIV, 70-200mm at 75mm, ISO 400, f/2.8 @ 1/4000 sec.

VIEW TIME
Year-round

IDEAL TIME OF DAY
Sunrise to sunset

VEHICLE
Any

HIKE
Easy to moderate

Although the Grand Canyon's incredibly beautiful vistas and enormous geological features are impossible to overlook, the Grand Canyon National Park offers even more to the wildlife photographer. With over 355 species of birds, 89 mammals, 47 reptiles, and 9 amphibians, the park's animal diversity would keep any shutterbug busy for years. However, three creatures stand out as outstanding photographic experiences: Rocky Mountain mule deer, Rocky Mountain elk, and the endangered California condor.

Primarily due to the five million visitors and non-hunting status of the park, mule deer and elk roam the oak and pinion forests seemingly oblivious to visitors and passing traffic—including photographers. This does not mean they are easy to photograph or not dangerous. Park rules make it illegal to approach or feed wildlife. In other natural settings, these rules would make it difficult to photograph these animals. However, at the canyon, most wildlife is accustomed to visitors so you do not need to

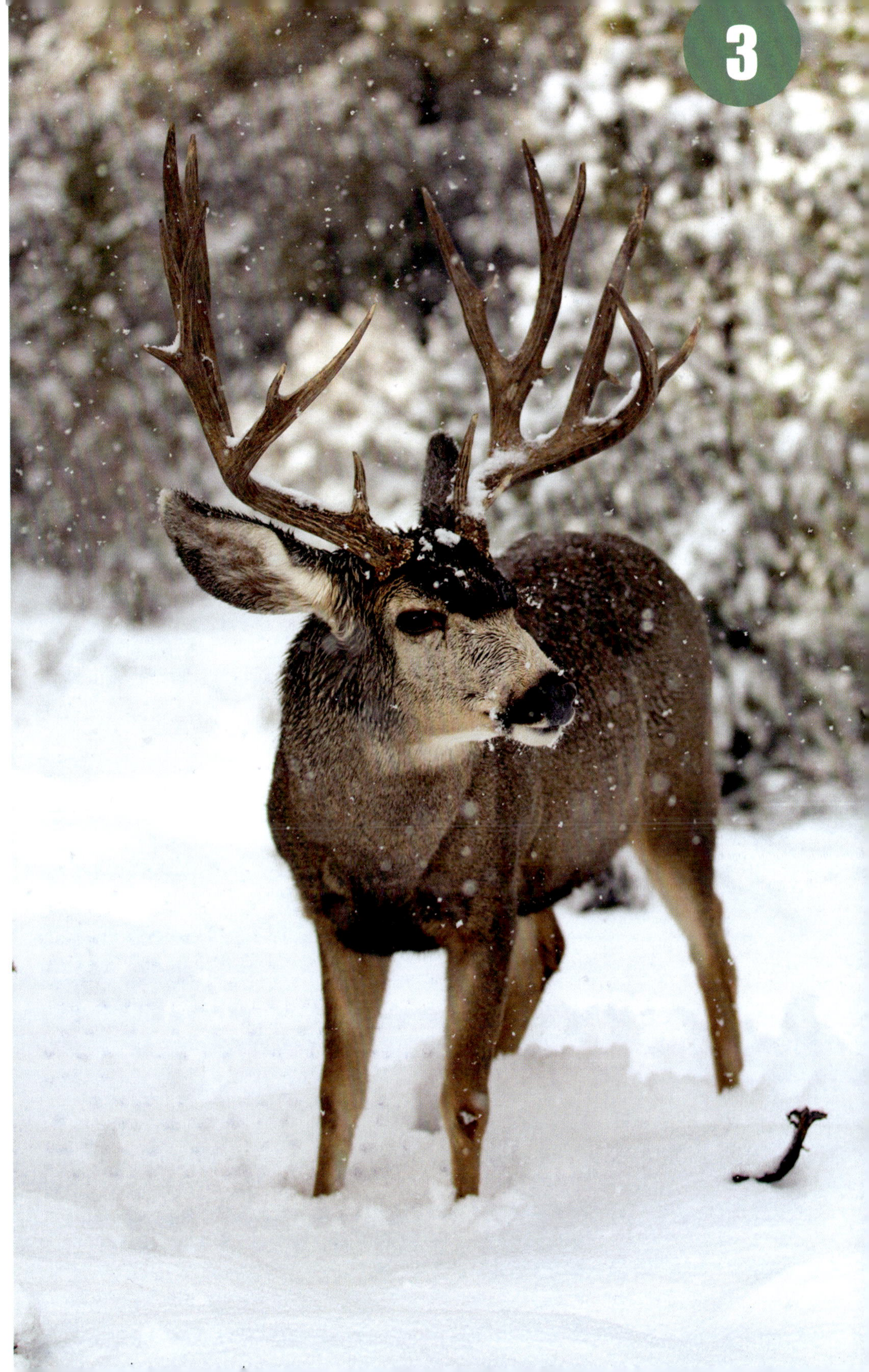
3

SOUTH RIM
Grand Canyon National Park

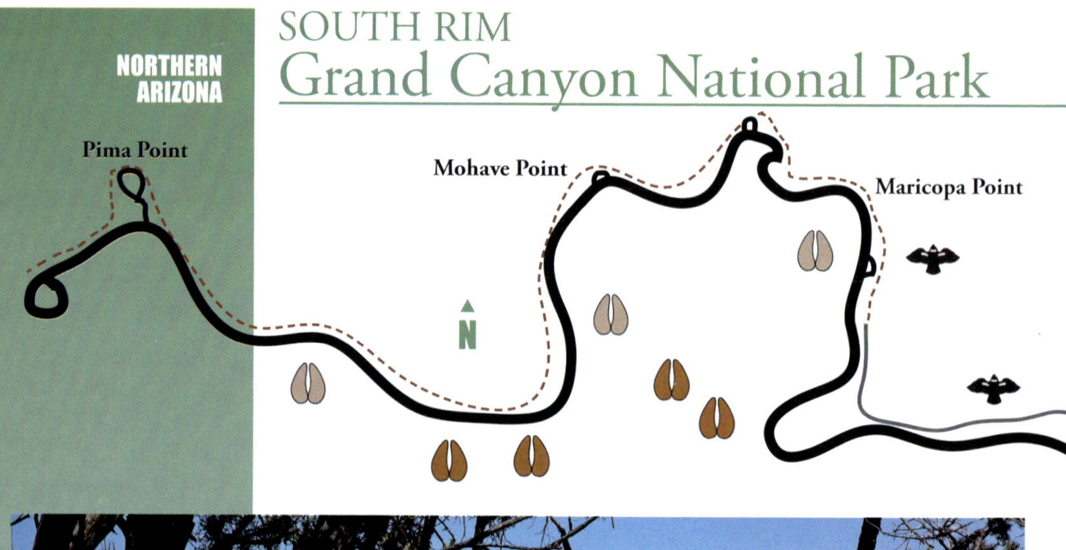

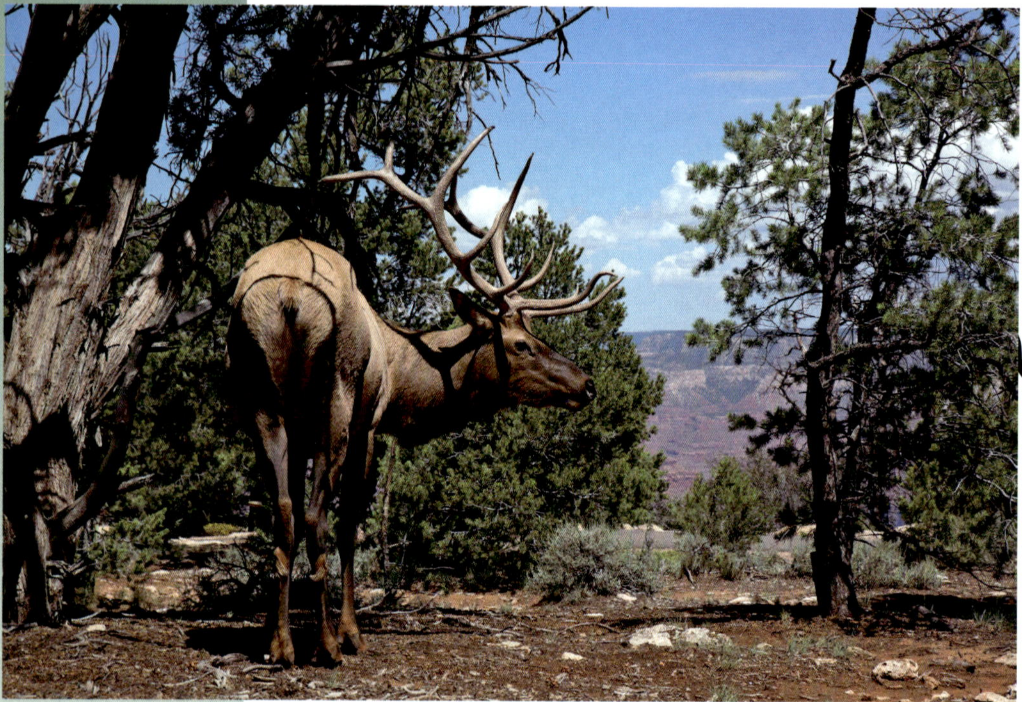

Rocky Mountain elk at the edge of the Grand Canyon. Canon 5DMIII, 24-105mm at 75mm, ISO 400, f/11 @ 1/800 sec.

"approach" wildlife in order to get close enough for a great image.

Rocky Mountain mule deer frequently browse between Mohave Point and Grand View Point. A particularly large group of deer hangs out at the intersection of South Entrance Road and Desert View Drive, while another concentration enjoys the area around Village East.

Drive along these park roads pre-dawn and scan for deer on either side of the road. When you locate them, find a legal pullout to park. Use a **telephoto lens** on a sturdy **tripod** with a **ball or gimbal head**. Take your time and allow the deer to go about their business.

Although easy to photograph at any time of the year, November and December offer the chance to see the deer in rut (breeding behavior).

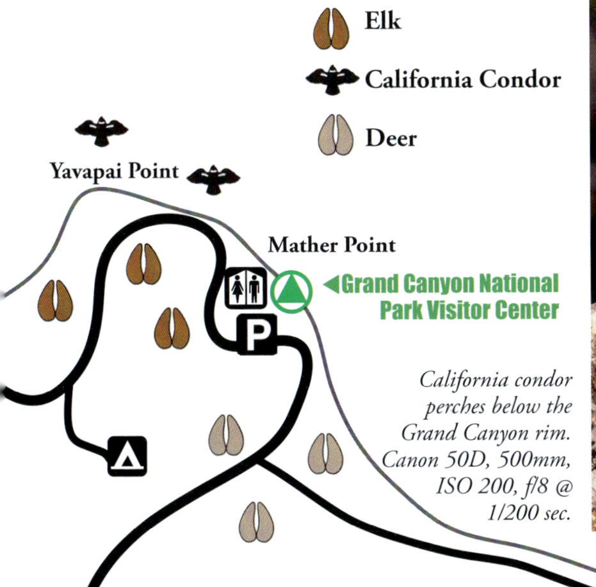

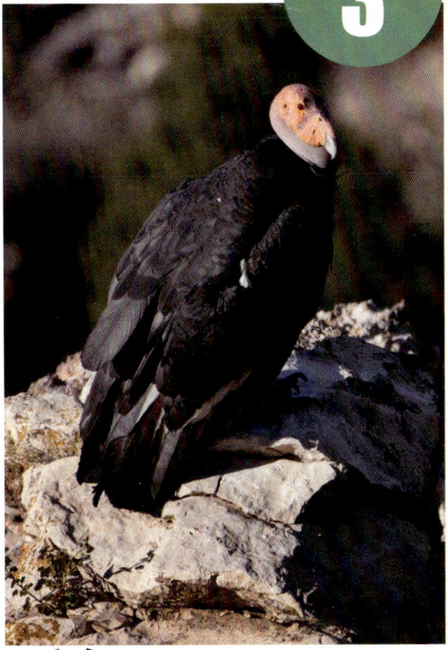

3

Elk

California Condor

Deer

Yavapai Point

Mather Point

Grand Canyon National Park Visitor Center

California condor perches below the Grand Canyon rim. Canon 50D, 500mm, ISO 200, f/8 @ 1/200 sec.

Grand View

180

64

DIRECTIONS:

From Flagstaff, travel 74 miles (119.1 km) north on Highway 180 to the entrance of Grand Canyon National Park.

Alternatively, from Williams (30 miles (48.3 km) west of Flagstaff on I-40), turn north on AZ 64 and travel 51.8 miles (83.4 km) to the southern park entrance.

The park charges an entrance fee. For information on the park visit **www.nps.gov/grca**.

During late June and July, fawns test out their new legs and are relatively easy to photograph. In winter, a fleeting snowstorm can transform the scene into a magical wonderland for the deer to pounce in the fluffy white stuff.

Begin looking for elk as soon as you enter the park's south entrance. Large herds dwell throughout the park from Hermit's Rest on the west end of the park to Desert View on the extreme east side. Drive slowly and search for this up to 700-pound (318-kg) herbivore. Do not get too close. Cow elk will aggressively defend their calves, while bulls may mistake you for a competitor and attempt to drive you from their territory. Therefore, stay at least 100 feet from elk, and use an appropriate telephoto lens for the image you want. Elk stay in the area year-round, but the rut (when the bulls bellow to attract females) occurs in September and October. During the rut, their behavior changes, and you may witness bugling, breeding, and general mayhem.

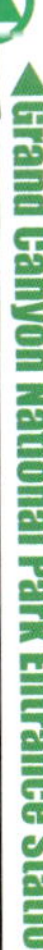

During the summer months, visitors can often see several condors either soaring in the updrafts of the canyon or roosting on the rock outcrops just below the Rim Trail. To find these special birds, look for crowds of visitors looking skyward, or ask one of the park rangers if there are any in the area. In a one- or two-day trip during the summer, you can expect to spot several condors. To record a flying or roosting condor, handhold a **400mm lens** on either a **cropped frame** or **full-frame sensor camera** (if shooting with a full-frame sensor camera, you may need a **1.4x** or **2.0x teleconverter** to fill your frame with the regal bird). Under sunny skies, start with an ISO speed around ISO 400 to keep your shutter speeds fast enough to freeze motion.

NORTHERN ARIZONA

Flagstaff Nordic Center

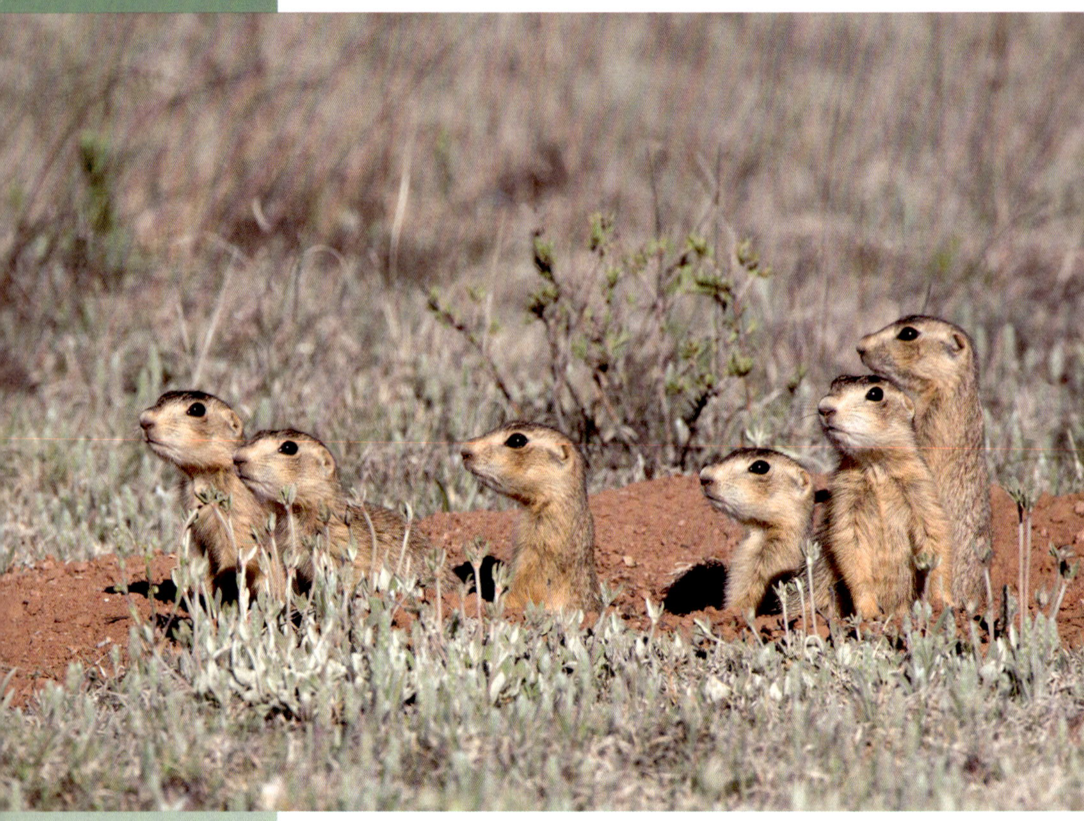

Family of Gunnison's prairie dogs peer from the safety of their den. Canon 50D, 500mm, 1.4x teleconverter, ISO 400, f/13 @ 1/800 sec.

VIEW TIME
May to October

IDEAL TIME OF DAY
Late morning to sunset

VEHICLE
Any

HIKE
Easy

Nestled among the ponderosa pine and aspen trees, the Coconino National Forest permits the Flagstaff Nordic Center to offer cross-country skiing and special events as a private concession. Within sight of the visitor's center, Gunnison's prairie dogs burrow in the soft soils surrounded by sagebrush. These dogs (which are related to North American ground squirrels) are accustomed to human traffic and often strike photogenic poses at close distances. Throughout the United States, prairie dogs are becoming rare, and this site gives photographers an opportunity to get close to a remnant of ancient Arizona.

Park at the visitor's center and listen; you will soon hear the squeaks of the adult prairie dogs as they communicate with the colony. Locate the active dogs, and set up your hunting blind about 20 yards (18.3 m) away and with the sun to your back. Sit down inside the blind, and quietly set up your **longest telephoto lens** on a sturdy **tripod**. Put the camera on high-speed continuous shooting mode and continuous autofocus.

Within approximately 30 minutes, the dogs will peer out of their burrows and give you a show. During late May and into June, the adorable young pups will begin to exit the burrow and frolic around the colony.

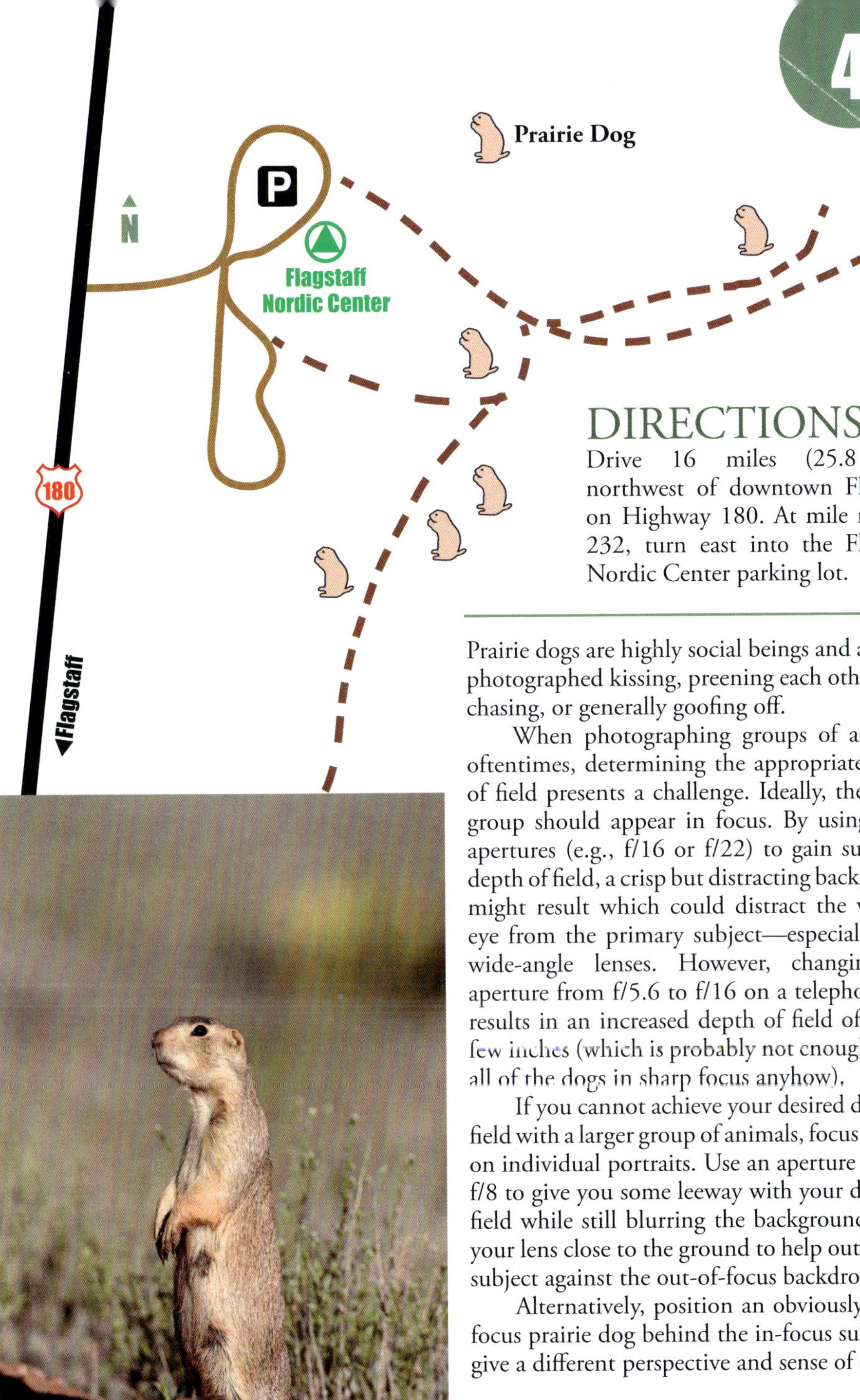

DIRECTIONS:

Drive 16 miles (25.8 km) northwest of downtown Flagstaff on Highway 180. At mile marker 232, turn east into the Flagstaff Nordic Center parking lot.

Prairie dogs are highly social beings and are best photographed kissing, preening each other's fur, chasing, or generally goofing off.

When photographing groups of animals, oftentimes, determining the appropriate depth of field presents a challenge. Ideally, the entire group should appear in focus. By using small apertures (e.g., f/16 or f/22) to gain sufficient depth of field, a crisp but distracting background might result which could distract the viewer's eye from the primary subject—especially with wide-angle lenses. However, changing the aperture from f/5.6 to f/16 on a telephoto lens results in an increased depth of field of only a few inches (which is probably not enough to get all of the dogs in sharp focus anyhow).

If you cannot achieve your desired depth of field with a larger group of animals, focus instead on individual portraits. Use an aperture around f/8 to give you some leeway with your depth of field while still blurring the background. Keep your lens close to the ground to help outline the subject against the out-of-focus backdrop.

Alternatively, position an obviously out of focus prairie dog behind the in-focus subject to give a different perspective and sense of depth.

NORTHERN ARIZONA

PAGE SPRINGS AND Bubbling Ponds Hatcheries

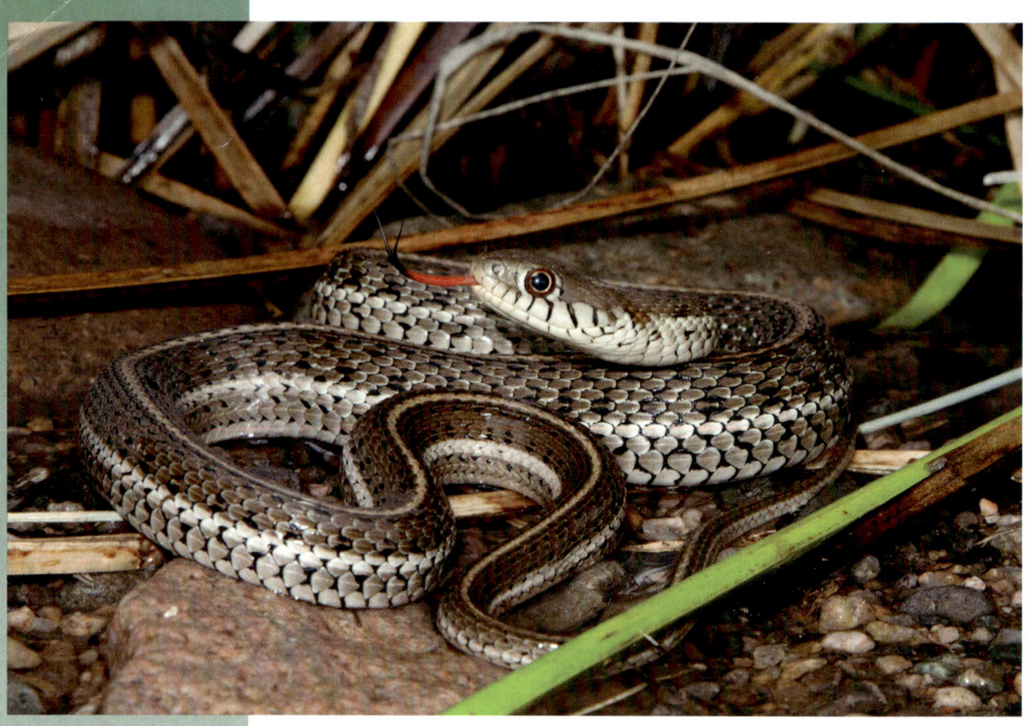

The Mexican garter snake is one of Arizona's rarest snakes. Canon Digital Rebel XTi, 24-105mm at 105mm, ISO 400, f/18 @ 1/200 sec., fill flash.

VIEW TIME
Year-round

IDEAL TIME OF DAY
Early morning to late afternoon

VEHICLE
Any

HIKE
Easy to moderate

Owned by the Arizona Game and Fish Department, this expansive 220-acre (89-hectares) riparian area houses a mix and diversity of species found in few places in Arizona. The property includes a huge cottonwood gallery, natural springs, and a section of Oak Creek which offer bird enthusiasts and photographers opportunities to view warblers, flycatchers, orioles, tanagers, herons, egrets, several species of hummingbirds, and ducks by the hundreds. Javelina, mule deer, raccoon, and the occasional river otter round out the mammals, while the rare Mexican garter snake is relatively abundant as well. During the summer months, loads of insects appear, including enough calling cicadas to dull ones ears.

Hike the well-maintained 1.5 miles (2.4 km) of developed nature trails. Carry either a **long telephoto zoom lens** or a **400mm, or longer, fixed telephoto lens**. Use either a **cropped frame camera** or a **teleconverter** to enlarge the subject within your final frame. Use a pop of **fill flash** to add a catch light to your subject's eye (see "Fill With Fill Flash" on page 88). If shooting in low light, bring a **tripod** along, despite the long walk. A good set of binoculars can help you find small birds in thick vegetation.

Set out in the early morning to take advantage of a peak activity period for both birds and mammals. During the spring, take your bird calls to entice your subjects to come a little closer. Although the birds and mammals here are somewhat accustomed to people, most are skittish and you will need to move slowly and quietly to get close.

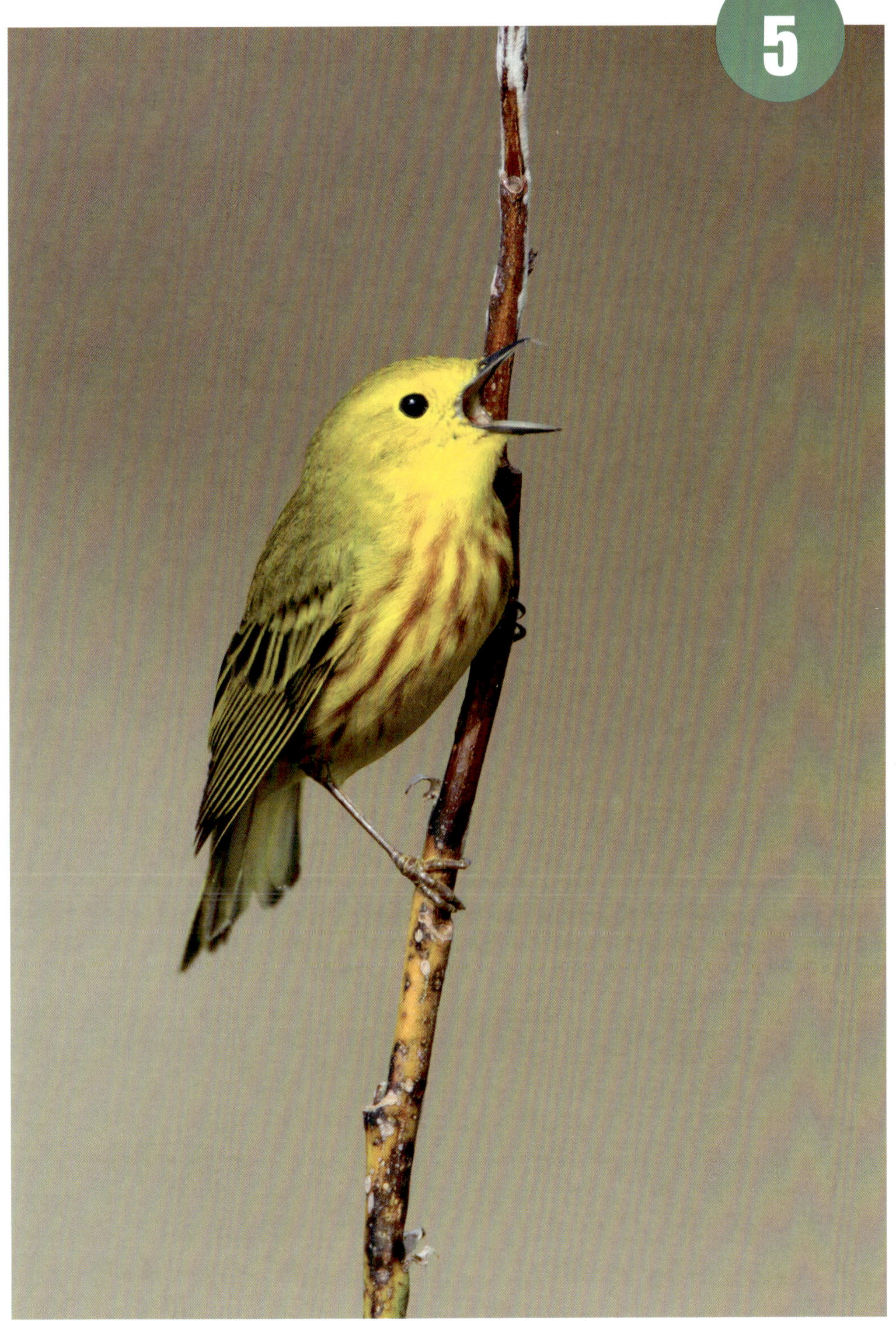

Yellow warbler sings to attract a female. Canon 7D, 500mm with 1.4x teleconverter, ISO 200, f/7.1 @ 1/500 sec.

PAGE SPRINGS AND Bubbling Ponds Hatcheries

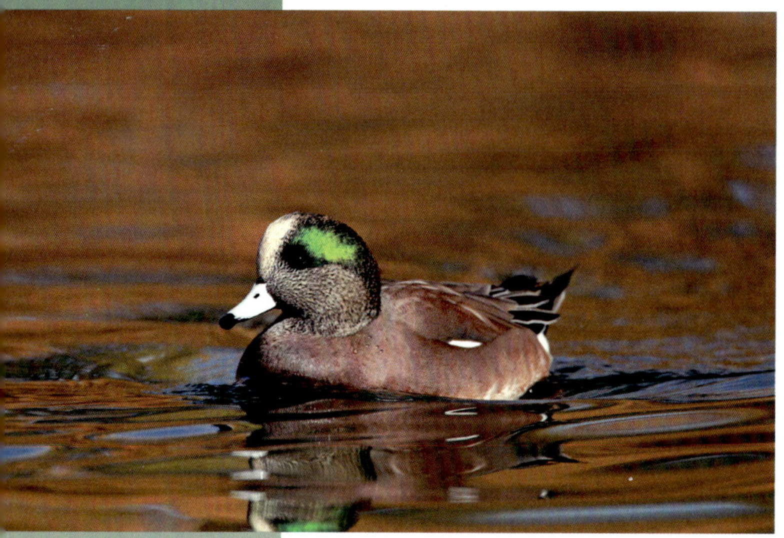

Photographing from a low angle with a wide aperture enabled a soft backdrop behind an American wigeon duck. Canon 1DMIV, 500mm, 1.4x teleconverter, ISO 800, f/5.6 @ 1/2000 sec.

During the winter months, ducks calmly paddle in the ponds of the hatchery searching for vegetation on which to feed. Year-round, small riparian birds on both sides of the river search for food and bathe at the water's edge. Herons stand in the shallow water waiting for an unsuspecting fish to swim past and become their next meal.

In summer, watch for the long, ribbon-patterned Mexican garter snake at the edge of the small ponds near the nature trail parking lot and along the river as it searches for food.

Do not attempt to handle this reptile as state and federal regulations protect it. However, if you are lucky enough to find one slithering around, use your telephoto lens to take close-up images.

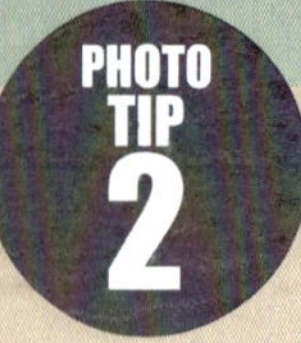

Ditch the Tripod or Monopod?

If there was ever a frequently asked question without a simple answer, it is "When do I use a tripod or monopod to support my camera?" Too many variables allow for a one-size-fits-all answer.

The photographer will almost always take a sharper image when using a camera atop a tripod. However, when a situation arises where a tripod or monopod is not practical, you may need to hand-hold a long, heavy telephoto lens.

When hand-holding your camera, adhere to the Inverse Lens Rule (which is more of a guideline). This rule states that one can generally hand-hold a lens and obtain a sharp image if the photographer sets a shutter speed faster than the inverse of the length of the lens (1/focal length of lens). For instance, if you use a 60mm lens, only hand-hold the camera with a shutter speed equal to 1/60 of a second or faster. Similarly, when hand-holding a 400mm lens, use a shutter speed of 1/400 of a second or faster.

Some caveats apply when using this rule. Is the photographer capable of holding the lens for prolonged time frames without tiring? Is the subject moving or sitting still? Photographing a moving subject requires a greater shutter speed and a steadier hand than photographing a stationary object.

DIRECTIONS:

From the intersection of AZ 89A and AZ 260 in Cottonwood: Travel 7.3 miles (11.8 km) northeast on AZ 89A and turn right onto Page Springs Road. The hatchery will be on the right in 3.1 miles (5 km).

From I-17: Take exit 293. Head west on Cornville Road for 8 miles (12.9 km) to Page Springs Road. Turn right onto Page Springs Road.

From Page Springs Road, travel north 4.4 miles (7.1 km) to the Nature Trail parking lot. Continue traveling another 0.5 miles (0.8 km) to the Bubbling Ponds Hatchery parking lot.

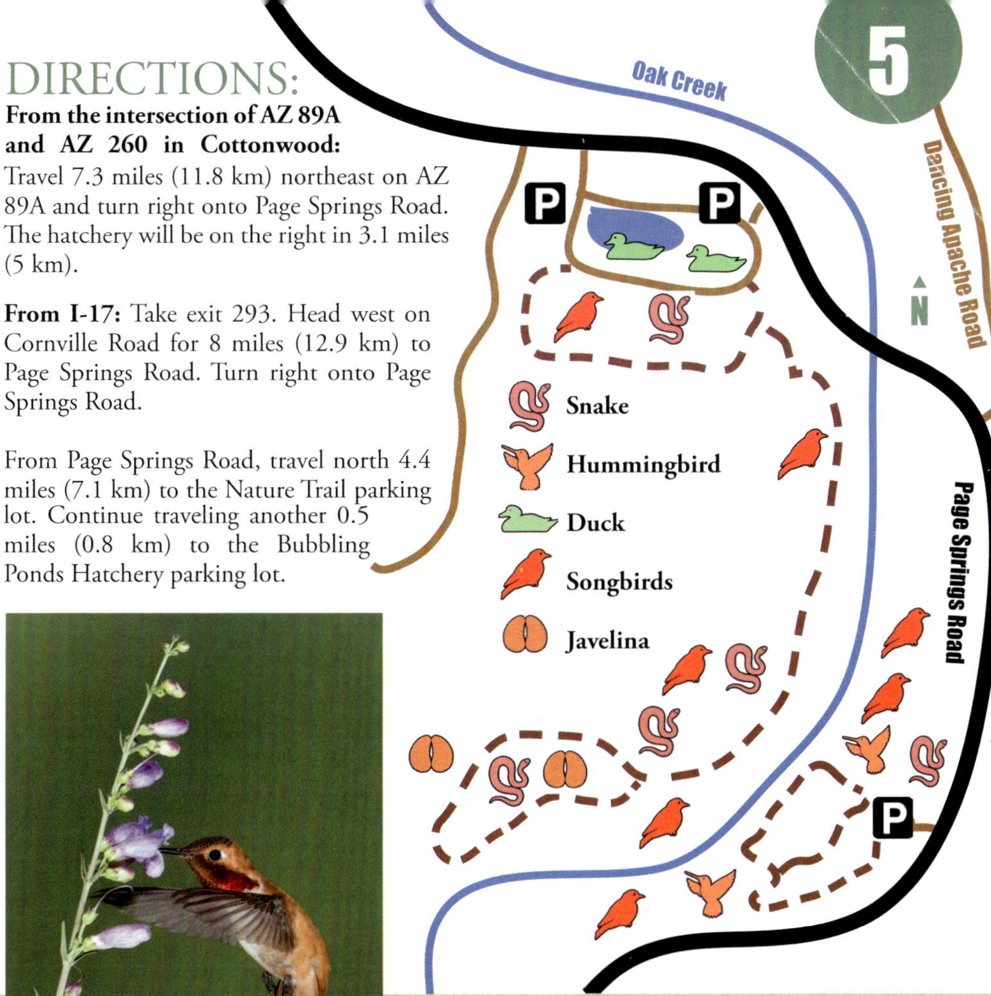

Rufous hummingbird feeds on a late summer penstemon. Canon 1DMIV, 70-200mm at 292mm, 2x teleconverter, ISO 250, f/16 @ 1/200 sec., high-speed hummingbird set-up.

When shooting at night, using flash as the only light source, hand-holding the camera and lens becomes a more practical and flexible solution. The flash normally fires between 1/500 and 1/1500 of a second. Since the flash is the entire light source, the effective shutter speed of the shot is the speed of the flash. For example, a flash duration of 1/500 sec. is the same as setting the shutter speed to 1/500 sec. At those speeds, hand-holding a short telephoto lens is reasonable.

Oftentimes, you may not be able to set up and use a tripod. Butterflies often rest in the middle of thick bushes where it is difficult to place a tripod without disturbing the plant and scaring away the occupants. Dragging a tripod with you while walking or crawling through thick brush is not worth the hassle. Also, tripods are not useful when lying on your stomach taking eye-level images of horned lizards.

Every situation (and every photographer!) differs, and one should make the decision to hand-hold or place the camera on a tripod (or monopod) on a case-by-case basis.

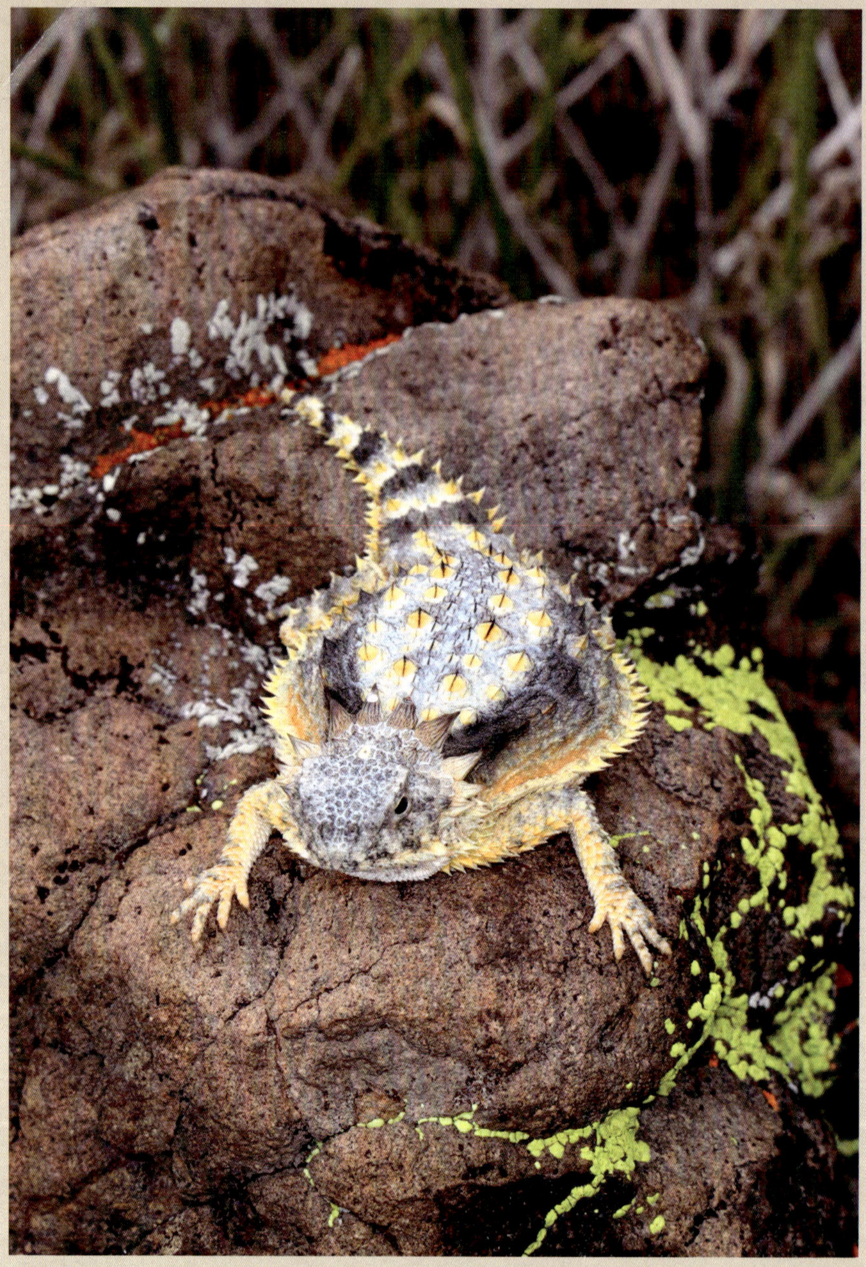

Regal horned lizard on a lichen-covered rock near Sunset Point along I-17. Canon 50D, 100mm macro, ISO 400, f/16 @ 1/80 sec., off-camera flash.

Wildlife of

WESTERN ARIZONA

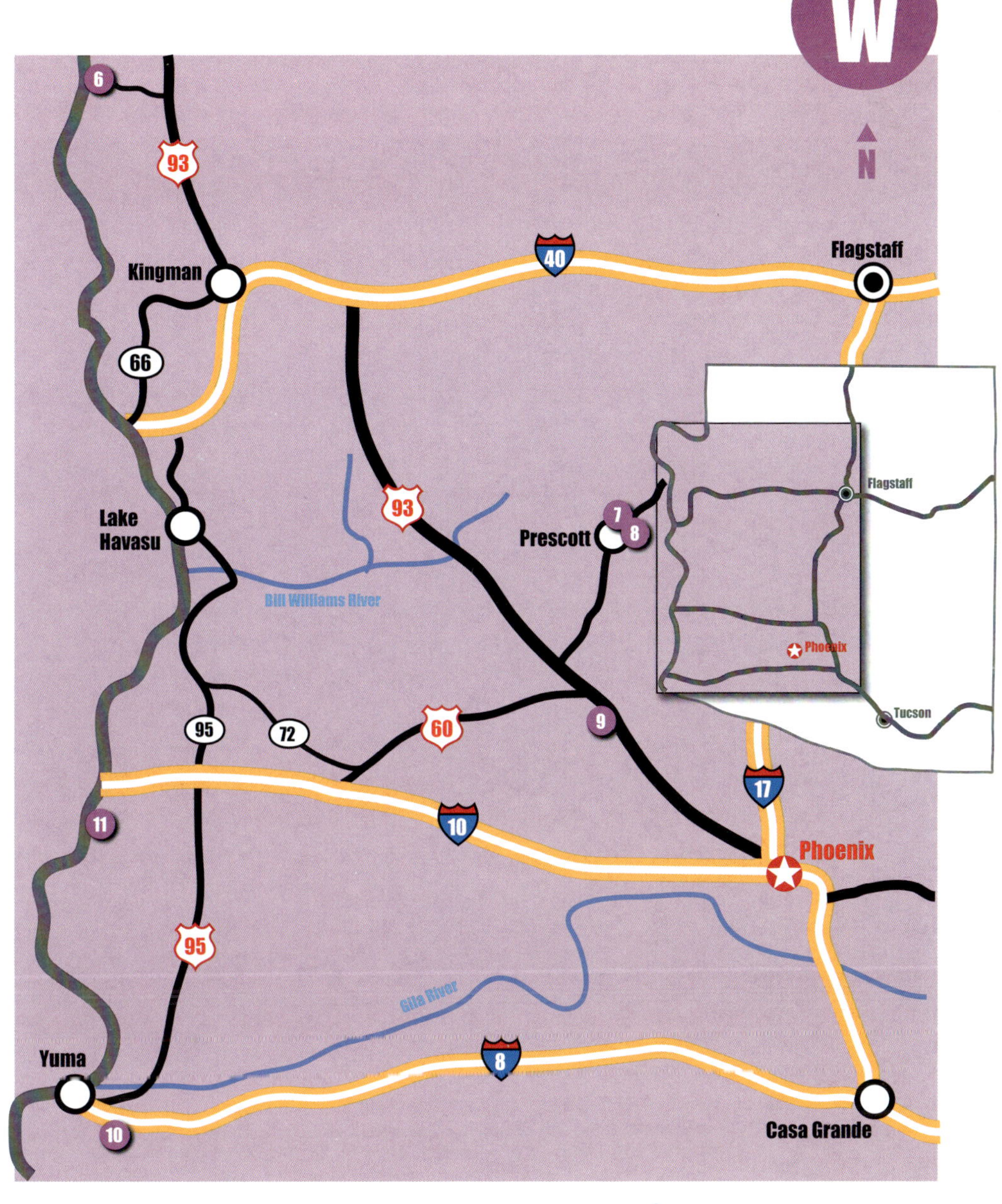
W
N
Kingman
Flagstaff
Lake Havasu
Prescott
Bill Williams River
Phoenix
Tucson
Gila River
Yuma
Casa Grande
93
40
66
95
72
60
10
17
8
6
7
8
9
10
11

Willow Beach

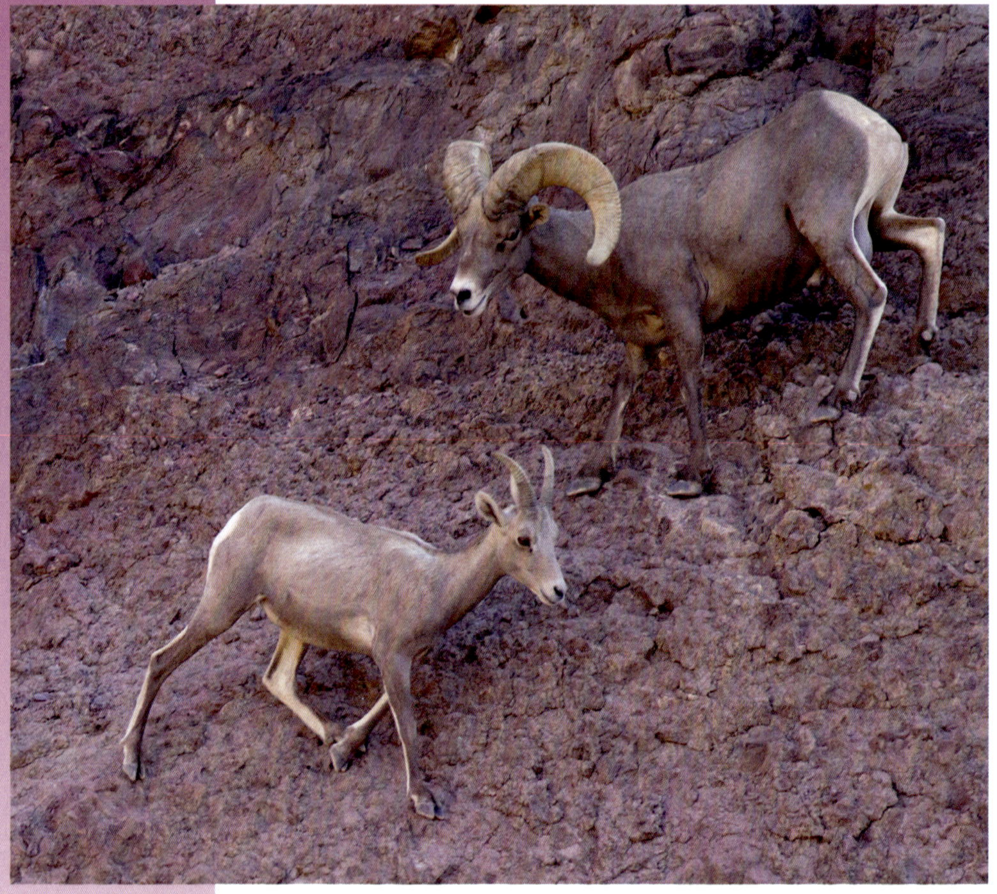

Ram and ewe desert bighorn sheep traverse the treacherous terrain along the Colorado River. Canon 1DMII, 400mm, ISO 500, f/5.6 @ 1/400 sec.

VIEW TIME
Late May to August

IDEAL TIME OF DAY
Early morning

VEHICLE
Any

HIKE
None; boat required

Across Arizona, iconic desert bighorn sheep elude even the most observant photographer. These shy, retiring beasts stand guard against predators and other dangers while inhabiting the hottest, driest, and most impenetrable mountains and hills in the Grand Canyon State. In late May through early August, when temperatures soar past 100 degrees F (38 C), desert bighorn sheep come down to the Colorado River to drink its cool, lifesaving water and pose for photographers.

This section of the Colorado River is just below Hoover Dam, and even though the dam "tamed" the river, the fast moving water in this area remains a hazardous environment. If you are uncomfortable with navigating a boat in this type of water, find a river guide or someone else to drive the boat. Two people can often do the job better than one, so bring a friend to help navigate the river while you photograph!

Do not get any closer than about one mile (1.6 km) downstream from the dam (dangerous currents and shallow rocks create many hazards here).

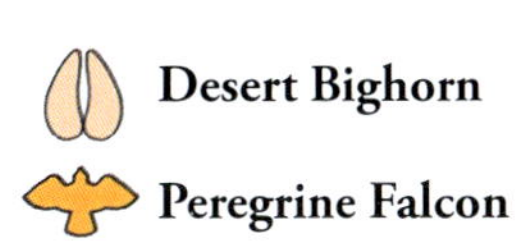

DIRECTIONS:

From Kingman, travel 56.2 miles (90.5 km) north on US93 and turn onto Willow Beach Road. Drive 4.2 miles (6.8 km) to the Willow Beach Marina.

An alternate route is to take US 93 for 21 miles (33.8 km) south out of Boulder City, Nevada to Willow Beach Road (where you turn right onto this road towards the marina).

Do not float more than a few miles downstream of the boat ramps. After that the river becomes quite wide and the sheep are difficult to locate. Take plenty of water and food with you on the river. The temperatures commonly exceed 100 degrees F (38 degrees C) so wear a hat and lather sunscreen on any exposed skin.

Not only are the sheep attracted to the water to drink, but they also live, breed, feed, and generally hang out in this area. Large rams vie for the affections of the ewes while the ewes attend to and feed their lambs. All are seemingly oblivious to the strange beings in boats.

Pack a **moderate telephoto lens (e.g., 300 or 400mm)** with image stabilization. Because you will photograph from a boat in constant motion (due to the current or rocking from waves of passing watercraft), set your camera to high-speed continuous shooting mode to quickly record bursts of four to six images at a time when the opportunity arises. Shoot with a wide aperture (such as f/2.8 or f/4) to keep your shutter speed as high as possible (e.g., 1/1000 of a second or faster). If needed, increase your camera's ISO speed to obtain the appropriate shutter speed.

As you meander along, stay close enough to the bank to see the sheep but far enough to keep from hitting rocks or other protuberances. Look for a large white rump (butt) patch among the large boulders. When you spot one or more of these elusive creatures, approach them slowly and keep the bow of the boat facing into the current. Stay in the boat! Watch patiently to see if the sheep do something interesting such as drink water, butt heads with their neighbors, chase a ewe, or feed their lambs.

Summer storms often bring lightning along the Colorado River. In case of lightning, get to the shoreline so that the boat is not in the middle of the river. Or, if you are close enough, go to the boat ramp and get out of the boat. In case of a passing shower, take rain gear to protect you and your camera equipment.

As an added treat, keep an eye out for the peregrine falcon. They breed around Willow Beach, so a keen observer can spot both adults and juveniles sitting on rocks, drinking the cool water, or flying past your boat.

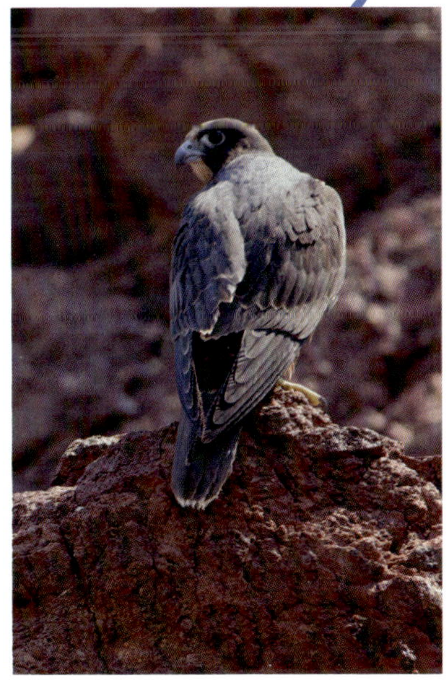

Peregrine falcon rests after a bath in the Colorado River. Canon 1DMIV, 70-200mm at 400mm, 2x teleconverter , ISO 640, f/6.3 @ 1/640 sec.

Making the Photo
1

Arachnophobes Beware!

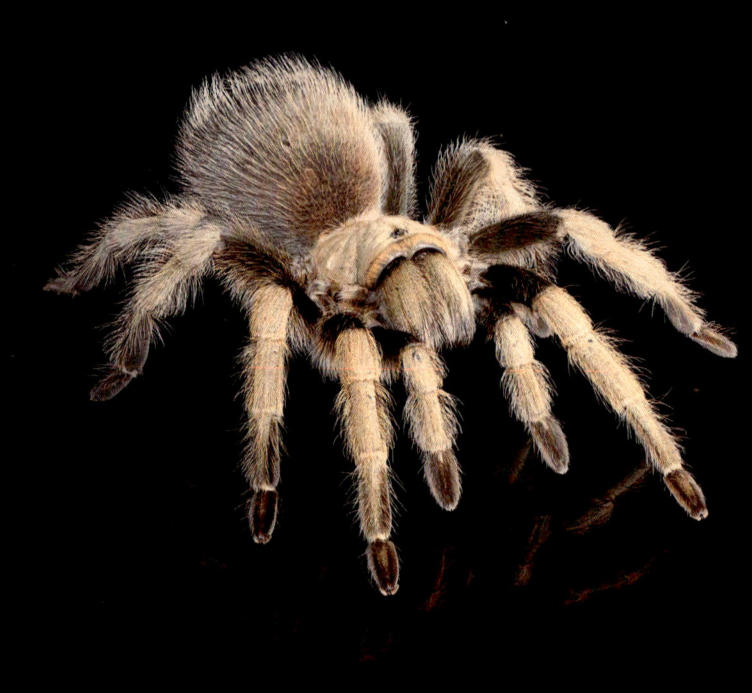

Stacked image of a tarantula from the Quartzsite area. Canon 5DMIII, 100mm macro, ISO 500, f/16 @ 1/200 sec., off-camera flash.

I love the under-appreciated animals—those often referred to as creepy, crawly critters. Snakes, frogs, scorpions, and yes, the lowly, hated spiders are among my favorites. I am not sure why, but since childhood, spiders have not frightened me. Their complicated life histories, mechanisms of killing their prey, and beautiful webs fascinate me.

Arizona has its share of these eight-legged, eight-eyed hunters. Most live quiet lives hidden in the vegetation that surrounds us. No, the bites you get at night in bed are not from spiders but from some other loathsome insect that has an interest in your body fluids. In fact, unless we try hard, it is difficult to get bitten by a spider (with the possible exceptions of black widows or the brown recluse).

One of my favorite spider photographic subjects—and the star of several classic sci-fi movies—is the tarantula. As large as an out-stretched hand, tarantulas live underground and come out at night to hunt their prey. After capturing an unlucky victim, they inject them with venom and suck their body juices dry. This seems gory, but everything has to eat.

Although tarantulas are found across Arizona, my favorite place to photograph them is the area surrounding Quartzsite and the other hot desert roads and washes in the western part of the state—especially during the monsoon season or early winter. I either drive the dirt roads at night or walk dry washes using a flashlight to find my subjects. I carry a large plastic jar in which to place my specimen (yes, you have to capture them to make great images).

Although I have taken many great shots of tarantulas in the wild, my favorite images were taken at home on my back porch. I set up my tent diffuser (see "Using a Tent Diffuser" on page 212) and placed one piece of black, reflective Plexiglas in the bottom and one just above the spider. I positioned a small flash in the camera's hot shoe and pointed the front of the flash either to the side or slightly upwards. By bouncing light off of the interior of the diffuser, I obtained wonderful surround light on my subject.

Then, I gently removed the tarantula from its container and placed it on the reflective surface. It went nuts and ran all around, so I had to convince it to calm down—truly a Zen process! When the tarantula started throwing its body hairs (called urticating hairs) at me, I held my breath so as not to inhale them and looked away so they did not end up in my eyes. Because these hairs can cause damage to eyes or skin, wear gloves, safety glasses, and a cloth mask when handling these spiders. At this point, I told the spider to "Chill out!" I used a pencil to position the spider in the middle of the plastic sheet and did my best to face it a little (25-45 degrees) towards the camera lens.

I felt my macro lens would be too long for the scene I envisioned, and I could not get close enough to the spider while having the flash within the tent at the same time. Instead, I used my 24-105mm wide-angle lens to obtain frame-filling shots and allowed my flash to bounce off the tent diffuser walls. I used manual exposure on both the camera and flash. I also used a small aperture (e.g., f/16) to obtain the greatest depth of field.

Oftentimes f/16 will not give me a sharp image from front to back, so in this image, I took several exposures at different focus points and combined them with focus stacking software (see page 196). The results are, in my mind, beautiful. The entire tarantula can be seen without any distraction from competing elements. The reflection off of the black Plexiglas is unique and adds to the intrigue of this image. While the tarantula remained still, I made sure to take close up images of the eyes.

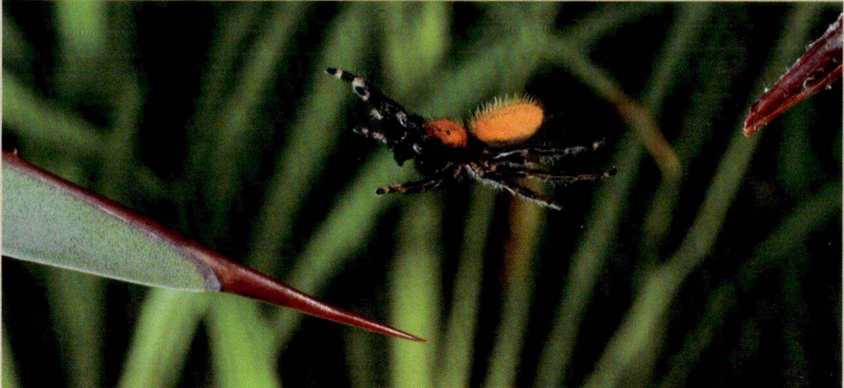

An Apache jumping spider "jumps" between two spines of an Arizona agave. Canon 7DMII, 100mm macro, ISO 400, f/16 on Bulb mode, Cognisys high-speed shutter, several remote flashes, Phototrap infrared trigger.

WESTERN ARIZONA

Fain and Williamson Valley Roads

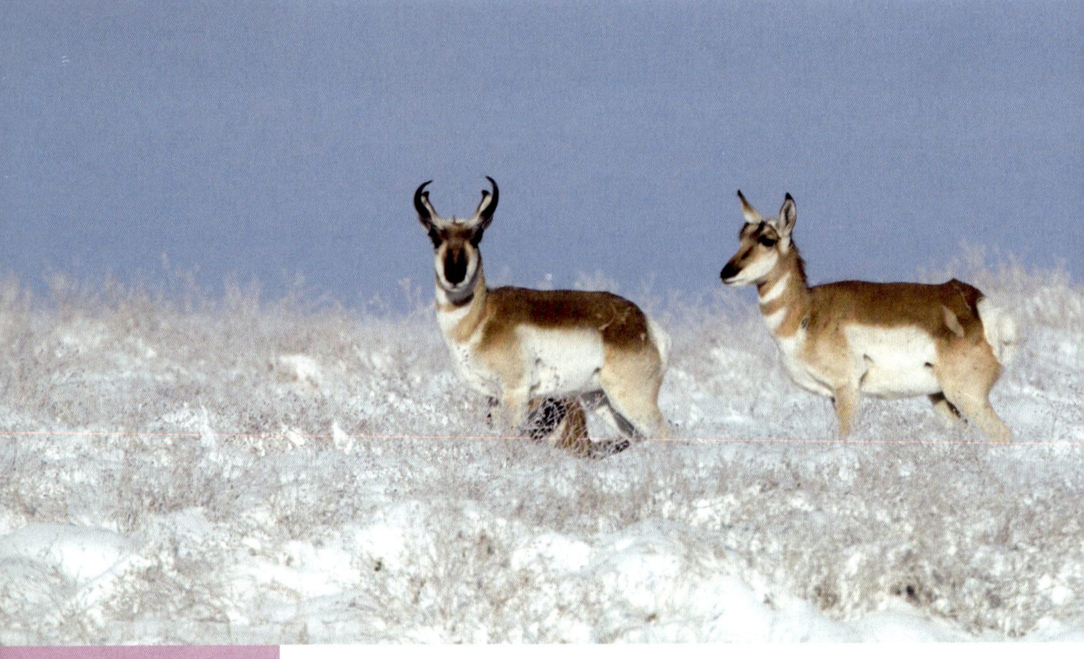

ABOVE: A buck and doe pronghorn after a snow storm in Prescott Valley. Canon 7DMII, 500mm, 1.4x teleconverter, ISO 400, f/8 @ 1/2500 sec.
RIGHT: Two three-month-old pronghorn twins. Canon 70D, 500mm, 1.4x teleconverter, ISO 200, f/5.6 @ 1/500 sec.

VIEW TIME
Year-round

IDEAL TIME OF DAY
Early morning

VEHICLE
Any

HIKE
Easy

As the fastest land animal in North America, American pronghorn (also incorrectly called antelope) require vast open areas to supply food and shelter. From birthing their young to spotting danger at several hundred yards, open areas provide key components for the survival of these mammals. Historically, large herds of these pronghorn roamed the vast high-elevation grasslands around Prescott and Prescott Valley. They continue to do so today, specifically off Fain Road in Prescott Valley and Williamson Road to the north of Prescott.

Pronghorn are used to vehicles driving on the roads adjacent to their homes, but they consider a photographer on foot to be a dangerous predator, so photograph them from your car. When you spot a pronghorn, slow down and find a safe place to pull over. Shut the engine off to eliminate any vibrations. Roll down your window, and either use a window mount or a bean bag to stabilize a **long telephoto lens** (the longer the focal length, the better). Use high-speed continuous shooting mode to reduce image blur.

Photographing these beautiful creatures offers a few unique challenges (primarily wind and heat related). Cameras on long telephoto lenses, and especially those mounted high on a tripod or on the window of a vehicle, will shake under even slight wind currents. Add the impact of a cropped frame and 500mm telephoto lens with 1.4x teleconverters and the result is the equivalent of a 1,120mm lens (similar to a 22x pair of binoculars!). At

FR 21
Chino Valley
5
Pronghorn
Williamson
Outer Loop Road
N
Prescott Municipal Airport
89A
Fain Road
89
Prescott Valley
69
Prescott

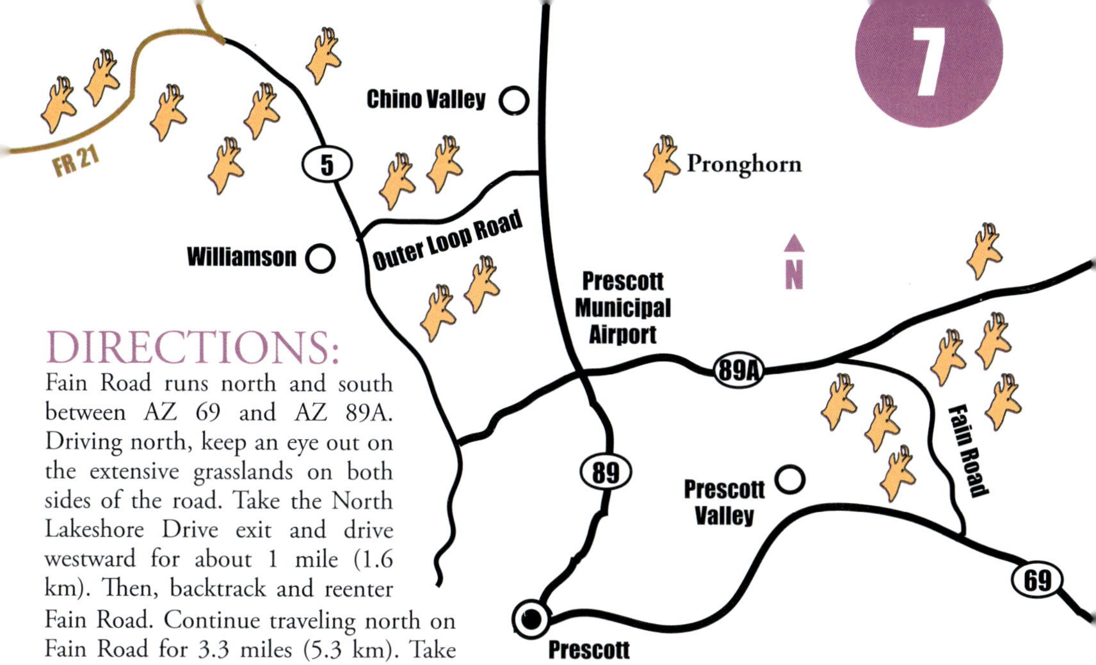

DIRECTIONS:

Fain Road runs north and south between AZ 69 and AZ 89A. Driving north, keep an eye out on the extensive grasslands on both sides of the road. Take the North Lakeshore Drive exit and drive westward for about 1 mile (1.6 km). Then, backtrack and reenter Fain Road. Continue traveling north on Fain Road for 3.3 miles (5.3 km). Take a right (east) on AZ 89A. Travel 4 miles (6.4 km), scanning once again both sides of the road. Turn around and repeat the Fain Road drive, except travel south this time. Because of the rolling hills, herds of pronghorn are difficult to spot if you only travel one direction.

Make one more trip north on Fain Road to AZ 89A and turn right (east). Travel for 5 miles (8.1 km) and keep an eye on both sides of the road, especially around the housing developments. When you arrive at the bridge just before traveling uphill towards Jerome, turn around and go back to Fain Road.

At the Fain Road intersection, continue on AZ 89A for 6.6 miles (10.6 km) to AZ 89. Turn right (north), and travel 6.3 miles (10.1 km) to Outer Loop Road and turn left. Pronghorn can appear along the entire length of Outer Loop Road.

Drive 5.4 miles (8.7 km) on Outer Loop Road and then turn right (north) onto Williamson Valley Road. Once you pass Fair Oaks Road (after about 8 miles (12.9 km)), start looking again for pronghorn along the next 10 miles (16.1 km).

---

these magnifications, even the slightest lens movement can cause softness in the image. Use as fast a shutter speed as possible and photograph early in the morning when the wind is calmer than any other time of day.

When looking across an expansive field (or a lake) during a hot summer day, you may notice "heat waves." This mirage results from differences in air temperature impacting the refraction of the light. These heat waves will cause the image to have an out-of-focus look, similar to using a blur filter when taking human portraits. No matter your camera, lens, or settings, there is no cure for this, except to get out early in the day to avoid it in the summer months.

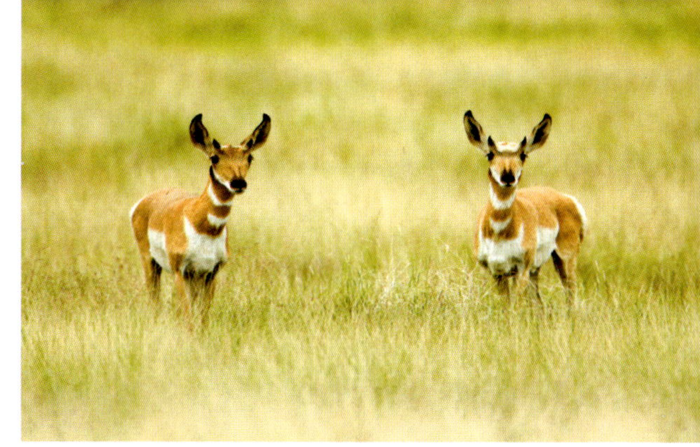

Watson Lake

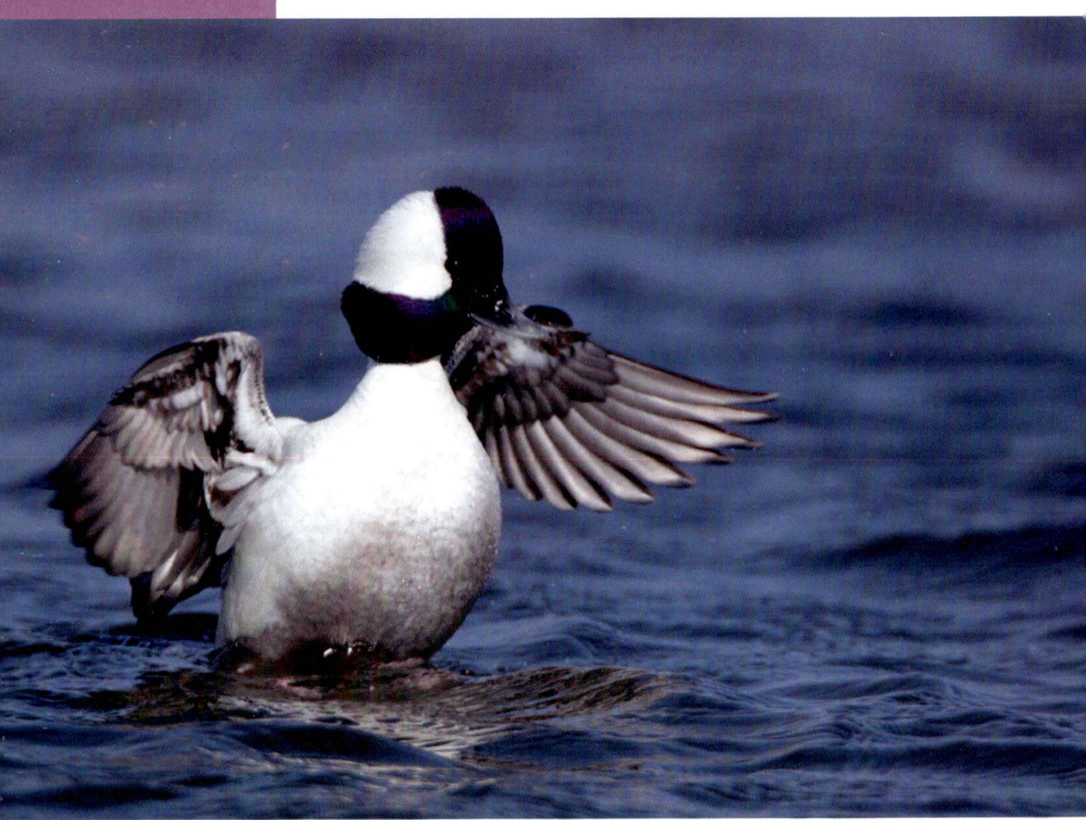

Bufflehead duck feels his oats. Canon 50D, 500mm, 1.4x teleconverter, ISO 200, f/6.3 @ 1/200 sec.

VIEW TIME
October to May

IDEAL TIME OF DAY
Early afternoon

VEHICLE
Any

HIKE
Easy

In the early 1900s, the Chino Valley Irrigation District created Watson Lake when they built a dam on Granite Creek. Although many outdoor photographers visiting Watson Lake focus on the unique jumbo rock structures bordering and emerging from the water, migrating birds from northern cold climates flock to this watering hole and take center stage. The open water, abundant aquatic vegetation, and relatively warm climate attract Canada geese, American coots, and several species of ducks.

Although easy to photograph at several locations in the state, Canada geese are especially approachable at Watson Lake. After arriving, use your binoculars to see where these giant birds hang out. No need for a blind. Carry your gear, sit down near the water's edge, and fire away when you spot them. Use a relatively high shutter speed (e.g., 1/500 sec.), open aperture (e.g., f/8), and high-speed continuous shooting mode. Use a **telephoto lens** to watch for interesting behavior such as fighting, calling, flying, or sexual advances. Prepare to stay a long time to let the geese get used to your presence and go about their normal, quite active lives.

Not necessarily a rare bird, but under most circumstances a very unapproachable one, the bufflehead duck commonly appears near the pier at Watson Lake. Again, get out the binoculars and look for this diminutive

8

Watson Lake Park

Watson Lake Park Road

N

89

Watson Lake

DIRECTIONS:

From the intersection of AZ 69 and AZ 89 in Prescott, travel north on AZ 89 for 3.8 miles (6.1 km) to Watson Lake Park Road. Turn right at the roundabout and proceed into Watson Lake Park. Continue about 500 feet (152 m). Pay your parking fee, and then turn right on the unmarked paved road to the fishing pier. Travel 0.2 miles (0.3 km) to the parking lot. For more information, visit **www.cityofprescott.net/services/parks/parks/index.php?id=24**.

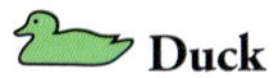 **Duck** 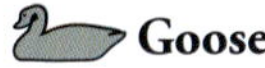 **Goose**

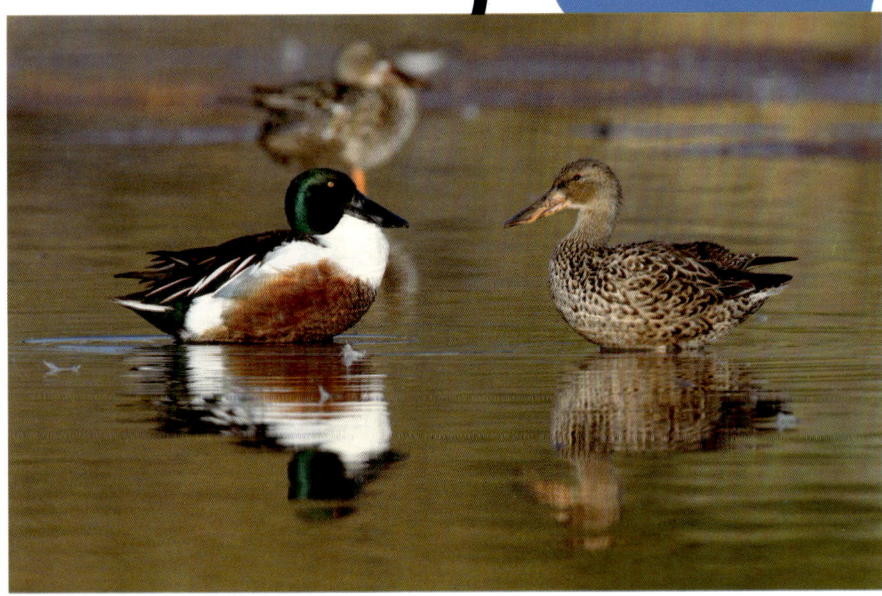

Pair of northern shoveler ducks. Canon 50D, 500mm, 1.4x teleconverter, ISO 200, f/10 @ 1/400 sec.

white-bodied and multicolored-head duck as it alternatively dives and appears at the water's surface. Buffleheads eat fish and constantly dive for their dinner.

After noting the location of the hungry ducks, set up your blind at the water's edge as close as possible to their feeding area. The ducks will leave at your approach, but do not worry, within about 30 minutes of entering your blind, they should reappear and continue to dive for food.

Place your camera on high-speed continuous shooting mode and continuous autofocus. Use a high shutter speed and wide aperture to isolate your primary subject—and freeze their motion. With patience and persistence, great photographs of these nervous little beasts await.

PHOTO TIP 3

Get Close!

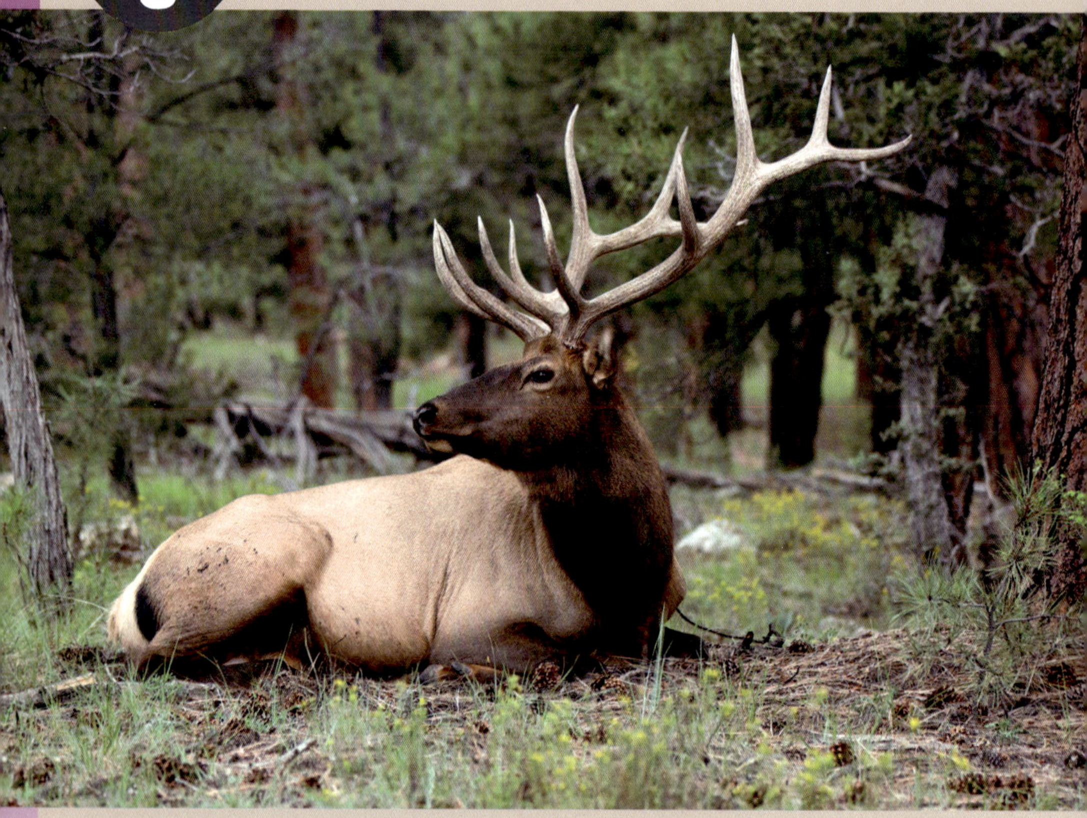

A large bull Rocky Mountain elk relaxes in the forest in the Grand Canyon National Park. Canon 5DMIII, 70-200mm at 170mm, 2x teleconverter, ISO 1600, f/5.6 @ 1/160 sec.

Wildlife photographers must not only act proficiently with their photography equipment, but they also need to approach their subjects closely in order to fill their frame with the animal. Seeing an image of a kestrel flying into its nest with food in its talons, a glare in the eye of an elk, and the texture of a hummingbird's feather has greater impact if these critical details appear large in the photograph. If these elements fill only a small part of the frame, they could disappear among the vast real estate and yield no impact at all.

Before wildlife photographers can get close to their subjects, they must know them. Although this book and others like it (see the Additional Resources on page 227) go a long way to getting you in the right place at the right moment, the more time you spend researching an animal's behavior and habitat, the more likely you will be ready when that moment occurs. Therefore, find out where the animal lives, breeds, and whether it migrates. Gain insight to your critter of choice by reading the works of renowned wildlife photographers, talk to biologists and naturalists, or if the experts are unavailable, visit the local library to read about your target animal. Join your regional Audubon Society, bird watching club, and field photography organizations. Once you have a feel for the basics, head out to the field and scout your quarry with your new knowledge.

Take a pair of binoculars and leave your camera in the car. Walk around to get comfortable with the area so that, when you return, you do not waste precious time looking for a vantage point from which to set up your gear. Take notes (on paper or your smartphone) on what you see, especially the time of day and exact locations you spot wildlife. If you are comfortable with using trail cameras or video equipment, set them out to gather data when you cannot be on site. Check first with the managing agency for any legal issues when setting trail cameras on public property. Armed with the confidence that comes with knowing your subject, you can get close.

To test your comfort level with getting close animals, try photographing at sites where they are more accustomed to (and not afraid of) people. National parks, boardwalks constructed in areas where access is not normally gained, urban wildlife sanctuaries, and private ranches offer some of the best places to visit.

Even at these locations, though, a little attention to fitting in with the surroundings goes a long way towards getting close. Be quiet and move slowly. Wear clothing that fits in with the colors and tones of your location. Wear a hat to shade your shiny face, and get in front of a bush or tree so that you do not silhouette yourself on the horizon. Sit in the shade, if possible, and have your camera on a tripod and pointed in the direction of your subject. A slight movement from repositioning your camera can easily scare off a potential subject.

When visiting locations with more cautious wildlife, employ more drastic techniques. Arrive at your location a couple of hours before sunrise so that you can outfit and position yourself well before the action begins (sleep deprivation is, in some strange way, a friend to great wildlife photographers). Dress yourself in camouflage and hide before your subjects arrive which will allow you access to the animals' private lives. If needed, crawl into your favorite blind and get your camera set up early.

Although difficult, you can try to sneak up on your quarry. Wear clothing that matches the colors of the surrounding vegetation. Locate the animal you want to follow and slowly move diagonally towards it. If you approach from an angle (and not directly at an animal), the critter will not feel as threatened and is not so inclined to leave. Do not look directly at your subject, but watch it out of the corner of your eye. Even if the subject is too small in the frame, take a few images as soon as you are able to do so. This assists in obtaining the correct exposure and at least gives you one image in case the critter runs away before you are in full frame distance. Then, move cautiously and diagonally to get a little closer. Take a few more frames until you are close enough for a frame-filling image.

If you are sneaking up on a bird during the breeding season, use a playback of its voice. Place the playback call close to the bird and in a place that you want the bird to move to.

Do not—under any circumstance—sneak too close to large mammals. If they let you get close enough for you to make images, they are likely not afraid of you and, if they become uneasy, they might charge. Instead, photograph large and potentially dangerous mammals using a telephoto lens or by working from a blind set-up at a safe distance.

WESTERN ARIZONA

Hassayampa River Preserve

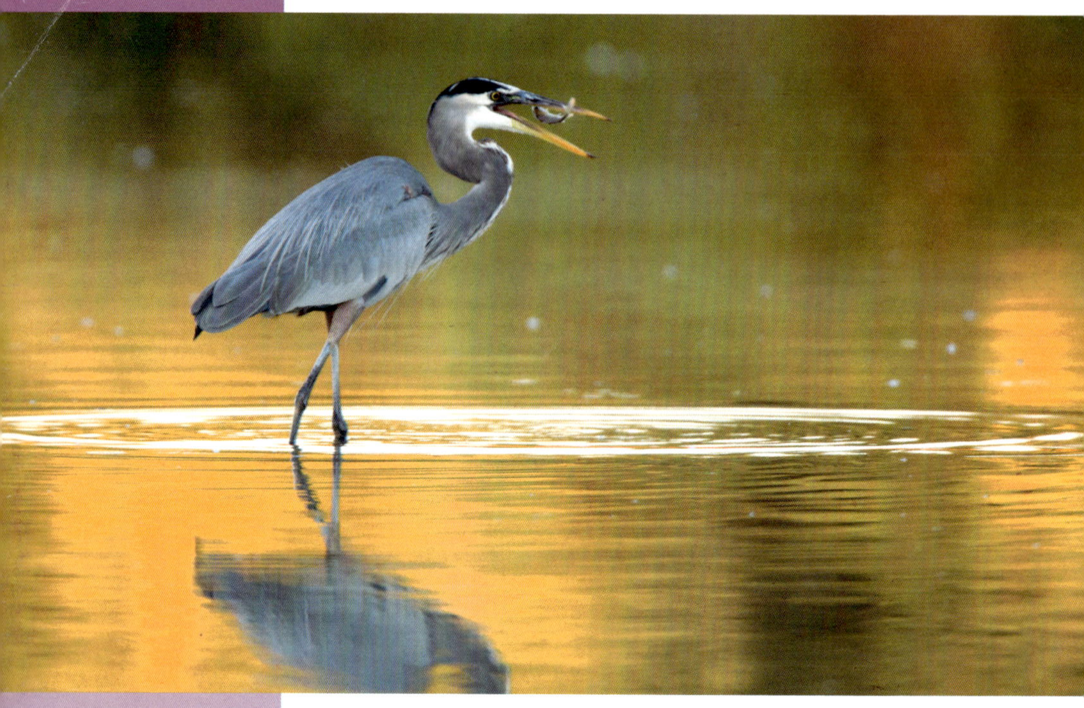

Great blue heron eats lunch. Canon 5DMIII, 500mm, 1.4x teleconverter, ISO 800, f/5.6 @ 1/250 sec.

VIEW TIME
November to June

IDEAL TIME OF DAY
Early morning

VEHICLE
Any

HIKE
Easy to moderate

For most of its 100-mile (161-km) journey through Arizona, the Hassayampa River flows underground. Then, just downstream from the Wickenburg Bridge, subsurface rock layers force the water to the surface for several miles at the Hassayampa River Preserve. Here the year-round water supply creates a lush riparian ecosystem (including hundreds of desert fan palm trees) packed with birds, lizards, fox, coyote, deer, and many of Arizona's more interesting wildlife. Once part of the Frederick Brill Ranch, The Nature Conservancy purchased the property in 1986. Today, the preserve manages this unusual Arizona habitat.

From late November to February, resident gray foxes feed on fan palm seeds. Sit in the small picnic area early in the morning or stroll the northernmost trails (Palm Lake Loop and Mesquite Meander) that circumnavigate the property while waiting for a fox sighting (enjoy the birdlife while you wait!). Although these are not as shy as most fox, any loud noises or sudden movement will scare them off. Go during the weekdays when human traffic is lower to increase your chances for a sighting.

Because of the thick vegetation, a **telephoto zoom lens** works best. Use a relatively high shutter speed (e.g., 1/250 of a second or faster) and relatively wide aperture (e.g., f/4.5 or f/5.6).

The Nature Conservancy sets out feeders for the hummingbirds, but other hungry birds (such as verdin, ladder-backed woodpecker, and Gila woodpecker) take advantage of the free meal, making them easy to photograph at close range. Verdin, cactus wren, curve-billed thrasher, Abert's towhee, Gila and ladder-backed woodpeckers, and great blue herons are

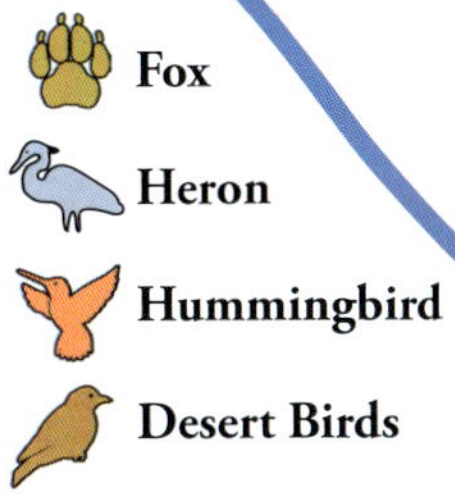

9

Fox

Heron

Hummingbird

Desert Birds

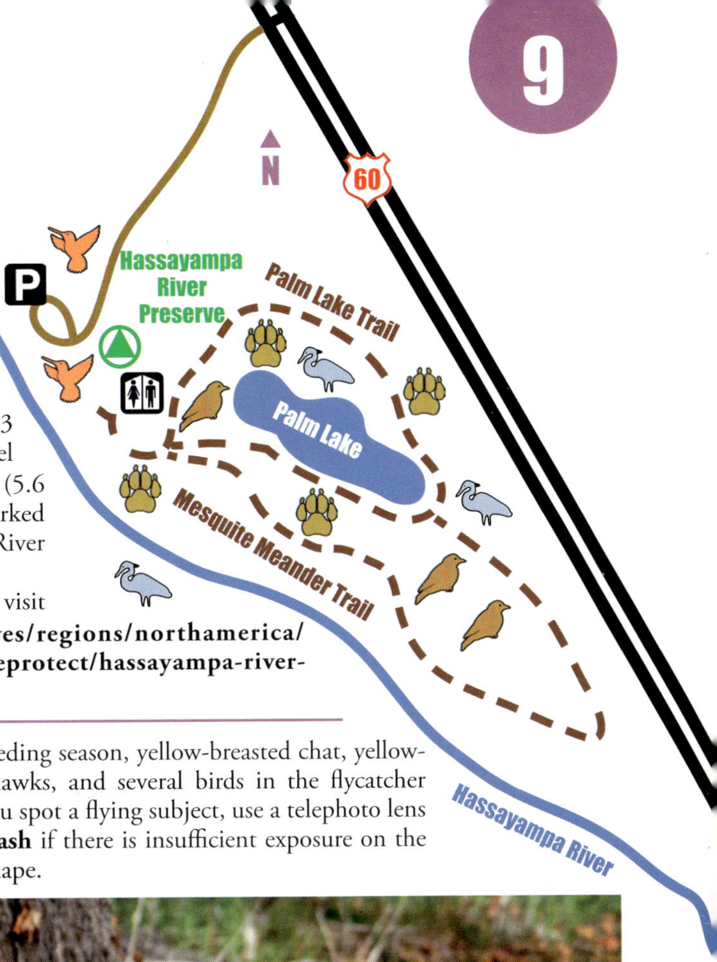

N

60

Hassayampa River Preserve

Palm Lake Trail

Palm Lake

Mesquite Meander Trail

Hassayampa River

DIRECTIONS:

From the intersection of AZ 93 and AZ 60 in Wickenburg, travel southeast on AZ 60 for 3.5 miles (5.6 km). Turn right at the well-marked entrance to the Hassayampa River Preserve (near mile marker 114).

For more information, visit **www.nature.org/ourinitiatives/regions/northamerica/unitedstates/arizona/placesweprotect/hassayampa-river-preserve.xml**.

year-round residents. During breeding season, yellow-breasted chat, yellow-billed cuckoo, gray and black hawks, and several birds in the flycatcher family nest here as well. When you spot a flying subject, use a telephoto lens mounted on a **tripod**. Use **fill flash** if there is insufficient exposure on the bird to illuminate its color and shape.

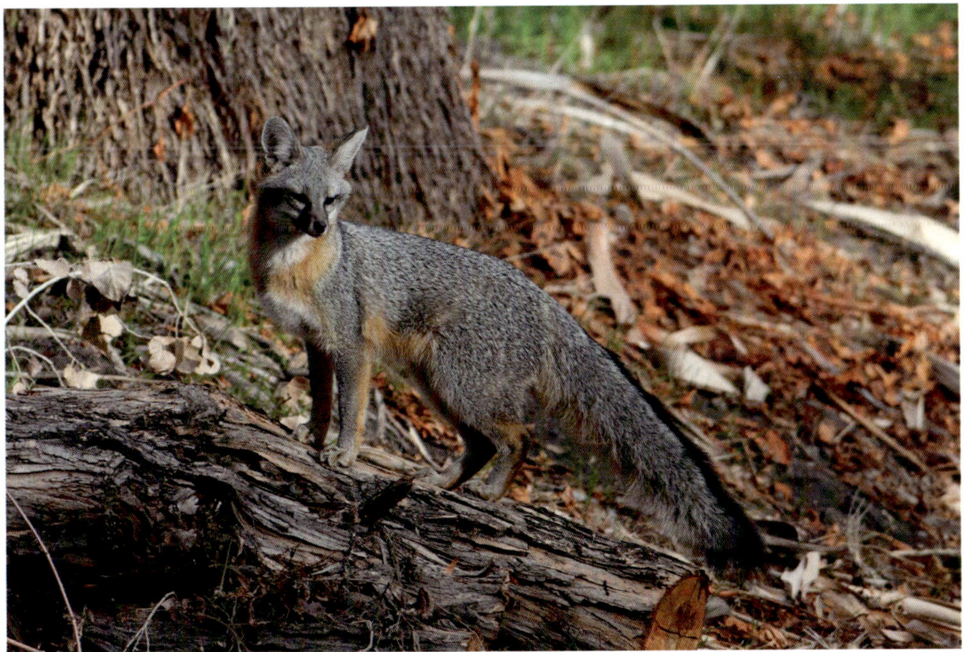

A gray fox poses for a portrait at the Hassayampa River Preserve. Canon 40D, 400mm, ISO 800, f/5.6 @ 1/500 sec.

Barry M. Goldwater Range

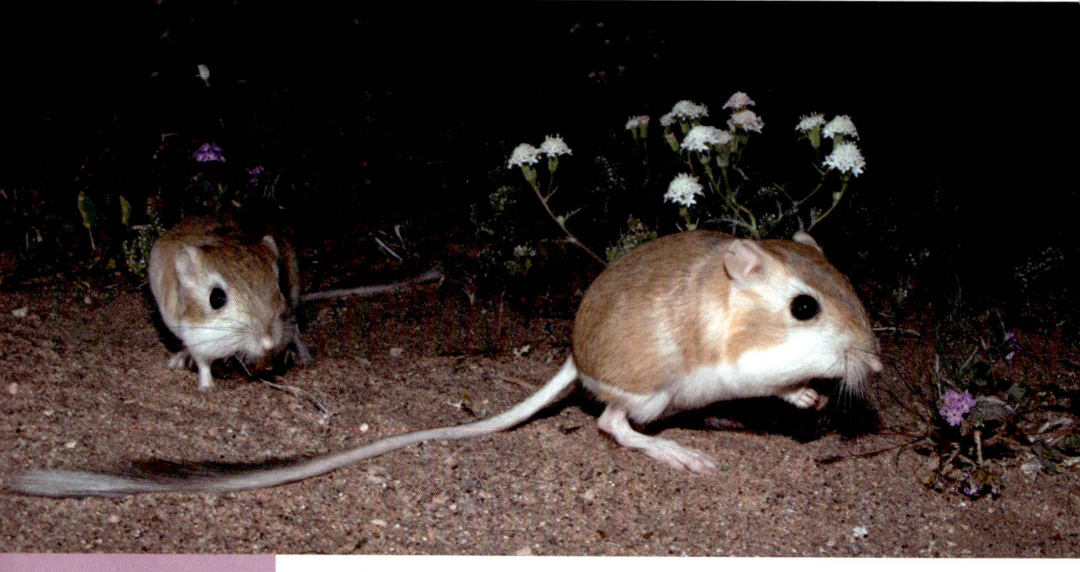

Desert kangaroo rats during spring flower bloom. Canon Digital Rebel XSi, 24-104mm at 24mm, ISO 400, f/22 @ 1/200 sec., on-camera flash.

VIEW TIME
December to May

IDEAL TIME OF DAY
Early morning, late afternoon, and night

VEHICLE
4WD high-clearance

HIKE
Easy to moderate

Covering over one million acres (404,685 hectares), the Barry M. Goldwater Range contains some of the most pristine Sonoran Desert habitat in the state. Used mainly as a training range for the United States Air Force, several locations remain open to the public. The range showcases expansive areas of open desert with intermittent washes. Expect to have frequent visits from the U.S. Border Patrol as they move about the range. However, for those dedicated photographers, the range offers photographic opportunities found nowhere else in Arizona. Key species typically sighted include the Le Conte's thrasher, loggerhead shrike, kit fox and desert kangaroo rat.

Scan the desert for green crucifixion thorn bushes which grow about three to seven feet (0.9 to 1.2 m) high and are full of thorns. Look for Le Conte's thrashers and loggerhead shrikes at the tops of these bushes.

Once you locate a bird, use your bird calls to get them to defend their territories (this technique works especially well in December and January when the birds start setting up their territories). Walk slowly and try to get close enough for an image. In early February, investigate each crucifixion thorn bush for bird nests. When you find a nest, set up a blind approximately 20 yards from it and wait for the adults to return to either incubate any eggs or feed their young. This technique requires several visits to the site and a lot of patience. Use a **telephoto lens** on a sturdy **tripod** and be extremely quiet.

At the intersection of County 9th Street South and South Avenue 46th East, look for desert kangaroo rat nests among the sand and sparse vegetation. Travel about one half mile (0.8 km) in any direction at this intersection, scanning the sandy areas adjacent to the road for mounds of

10

East Country 9th Street

S. Ave. 46 E.

Mohawk Wash Road

Desert Kangaroo Rat

Loggerhead Shrike

Le Conte's Thrasher

N

DIRECTIONS:

From Yuma, travel east on I-8 for 39.5 miles (63.6 km). Take exit 42 (Avenue 40 East) for Tacna. Turn south and go east on County 9th Street South (just a few yards south of the eastbound exit ramp). Travel east on County 9th Street South for 6.1 miles (9.8 km) to South Avenue 46 East/Mohawk Wash Road and turn right. The sandy road continues for another 17.6 miles (28.3 km). Park wherever the road widens, but watch for soft sand.

The range requires all visitors to secure a permit prior to entry. For permit information, visit **www.mcasyuma.marines.mil/Portals/152/Docs/Range/Range%20Guide-Map.pdf**

---

sand with several four-to-six inch (10.2-to-15.2 cm) holes in them. At the holes, watch for footprints or tail drags. Mounds with the most prints and tail drags indicate high activity.

Kangaroo rats like to scamper around after dark and move in and out of the nest mound searching for seeds or mates long into the night. Set up your camera on a **tripod** within a few feet of an active mound. Use a **short zoom lens, such as a 24-105mm,** to frame an area outside of a hole. Place some brown rice seeds in a small mound about one foot from the hole.

When photographing at night, visit during a full moon or use a very low power flashlight with a red filter in order to see the subject. Place one **flash** on each side of the camera at a distance of about three to four feet (0.9 to 1.2 m). Set the flashes to ETTL and use remote flash triggers to fire during the exposure. Pre-focus your lens using manual focus on the pile of rice about one foot (0.3 m) from the hole. Set your camera to a moderately wide aperture (e.g., f/8 or f/11), flash sync speed (the pre-defined shutter speed setting that allows the shutter curtain to move in front of the sensor before the flash stops), and single shot drive.

Once the kangaroo rats come out of their nest holes and find the seed pile, take a few frames on single-shot shooting mode. The light of the flash will scare them back into the hole, but be patient; they need the seeds for survival and will return for more in a few minutes.

Since the nearest services are at least 40 miles (64.4 km) away in Yuma, you might consider packing dinner, extra water, emergency supplies, extra batteries, and memory cards.

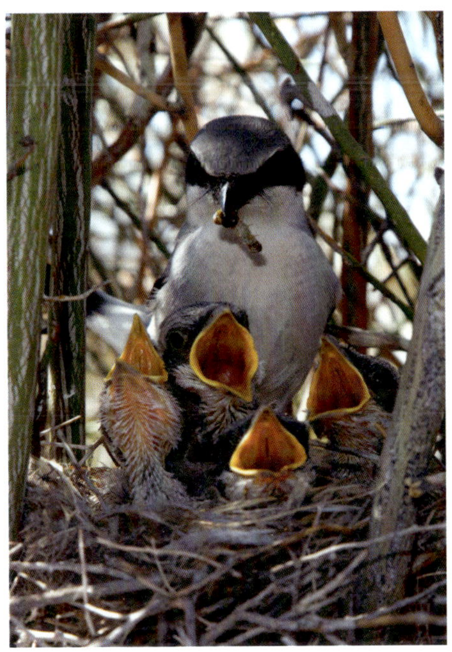

A camouflaged camera and remote release enabled me to photograph a loggerhead shrike feeding its young. Canon Digital Rebel XSi, 24-105mm at 97mm, ISO 400, f/10 @ 1/200 sec., on-camera flash @ -1/3 FEC.

Cibola National Wildlife Refuge

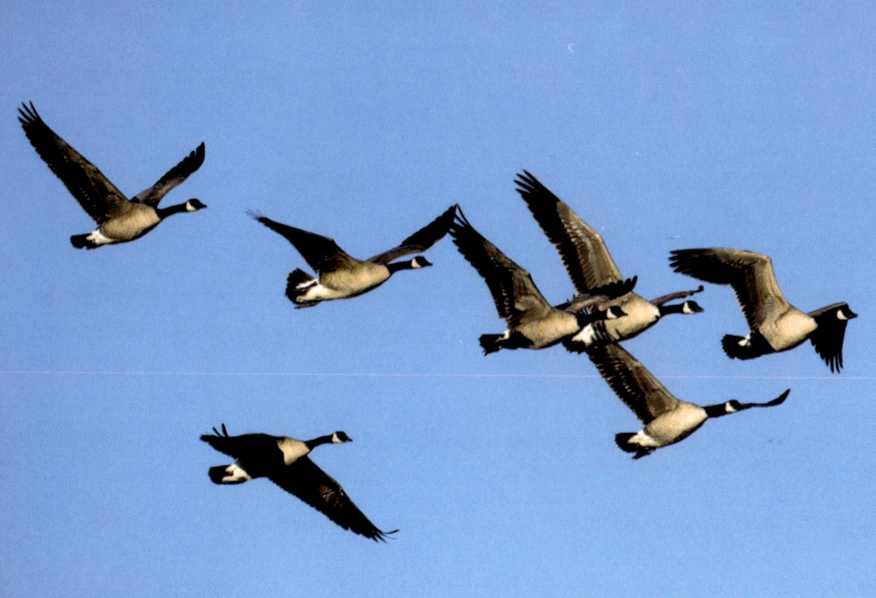

Flock of Canada geese fly to one of the many cultivated fields to feed. Canon 7DMII, 100-400mm at 400mm, ISO 400, f/7.1 @ 1/2000 sec.

VIEW TIME
November to March

IDEAL TIME OF DAY
Sunrise and sunset

VEHICLE
Any

HIKE
Easy

The U.S. Fish and Wildlife Service purchased the 18,444-acre (7,464-hectare) Cibola National Wildlife Refuge in 1964 to protect the unique riparian habitats along the Colorado River and to provide wintering grounds for sandhill cranes, Canada geese, all sorts of ducks, and other waterfowl. The lush green vegetation seems out of place in a land where rainfall averages less than four inches (102 mm) per year and summer temperatures rise in excess of 120 degrees F (49 degrees C). Beautiful mountain vistas and outstanding desert views frame the river corridor. Although most of the refuge falls in Arizona, part of it lies in California. Consequently, the best way to access the refuge is from the California side.

The three-mile (4.8 km) auto tour loop (also known as Canada Goose Drive) gives photographers access to a large part of the refuge and offers the best photographic opportunities. The refuge requires the visitors to remain in their vehicles while driving the loop except at the designated parking lot at the restroom. Refuge managers want to ensure that visitors wandering around the property do not flush wildlife from their feeding sites. The drive follows a somewhat rectangular loop. Along the way, you can observe birds loafing at several ponds, feeding in nearby planted fields, or moving between the two. Hundreds of sandhill cranes, thousands of Canada geese, and as many ducks winter at the refuge.

Photograph sandhill cranes and Canada geese when they move between the ponds and feeding areas as well as when they fly to their roost in the

DIRECTIONS:

From Blythe, California, travel 5 miles (8.1 km) west on I-10 to the Neighbors Boulevard/Highway 78 exit. Travel south on Neighbors Boulevard for 17 miles (27.4 km) to the Cibola Bridge. Continue for 3.5 miles (5.6 km) to the headquarters. At the headquarters, obtain a free map for the auto loop.

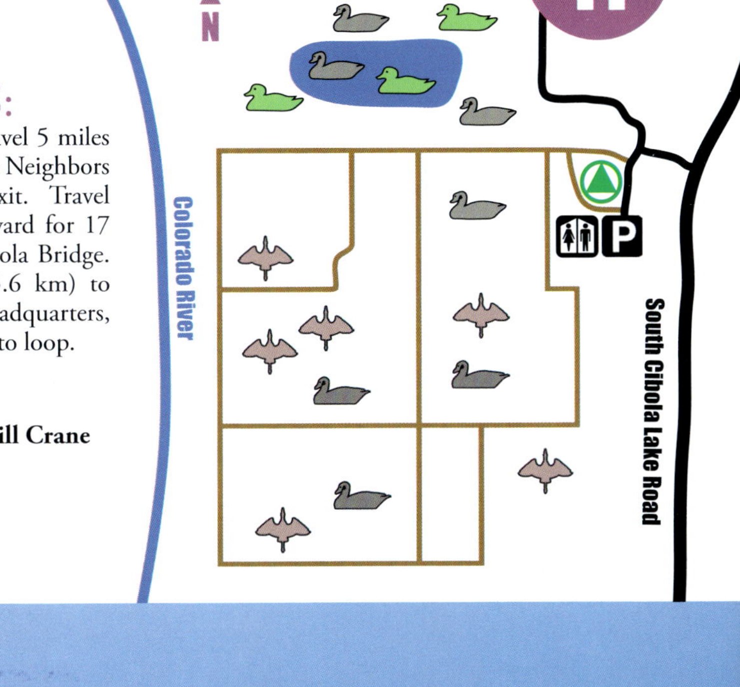

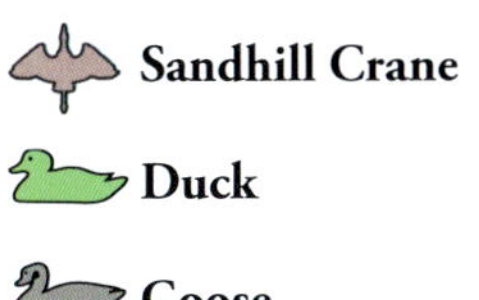

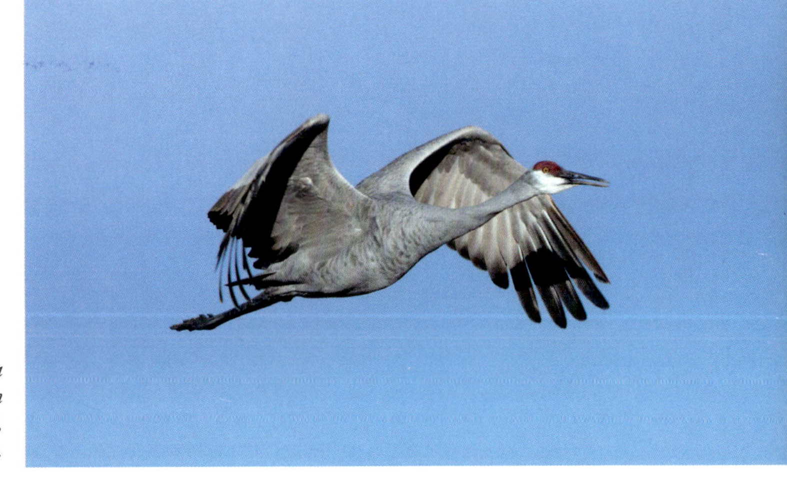

Sandhill crane flies to a nighttime roost. Canon 40D, 500mm, ISO 400, f/6.3 @ 1/3200 sec.

evening. Movement times vary, but expect the cranes to arrive in the fields around daybreak and leave to roost about one hour or so before dusk. The geese prefer to feed at night and rest in the ponds for most of the day. About one hour before sunset, geese in large numbers will move from the water to the planted fields.

Because of the vehicle restrictions, hand-holding your camera and lens for flight shots is necessary. Window mounts can assist in supporting your rig, but limit your ability to shoot flying birds. Set the camera on continuous autofocus, high-speed continuous shooting mode, and a relatively high ISO speed. Turn on the lens' image stabilization while using a **long telephoto lens (400mm or longer)**. Shoot with a wide aperture to help keep the shutter speed above 1/1250 of a second. If possible, keep the sun to your back when photographing the flying birds.

A day or two of scouting here will benefit you. Since the refuge is large and the birds move around to find the best feeding and loafing area (but tend to stay in the same places from day to day), scout the area for a day or two if you have time. On the day of your shoot, arrive early, and drive your vehicle to where you noticed activity the day prior.

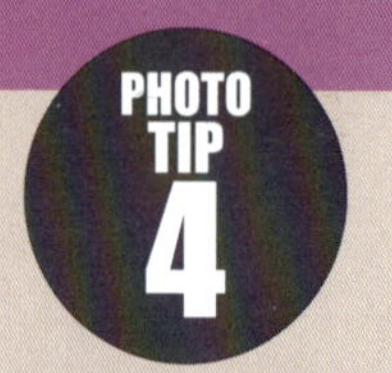

Bird Calls

Birds attract humans through their melodious voices (and of course, their beautiful colors). For centuries, song lyrics and poems have wooed us with ethereal descriptions of doves calming coos and mockingbirds midnight trills.

These voices developed to allow birds to communicate their feelings, moods, locations, and desires. Birdwatchers and scientists trained themselves to imitate certain bird songs to attract the birds and enable observers to get close enough for study. During the 1940s, when transportable sound recording equipment became available, scientists and bird watchers recorded bird voices, making it easier to identify and locate them. Today, smartphone applications like iBird Pro and Bird Tunes feature complete sets of easy-to-use bird calls.

Also known as "playback calls," bird calls work best during the breeding and nesting season. When a photographer uses a call, a male bird that has already set up (or is in the process of setting up) a territory will approach your area. In doing so, the eager bird aims to determine if there is a rival male trying to co-op his space. The defending bird will sing to let the interloper know that he is boss and that the new bird (imitated by your call) should leave. Alternatively, if a female in the area is looking for a mate, she may approach the call to see if the new bird on the block is attractive enough to gain her attention.

Bird calls serve two main goals. First, if you arrive at an area and do not know if the bird you wish to photograph is present, play its call and listen for a response. For instance, I use owl calls quite a bit when out at night. Obviously, I cannot see the owls, so I play their calls and wait a few minutes to see if they respond. The second goal of using bird calls is when you already know that the bird is in the vicinity and you wish to attract the bird close enough to take a frame-filling image.

Find a good place to take photographs and set up your blind (or camouflage yourself in some other manner—see "Hide and Seek: Camouflage" on page 18). Then, place a Bluetooth speaker below the perch that you want the bird to sit upon when you create your photographs. Call for a few seconds. If a bird shows interest, he/she approaches to look for the other bird in its territory.

Only play a call long enough to keep a bird's interest. If you play the call too long, your photographic subject may become afraid or aware they are being fooled and fly away. When abused, calls can negatively affect the birds. Too much calling, or calling during the breeding season, can potentially scare a territorial male and drive it away. During migration, some scientists believe that if bird calls are used in excess, the intended photographic subject uses up a tremendous amount of energy defending itself against a false enemy (the call) and may die before it reaches the breeding grounds.

Wildlife photographers have a moral responsibility to protect and preserve our animal subjects. For further information on the best and proper use of bird calls, visit **www.sibleyguides.com/2011/04/the-proper-use-of-playback-in-birding**. If photographers use these guidelines, we will do no harm to the wildlife we enjoy photographing.

RIGHT: Common yellow-throated warbler calls to defend its territory in response to an imitation bird call. Canon 40D, 500mm, 1.4x teleconverter, ISO 200, f/9 @ 1/400 sec.

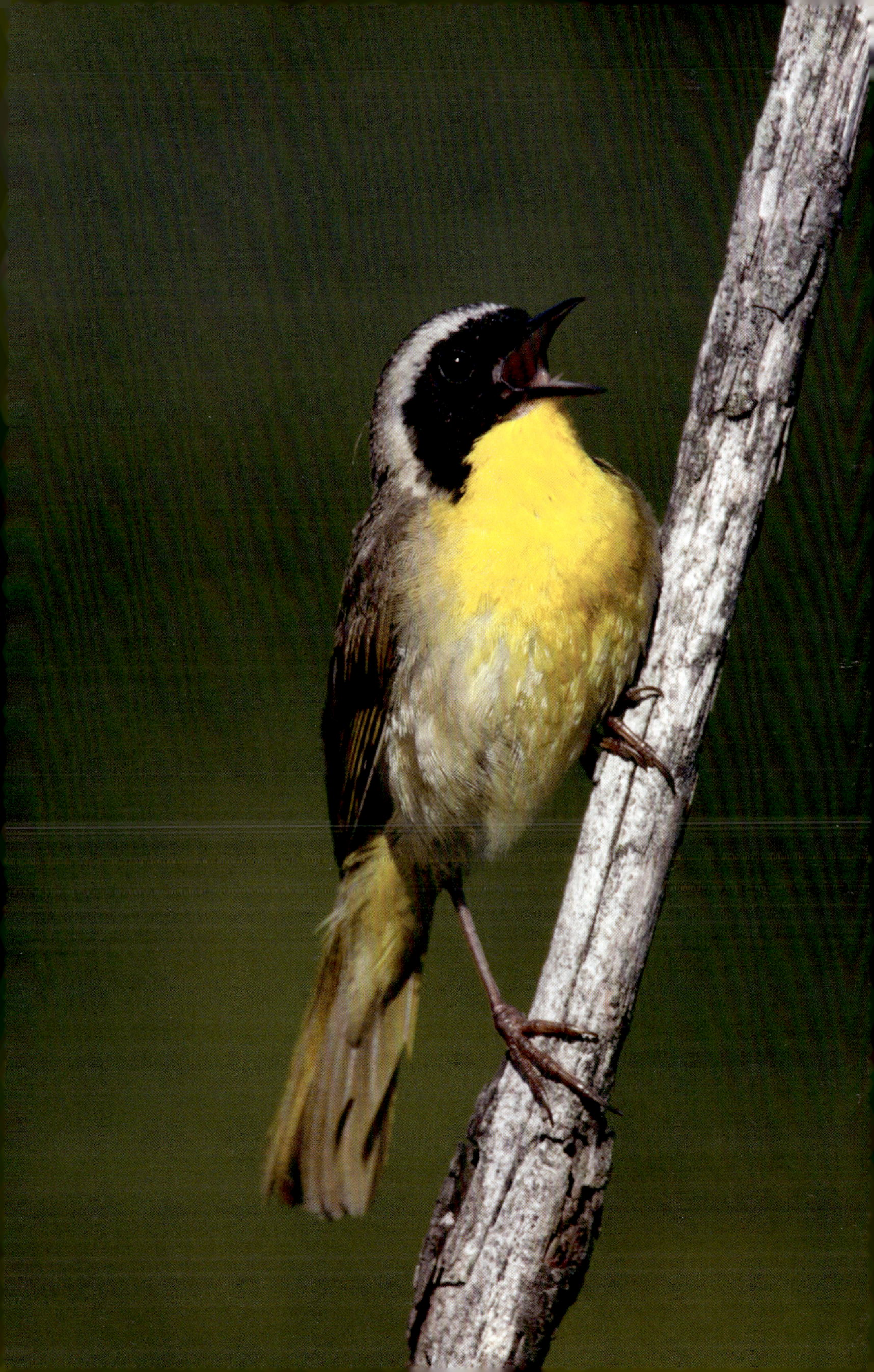

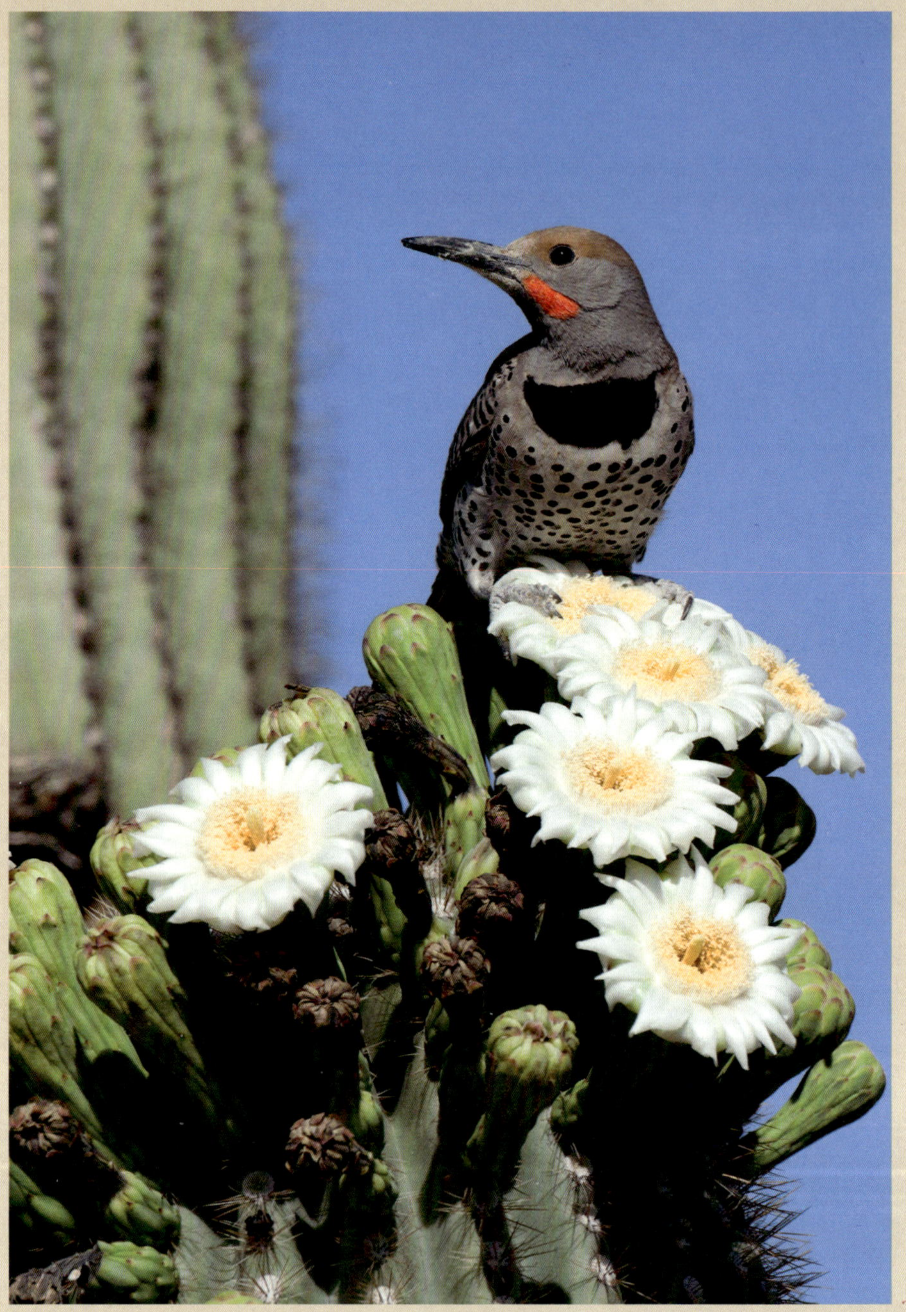

Gilded flicker sits on saguaro cactus flowers. Canon Digital Rebel XTi, 500mm, ISO 200, f/10 @ 1/500 sec.

Wildlife of

CENTRAL ARIZONA

C

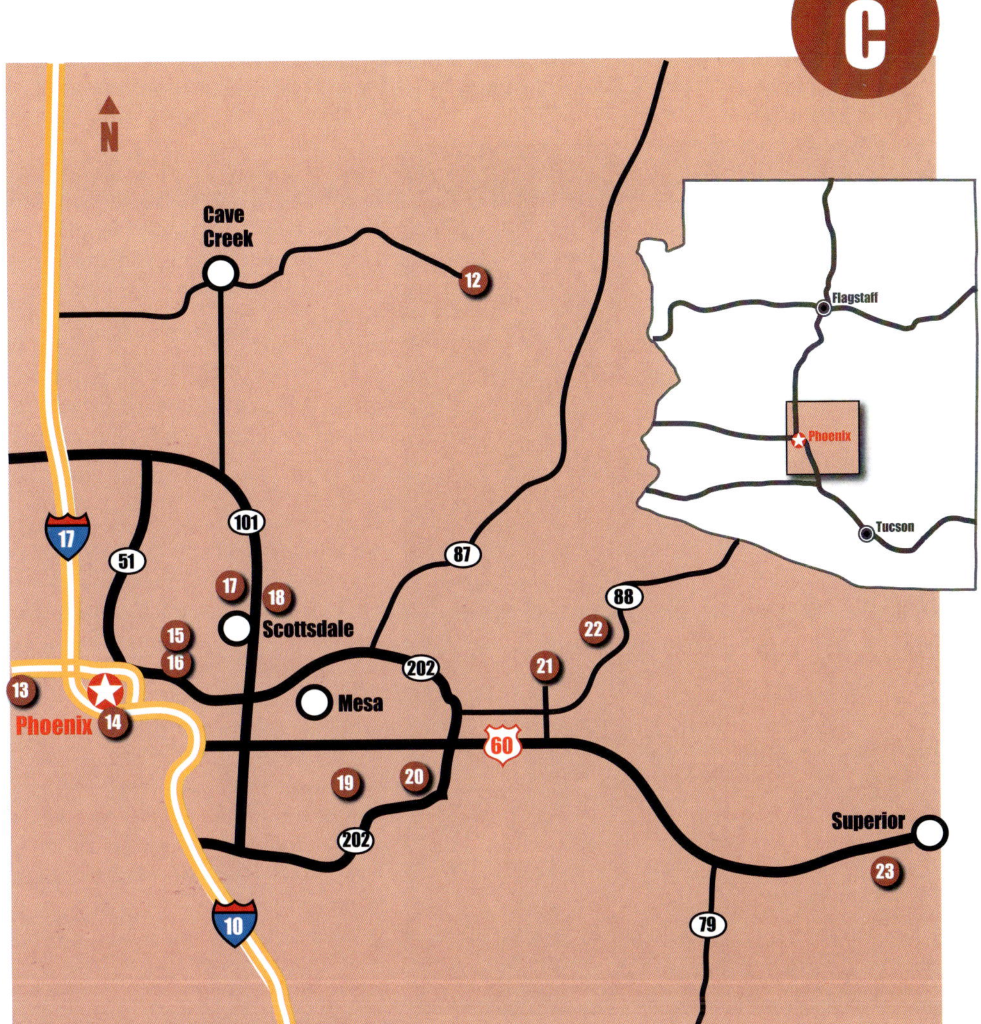

N
Cave Creek
12
Flagstaff
Phoenix
Tucson
101
17
51
87
17
18
Scottsdale
88
15
22
16
202
21
13
Mesa
Phoenix
14
60
19
20
Superior
202
23
10
79

Bartlett Dam Road

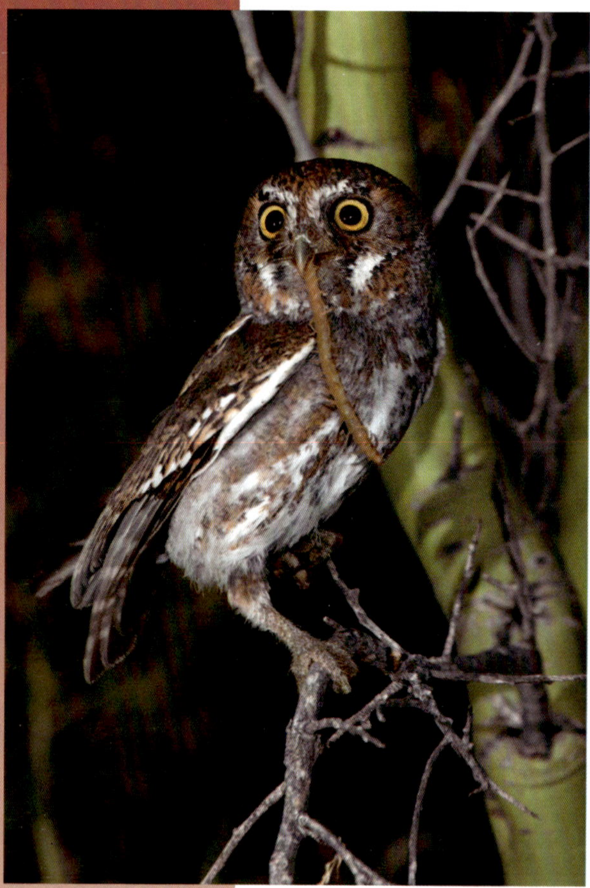

Elf owl with centipede gets ready to feed its babies. Canon 40D, 100-400mm at 400mm, ISO 200, f/10 @ 1/250 sec., off-camera flash.

Along Bartlett Dam Road (east from North Cave Creek Road to Bartlett Lake), giant saguaro cacti, palo verde trees, ironwood trees, and smaller cacti thrive in some of the richest Upper Sonoran Desert habitat found in Arizona. The gentle granitic rolling hills host two of Arizona's most abundant owls. Western screech owls are year-round residents, while elf owls, the smallest owl in the world, migrate from Mexico to the U.S. to breed. Both species breed in woodpecker-made cavities in the saguaro cacti and ironwood trees. Although elf owls spook more easily than screech owls, both are abundant and easy to find (with some patience and artificial calls) from late March until early July.

Photographing these diminutive owls requires different skills and equipment from that used to photograph birds during the daytime. Use a **moderate telephoto zoom lens (e.g., 100-400mm)** on a **cropped frame camera**. Leave your tripod behind. Bring a powerful flashlight to not only navigate in the desert, but also to shine on the owls so that your camera's autofocus will work in the low lighting conditions. Alternatively, a headlamp frees both hands to handle the camera, flash, and other equipment.

VIEW TIME
Late March to early July

IDEAL TIME OF DAY
Night

VEHICLE
Any

HIKE
Moderate to strenuous

Begin at dusk about three miles (4.8 km) east of the intersection of Cave Creek Road and Bartlett Dam Road. Find a safe place to pull off to the side of the road and step outside. Listen carefully for about five minutes to see if you can hear either species calling. If you cannot, play their calls for about two minutes, and then listen for another five minutes. Play the calls again for two minutes and listen for five minutes. If you receive no response to either species call, drive one mile (1.6 km) further east and repeat the same sequence.

When an owl responds to the call, note the direction and approximate distance. Turn on your flashlight or headlamp and walk towards the origin of the call. Stop after a short walk, turn off the headlamp, and listen for the owl. If the owl continues to respond, keep walking towards it. If the owl stops responding, play the bird's call again for a few minutes. Keep listening, playing the call, and walking towards the owl. Shine your flashlight or headlamp in the surrounding vegetation to look for it. If the owl flies away

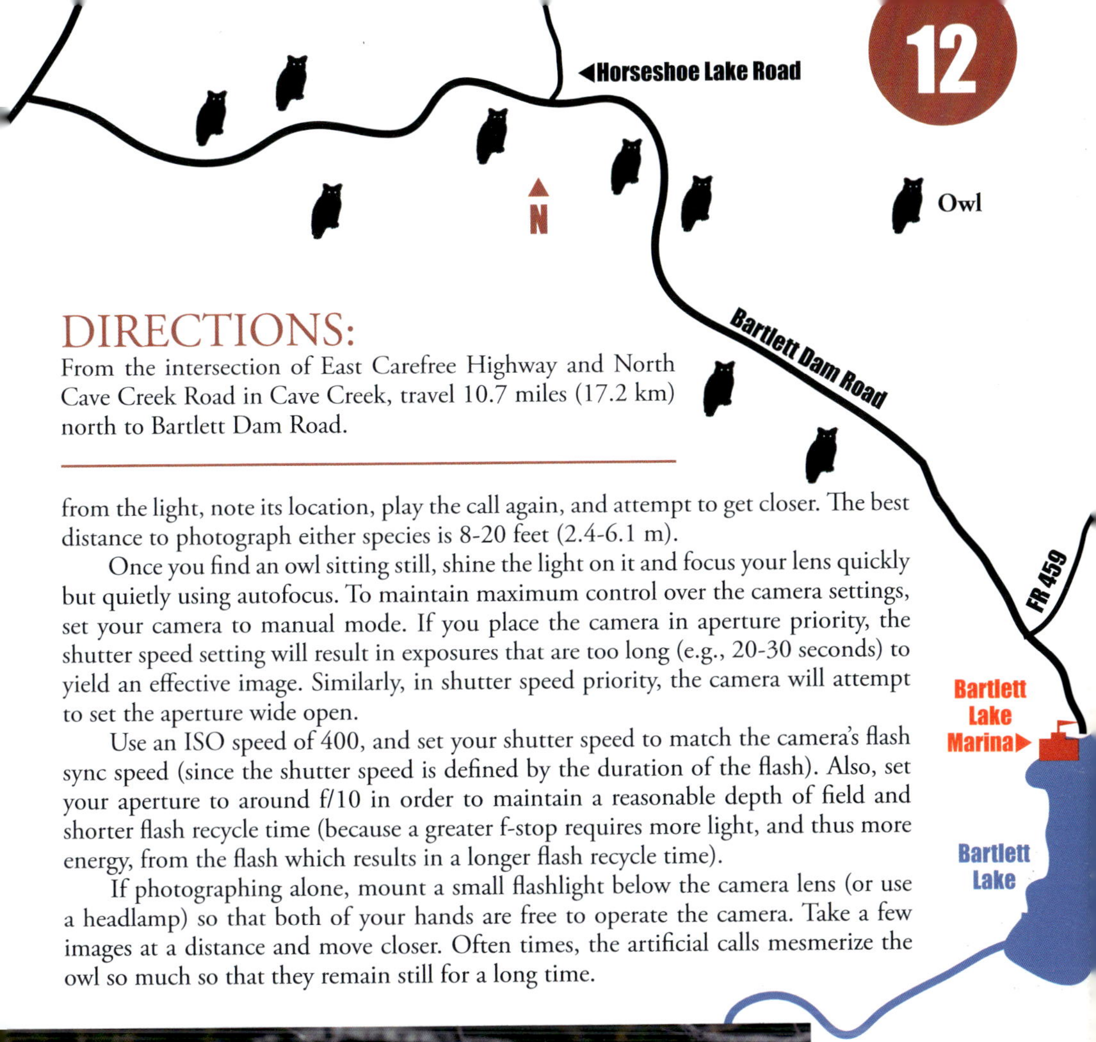

12

◄Horseshoe Lake Road

N

Owl

Bartlett Dam Road

FR 459

Bartlett Lake Marina►

Bartlett Lake

DIRECTIONS:

From the intersection of East Carefree Highway and North Cave Creek Road in Cave Creek, travel 10.7 miles (17.2 km) north to Bartlett Dam Road.

---

from the light, note its location, play the call again, and attempt to get closer. The best distance to photograph either species is 8-20 feet (2.4-6.1 m).

Once you find an owl sitting still, shine the light on it and focus your lens quickly but quietly using autofocus. To maintain maximum control over the camera settings, set your camera to manual mode. If you place the camera in aperture priority, the shutter speed setting will result in exposures that are too long (e.g., 20-30 seconds) to yield an effective image. Similarly, in shutter speed priority, the camera will attempt to set the aperture wide open.

Use an ISO speed of 400, and set your shutter speed to match the camera's flash sync speed (since the shutter speed is defined by the duration of the flash). Also, set your aperture to around f/10 in order to maintain a reasonable depth of field and shorter flash recycle time (because a greater f-stop requires more light, and thus more energy, from the flash which results in a longer flash recycle time).

If photographing alone, mount a small flashlight below the camera lens (or use a headlamp) so that both of your hands are free to operate the camera. Take a few images at a distance and move closer. Often times, the artificial calls mesmerize the owl so much so that they remain still for a long time.

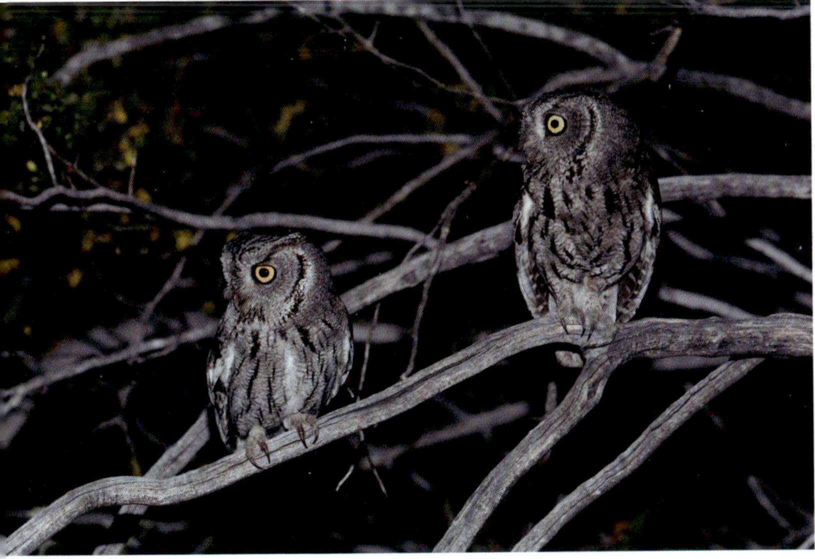

A pair of adult western screech owls. Canon 5DMIII, 70-200mm at 140mm, 2x teleconverter, ISO 500, f/10 @ 1/200 sec., on-camera flash.

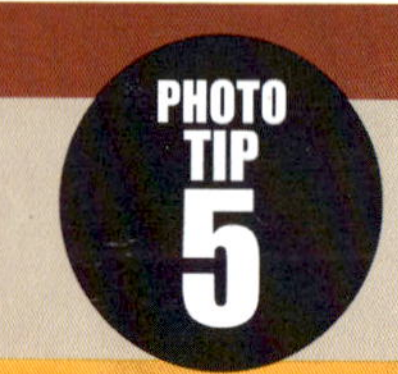

Chase the Light

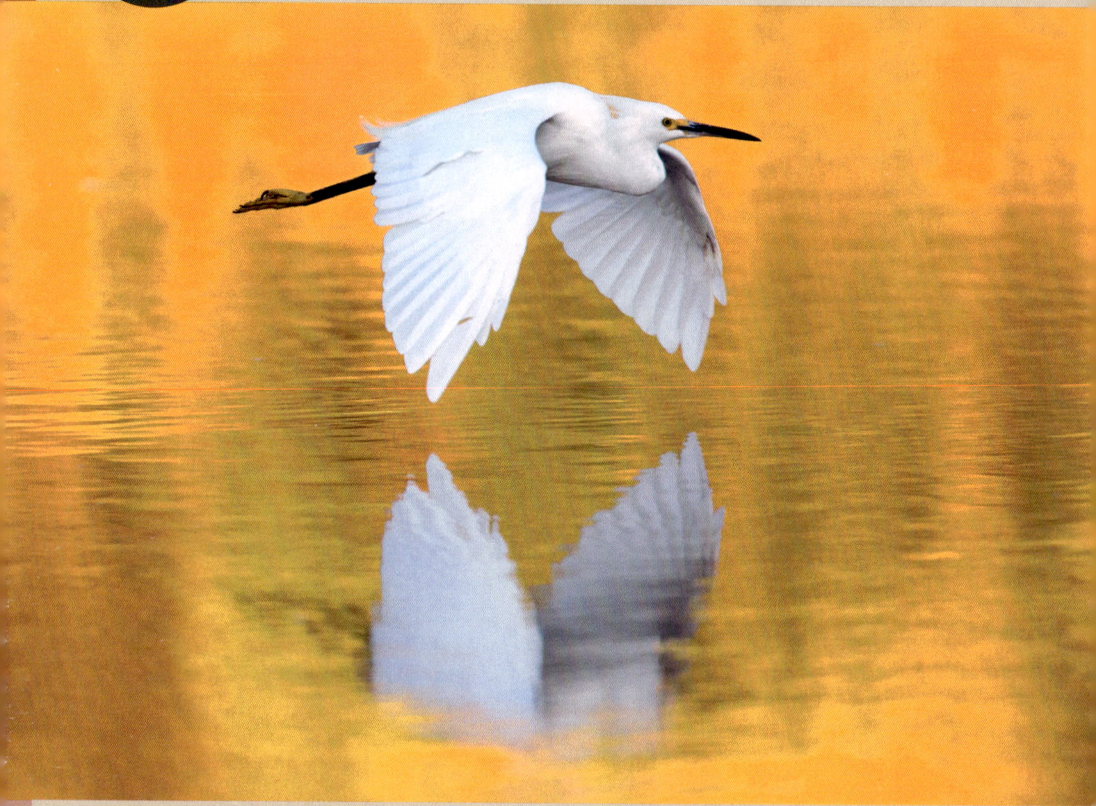

To prepare for this image, I scouted for a location where snowy egrets were active and where the early morning light bounced off nearby cottonwood trees and reflected into the calm waters. Canon 5DMIII, 500mm, 1.4x teleconverter, ISO 800, f/6.3 @ 1/800 sec.

Light provides the palette of colors photographers use to paint on our canvas. The final quality of an image is determined by not just the composition and subject matter, but also by the way in which light illuminates the scene. Along with an understanding of wildlife behavior and distribution, wildlife photographers must understand how light affects their photographs.

During a 24-hour day, the quality of light changes dramatically. In the early morning, as the sun cracks the horizon, the light travels through 50 or more miles (80.5 km) of atmosphere that might be dusty, moist, dry, clear, or a myriad of other air qualities. Calm air in the early morning immediately following sunrise produces warm red, yellow, and orange light. Called "The Golden Hour," this time frame offers ideal conditions to photograph wildlife, especially since the animals are more active in the early morning hours.

As the day progresses in Arizona, the sun begins to distance itself from the horizon and migrates slightly to the south. Now the light is moving straight down through less atmosphere. Colder, bluer light results due to the reduced dust and moisture in the air. The high angle of the sun causes dappled light to filter through vegetation and renders harsh shadows on the photographer's subjects.

Later in the afternoon, the sun moves closer to the horizon and, once again, begins to produce warmer colors. As the shadows become longer, the light becomes more evenly distributed. Although strong winds can affect the quality of light, late afternoon offers another great time to photograph wildlife.

As darkness approaches, the warm reds, yellows, and oranges become difficult to see, and the cold blues begin to dominate the color range. Within a couple of hours after sunset, all color fades into black, and wildlife photography enters the realm of artificial lighting.

Seeking the "best light" for your photography can pay off—or can produce predictable and boring results. If you wait for the "perfect" light, you will miss some wonderful opportunities. Relish all light in all weather conditions to add more drama and excitement to your images. Use falling snow to show additional motion in your frame. Tap into the soft and stimulating light following a rainstorm, regardless of time of day. Follow a subject throughout a cloudy day, where you will not need to worry about harsh shadows or using fill flash. As Alfred Stieglitz once said, "Wherever there is light, one can photograph."

Next to the quality of the light, how it falls on a subject can make or break the success of a photograph. Shoot with the sun behind you (called front-lighting), but not directly behind you. By having the sun a little off center on your subjects, the animal you are photographing will retain the texture of their feathers and fur while still catching a nice glint in their eyes. The colors also remain vibrant.

In addition, seek backlit subjects (where the sun falls behind the animal). Backlighting adds a thin halo of light around your subject to help them stand out from the background and gives a nice "pop" to your images. To avoid harsh shadows while keeping the soft colors, use backlighting during the early morning or late afternoon.

For mammals, seek side lighting to enhance the visual depth of their fur. For best results, use side lighting when a bright sun is throwing shaping shadows.

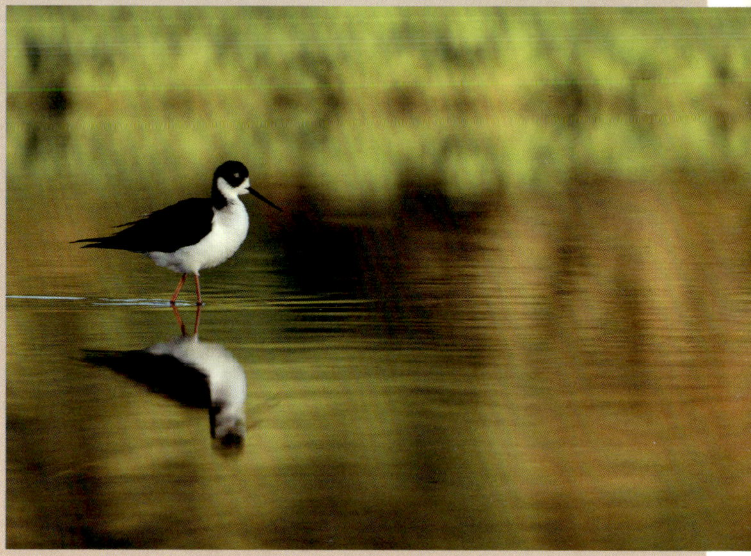

At dawn, the light reflects off the green vegetation and onto the water's surface to present a soft, saturated image of a black-necked stilt. Canon 1DMIV, ISO 400, f/4 at 500mm, 1.4x teleconverter, f/6.3 @ 1/640 sec.

CENTRAL ARIZONA

South Verado Way

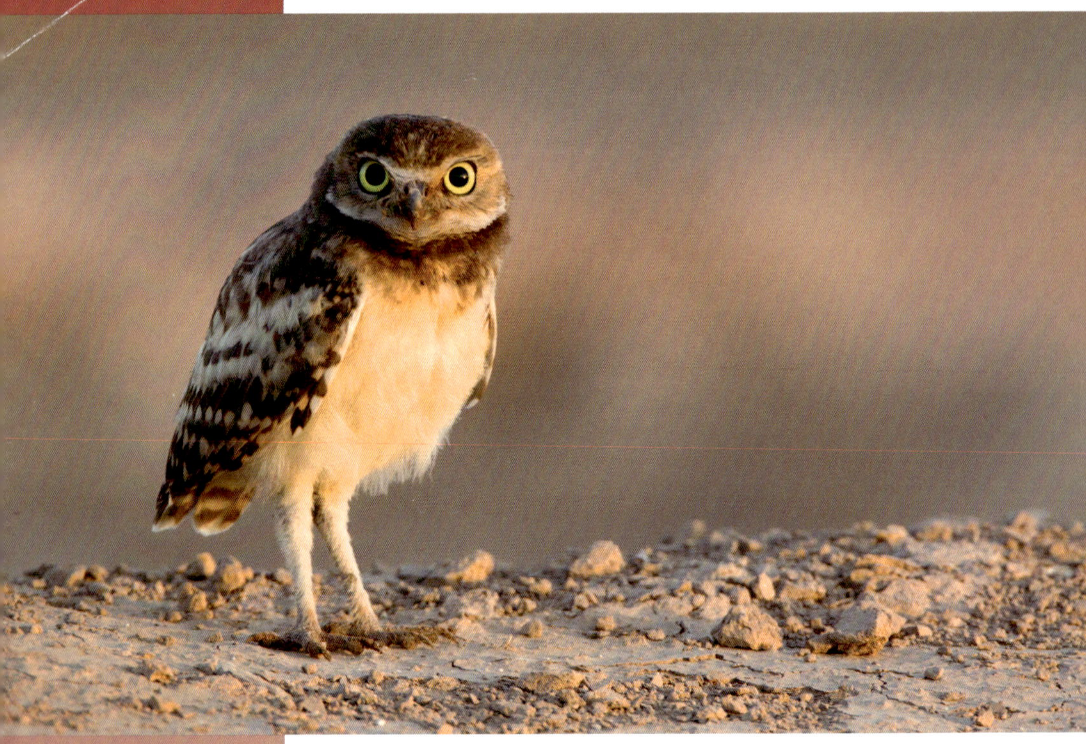

Baby burrowing owl in early morning light. Canon 1DMIV, 500mm, 1.4x teleconverter, ISO 400, f/6.3 @ 1/160 sec.

VIEW TIME
Year-round

IDEAL TIME OF DAY
Early morning; sunset

VEHICLE
Any

HIKE
Easy

The burrowing owl's large, yellow eyes, soft calls, and intelligent attitude pull at the heartstrings of many photographers. While these beloved creatures appear across much of Arizona, nowhere are they more abundant than near agricultural lands. The banks of irrigation canals offer the perfect place for these owls to excavate their nests while the nearby agricultural crops attract food for hungry owls and their young. Near the west end of Goodyear, in this 17-square-mile (44-square-km) agricultural zone, a very dense population of burrowing owls has created over 25 separate territories.

Ideal light occurs during the first three hours after sunrise—when the cooler temperatures encourage the owls to bask in the warm sun—or an hour or two before dusk. Bring a pair of binoculars, a **telephoto lens (in the range of 300-500mm)**, a **cropped frame camera** (or a **full-frame sensor camera** if you own a telephoto lens longer than 500mm), a bean bag, and plenty of memory cards.

Drive slowly along one of the north-south roads within the quadrant, stopping every so often and using your binoculars to look for owls. Owls will conspicuously stand outside of their burrows or on top of the canal berms adjacent to the road.

When you locate a likely subject, roll down the window (the one facing the owl), put your bean bag on the door jam, and slowly drive to within shooting range of the owls. Pull off the road a safe distance before stopping the car and shutting off the engine (if you leave the engine running, your

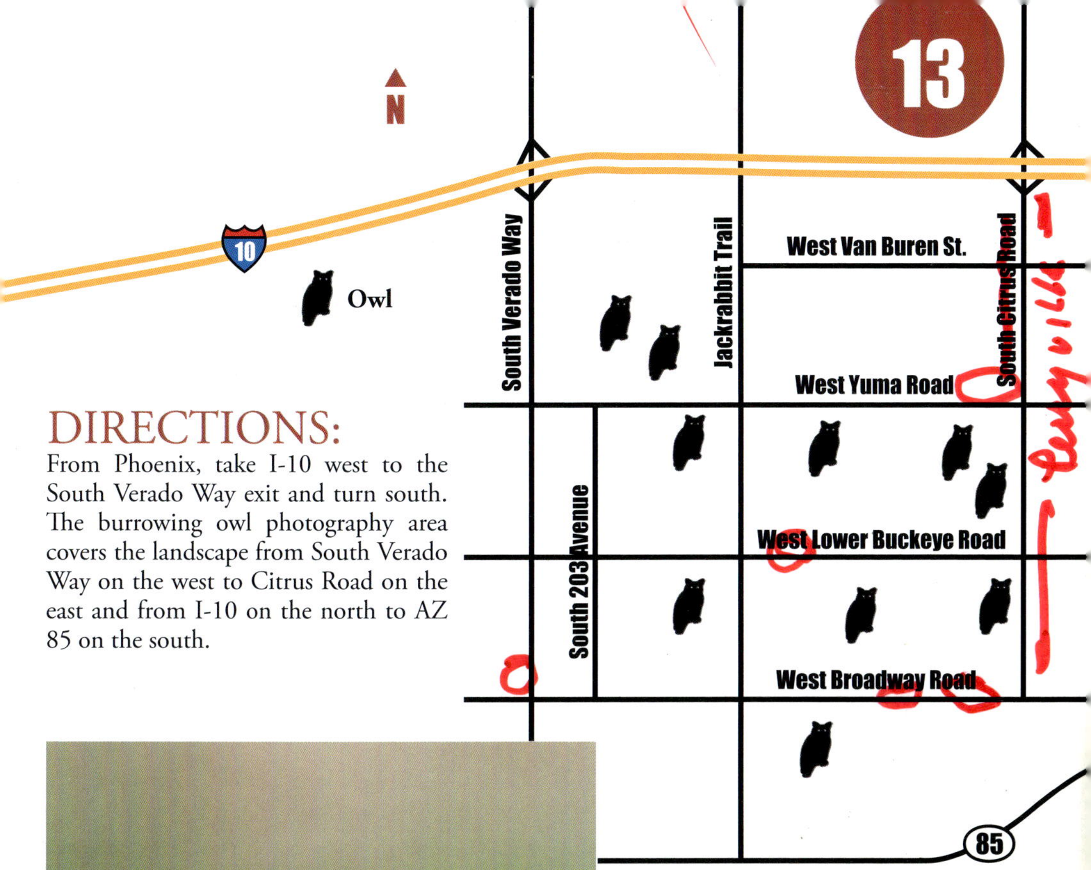

13

N
Owl
West Van Buren St.
West Yuma Road
West Lower Buckeye Road
West Broadway Road
South Verado Way
South 203 Avenue
Jackrabbit Trail
South Citrus Road
10
85

DIRECTIONS:

From Phoenix, take I-10 west to the South Verado Way exit and turn south. The burrowing owl photography area covers the landscape from South Verado Way on the west to Citrus Road on the east and from I-10 on the north to AZ 85 on the south.

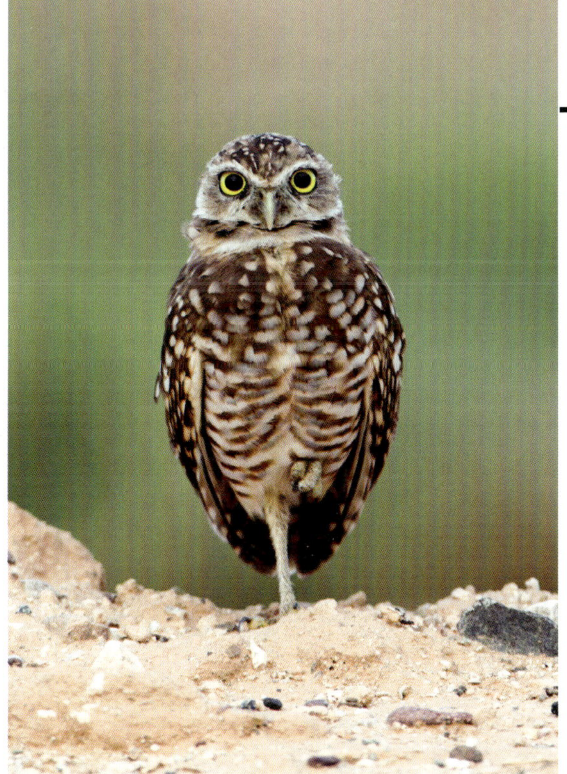

Adult female burrowing owl rests on one leg. Canon 1DMIV, 500mm, ISO 400, f/5.6 @ 1/320 sec.

photos may appear soft due to the engine's vibration). Use a moderately fast ISO speed (e.g., ISO 200 or 400), a moderately wide aperture (e.g., f/8), and a shutter speed of at least 1/500 of a second. Turn on image stabilization (or vibration reduction) mode, and enable your high-speed continuous shooting mode. If you are quiet and still, the owls will forget your presence and continue about their normal activities.

As you set up, look for front or back lighting. Find an owl standing on top of a berm so you can take an image with a colorful, out-of-focus backdrop. If you visit this area during the winter months, farmers may have planted a green crop that can provide a clean, rich, and color-contrasting backdrop for your photographs of these exciting owls.

Making the Photo

2

Ouch!

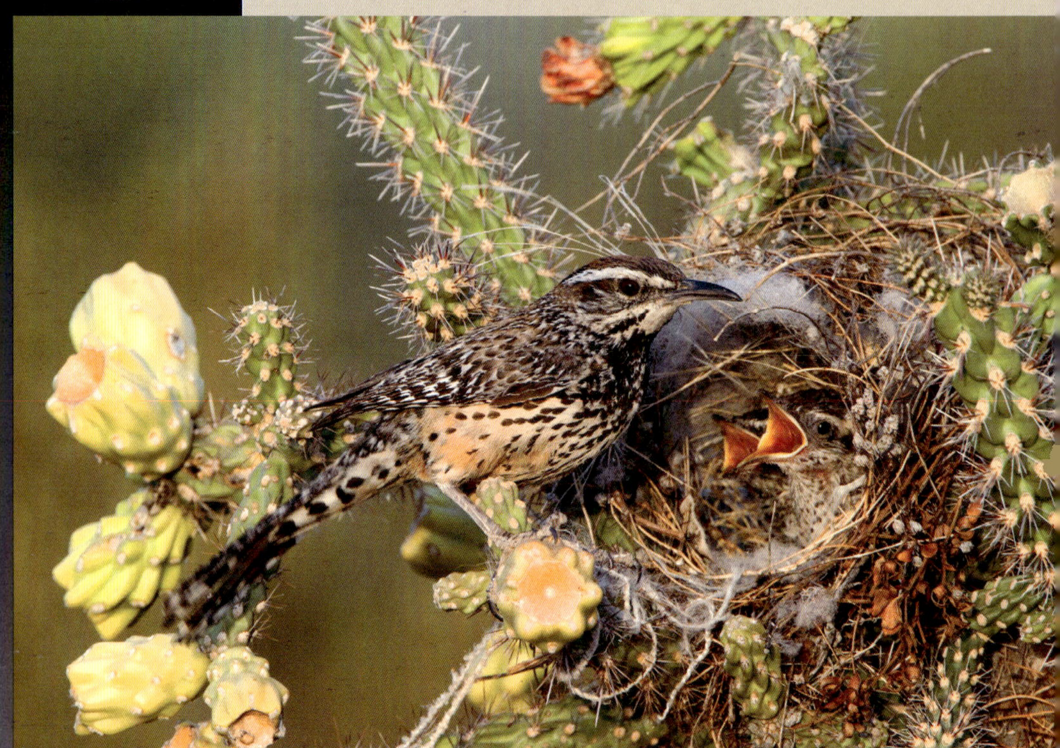

Cactus wren feeds its hungry babies. Canon Digital Rebel XTi, 500mm, 1.4x teleconverter, ISO 400, f/11 @ 1/125 sec.

When I arrived in the Grand Canyon State from the Midwest, one of my first revelations was the difference between the wet climate plants in Wisconsin, Massachusetts, and Illinois and the dry desert plants of Arizona. In the north, I could stroll through the countryside watching out for only the occasional briar or poison ivy plant. In the southwest, I was constantly picking prickly pear and cholla needles from various (and sometimes unmentionable) parts of my body. I learned to carry a comb with me so I could more easily extract the large cholla balls. I also became a believer that walking carefully and wearing large, leather boots was a "must," even in the heat of Arizona's searing summers. My wife began to complain about the small, and mostly unseen, sharp objects that I was bringing into the house and that ended up in her feet or clothing.

My appreciation for the wildlife that lives in the midst of these seas of cactus soon replaced my concerns about protecting myself from the "dangers" of hiking and hunting in Arizona. I found out from fellow biologists that the flower and fruit of the cholla and prickly pear provided important food for many of the deserts birds, mammals, and insects. As I tiptoed through cactus habitats during the spring, I noticed small birds darting out from cholla thickets. On closer inspection, I discovered bird nests in seemingly impenetrable and dangerously spiny cacti. I began to notice javelina and deer with cholla balls attached to their fur. Then it hit me!

In the wet climates of the northern United States, birds and other wildlife used the tall thickets of green-leafed plants, trees, and shrubs to live and hide from their predators. In the Sonoran Desert, predators could spot their prey from a long way in the relatively open habitat, but cannot often get to them because of the type of foliage. For example, predators could easily see birds that nest in a cactus. However, because of the cactus's sharp, spiny needles, hunters on the prowl could rarely reach the well-protected birds. Fallen cholla pieces cover the ground around each plant, making it a painful venture to travel there with unprotected feet. Even if a four-footed predator could find some access to the nest, there was no possible way for them to get their mouths on their young prey through the protection of the cactus spines.

As my appreciation for desert birds grew, my desire to photograph them nesting in the cholla groves also grew. Much to my surprise, it was easier than I had thought. I consulted with professional ornithologists and biologists to determine that most of the cholla nesting birds did so from February to about June. Fortunately for me, a short distance drive north from my home landed me in native desert habitat with plenty of cactus and desert birds.

My scouting consisted of driving close to potential breeding sites, using my binoculars to watch for nest building behaviors or baby bird feeding, locate the nest, and investigate further. It took me several weeks to get the hang of this observation process, but eventually I located a cactus wren in a cholla whose nest just happened to be facing west (and not very hidden from view). The west-facing nest meant that I did not have to return to this spot before work or on weekends but rather could photograph the site at a more leisurely pace in the evening after work.

The nest that I located sat in a piece of mostly cleared land on the north end of Phoenix. Thus I could drive my truck right up to the nest to photograph from my "truck blind." I recorded the wrens bringing food into the nest, but they always disappeared into the funnel-shaped nest ball before delivering the meal to their young. Then, a few days before the young cactus wrens were ready to fledge, the adult wrens opened up a larger opening in the nest, and I could see the babies begging for food when the adults delivered it. The adult birds brought food to the hungry and noisy young, offering an opportunity to photograph the lovely family. All of my hard work and pain had paid off!

To this day, almost every spring, I survey the local cactus patches looking for bird nests. If I find one that makes a great photographic study, I set up my gear, relax, and wait for the inevitable action. I still get stung by the pain of cholla spines and bring home sharp objects for my wife to complain about, but now I have a better appreciation for how the desert works.

CENTRAL ARIZONA

South Mountain Park and Preserve

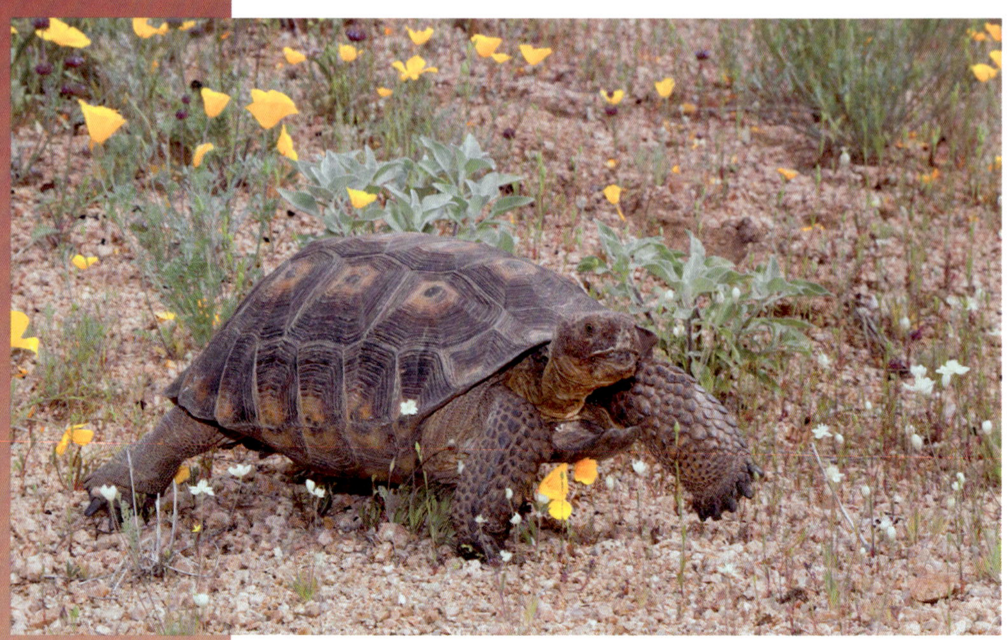

Desert tortoise strolls among the spring desert flowers at South Mountain Park. Canon 50D, 24-105mm at 58mm, ISO 250, f/11 @ 1/250 sec., on-camera flash.

VIEW TIME
February to June

IDEAL TIME OF DAY
Sunrise to late morning

VEHICLE
Any

HIKE
Easy to strenuous

South Mountain Park and Preserve encompasses 16,283 acres (6,589 hectares), making it the largest municipal park in the United States and one of the largest urban parks in the world. Rattlesnake, desert tortoise, chuckwalla, coyote, desert-adapted birds, and the giant saguaro cactus all call this metropolitan oasis home.

One of the more interesting animals found in the park is the common chuckwalla. The beautiful (in a lizard sort of way) black-bodied and carrot-red-tailed, 18-inch-long (45.7-cm) chuckwalla inhabits the larger boulder outcroppings of the park. They are most active from February to May when the males defend their territories and search for females.

Drive slowly on the roads near the top of the mountain and look for small, irregular bumps on the boulders indicative of chuckwallas basking in the sun and surveying their territories. If you spot a chuckwalla near the road, look for a safe pull-out and attempt to photograph the critter from your vehicle. As soon as you attempt to get out of your car, the lizards will probably run for cover beneath the behemoth rocks they inhabit. If you need to get out of the vehicle for a closer look, walk slowly and take several photographs as you approach them.

Use a **telephoto lens (e.g., 300 to 600mm)** to capture close-up portraits of this unique lizard. If you stay in your vehicle, sling a bean bag over your car door jam for stabilization. Otherwise, set the camera on a sturdy **tripod** and **ball head**. In either manual or aperture priority mode, use a fast shutter speed (e.g., 1/250 of a second or faster), an aperture of f/8 or less, and high-speed continuous shooting mode.

Maneuver your car (or body) so that the sun falls on your back and your shadow points towards your subject. Side or back lighting will cause

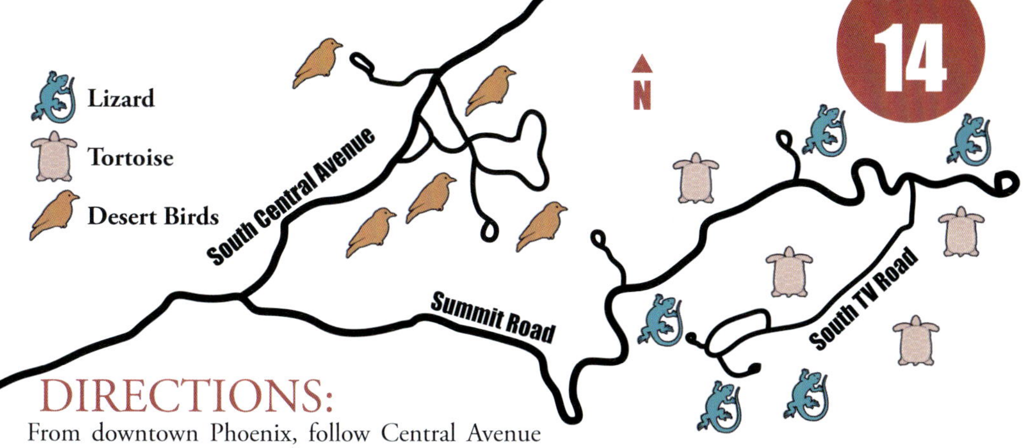

DIRECTIONS:

From downtown Phoenix, follow Central Avenue south to the entrance of South Mountain Park and Preserve.

For those traveling on I-17, note that no exit for Central Avenue exists. Instead, take Exit 195B for 7th Street. Turn south onto 7th Street and drive about 4.5 miles (7.2 km) to Dobbins Road. Turn right onto Dobbins Road and travel a half mile (0.8 km) before turning left onto Central Avenue. After about 1.5 miles (2.4 km), you will enter the park.

For information on the park's rules and regulations, visit **www.phoenix.gov/parks/trails/locations/south-mountain**.

---

deep, harsh shadows on the chuckwalla. A pop of **fill flash** can help with some of the shadows.

South Mountain Park also serves as home to many of the common (and some not so common) desert-adapted birds like the curve-billed thrasher, cactus wren, roadrunner, Gila woodpecker, gilded flicker, house finch, and Gambel's quail. In addition, the park provides one of the best places to photograph the Bendire's thrasher (a not-so-pretty, but certainly unique, species).

Arrive at the park at daybreak during the months of February through May. Find an open place with saguaro, prickly pear, and cholla cacti nearby. Sit down (watch where you sit!) and quietly survey the surroundings for bird movement. Listen for their calls. Soon you will see or hear bird activity. Note (in a notebook or mobile device) where the birds are and what behavior they demonstrate. Watch the nearby saguaro and cholla cacti to see if the birds are nesting or feeding young. Stay at this first location for about 30 minutes, and then move to another location at least 0.5 miles (0.8 km) away. Keep conducting your survey of bird locations until about 10 a.m. By this time, you should have noticed several bird territories and some nests.

To create frame-filling portraits of these special birds, use a telephoto lens on a sturdy **tripod** with a ball or **gimbal head**. You may consider hand-holding your gear, but your arms will tire over the course of an hour or so. If the birds appear comfortable with your presence, sit in a chair with the sun on your back and your shadow pointing towards the subject. When a bird arrives, use high-speed continuous shooting drive with a relatively high shutter speed (i.e., 1/250 of a second or faster). If the birds are not acting "normally," use a blind or ghillie suit to camouflage your presence.

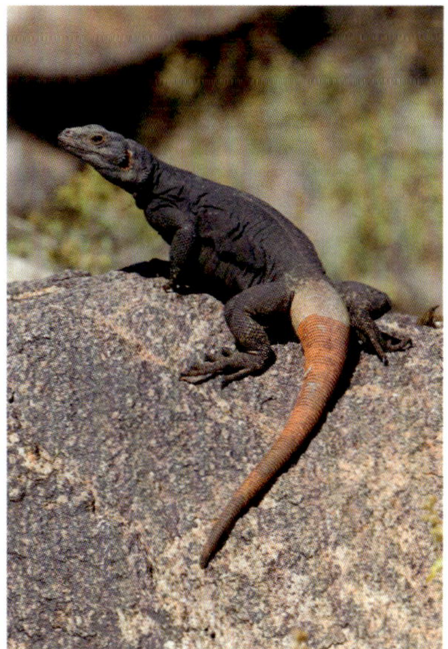

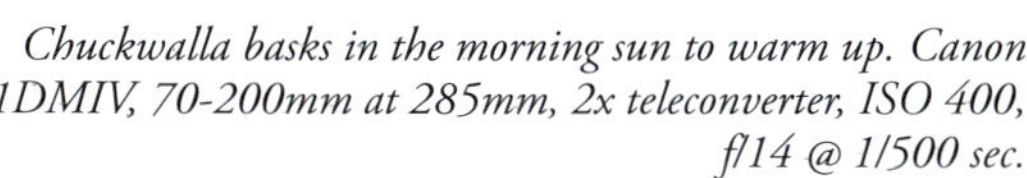

Chuckwalla basks in the morning sun to warm up. Canon 1DMIV, 70-200mm at 285mm, 2x teleconverter, ISO 400, f/14 @ 1/500 sec.

PHOTO TIP 6

Fill With Fill Flash

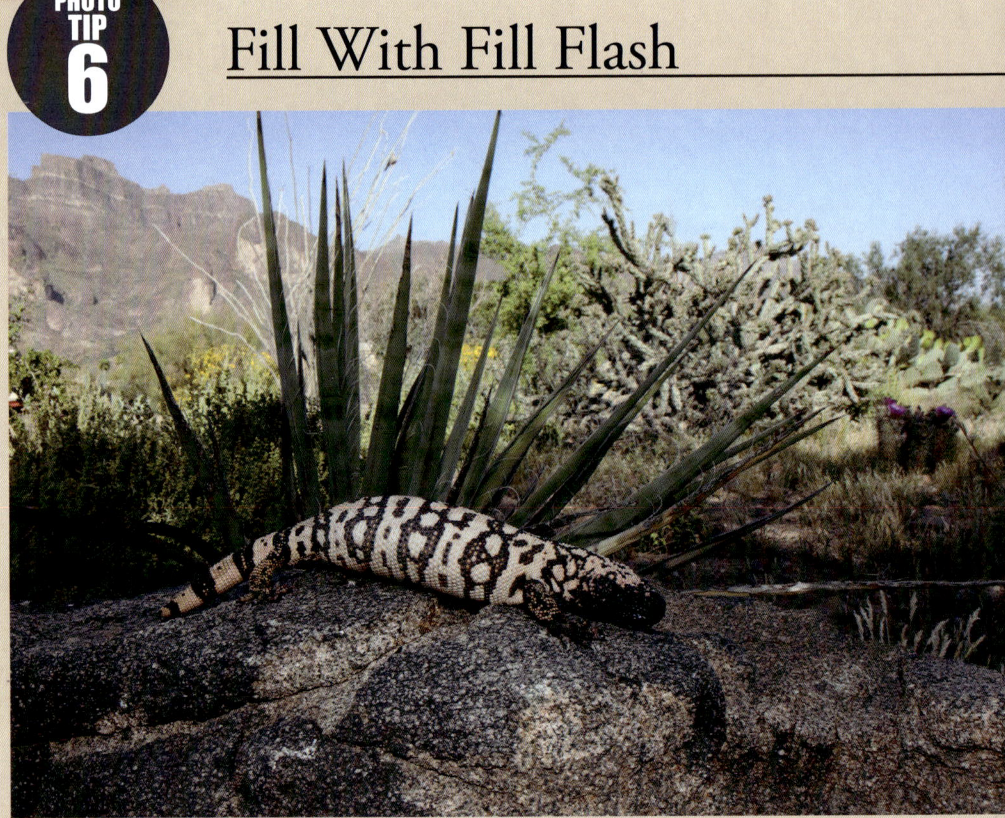

Gila monster. Canon 40D, 24-104mm at 24mm, ISO 400, f/14 @ 1/250 sec., on-camera flash @ -1 1/3 FEC.

Animals do not always do what you want them to do. After doing your homework and spending time scouting a site, you can typically place your subject in relatively effective light. At other times, squirrels scamper to the shade to keep out of sight from predators, lizards scoot from one bush to the next, and birds land in a location you could not predict. In these—and similar—instances, use flash to help obtain a proper exposure.

Flashes (also referred to as speedlites or strobes) are small sources of artificial light that either sit on the camera in the "hot shoe" or off-camera to add light to the scene. Modern day flashes use a mechanism called TTL ("through-the–lens" metering) where the flash and camera "talk" to each other to determine the correct amount of light based on the camera's settings for aperture and ISO speed as well as the subject's estimated distance.

In addition to TTL mode, manual flash mode allows the photographer to manually change the output power as needed. For the most part, the flashes output power can be halved, quartered, eighth, and so on until around 1/128 of the full output power.

By changing the output power, the speed of the flash changes. When set at full power, the flash may take about 1/1000 of a second for it to expel all of the stored energy. By reducing the power, the flash expels a lesser amount of energy, and at 1/128 power, the flash may only take 1/25,000 of a second to fire.

For the most part, only use the full power of the flash at night when it is your only source of light. At other times, employ the flash to add just enough light to assist you in obtaining an exposure where the subject is lit appropriately (appropriate exposure is a personal choice). Use the camera's flash exposure compensation function (which works like the exposure compensation feature, but for your flash) to vary the amount of flash output and give you the results you desire.

When you need to follow a moving subject to a new location, put a flash in the camera hot shoe and reduce the flash power by two stops (either in TTL or manual mode). Placing the flash on the camera affords you more mobility to follow the critters. Reducing the flash power still illuminates the subject without rendering a flat, overly bright look. If your subject is very close to you (in the case of a snake or lizard), use the pop-up flash (if available). When photographing animals more than two feet (0.6 m) away from the camera, a larger external flash unit will offer a more intense light source.

Even when using a powerful external flash, the animal may be so far away that the flash at full power cannot effectively reach the subject (depending on ISO and other factors, most unassisted flashes will not effectively illuminate subjects that are more than 50 feet (15.2 m) away). In this case, use a flash extender (such as the Better Beamer) to extend the light to great distances. These extenders attach to the flash unit and use a Fresnel lens to concentrate the flash's light into a smaller beam. Note that when using a telephoto lens of 300mm length or shorter, the beam is so narrow that it does not cover the entire scene (but works well enough to add an attractive catch light in the animal's eye).

On-camera flash and flash extenders expand the photographer's ability to take advantage of difficult lighting, but they can also cause either "red eye" or blue-colored "steel eye." Both of these unflattering effects result from the flash sitting too close to the lens, which causes the light from the animal's retina to reflect back to the sensor. To eliminate red eye or steel eye, reposition the flash away from the camera lens. Many photographic suppliers make a bracket that attaches to either the camera or tripod and raises the flash a few more inches above the camera's lens.

When working from a blind, or when you are not able to predict the exact location of your subject, set one or two flashes on flash stands several feet away from your camera. Not only does this placement eliminate the chance of red eye or steel eye, but this set-up also creates more natural-looking light which adds structural shading to fur, feather, and scales. Once you position your lights, use either wired or wireless flash remotes to trigger them.

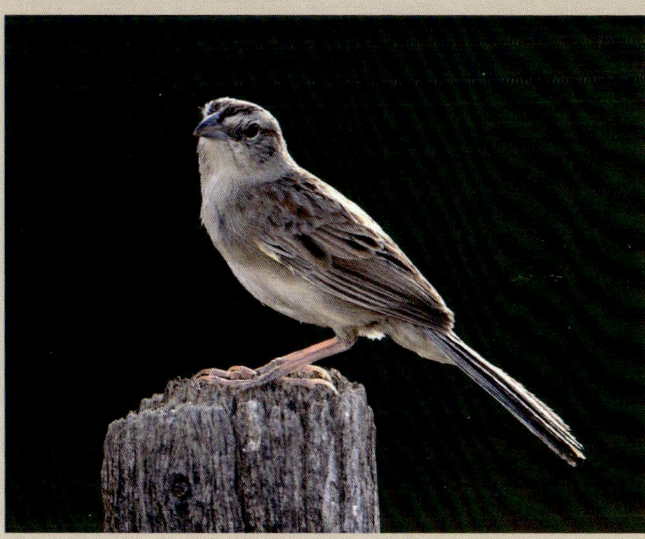

Botteri's sparrow. Canon XTi, 500mm, 1.4x teleconverter, ISO 200, f/5.6 @ 1/200 sec., on-camera flash @ +1 FEC.

Desert Botanical Garden

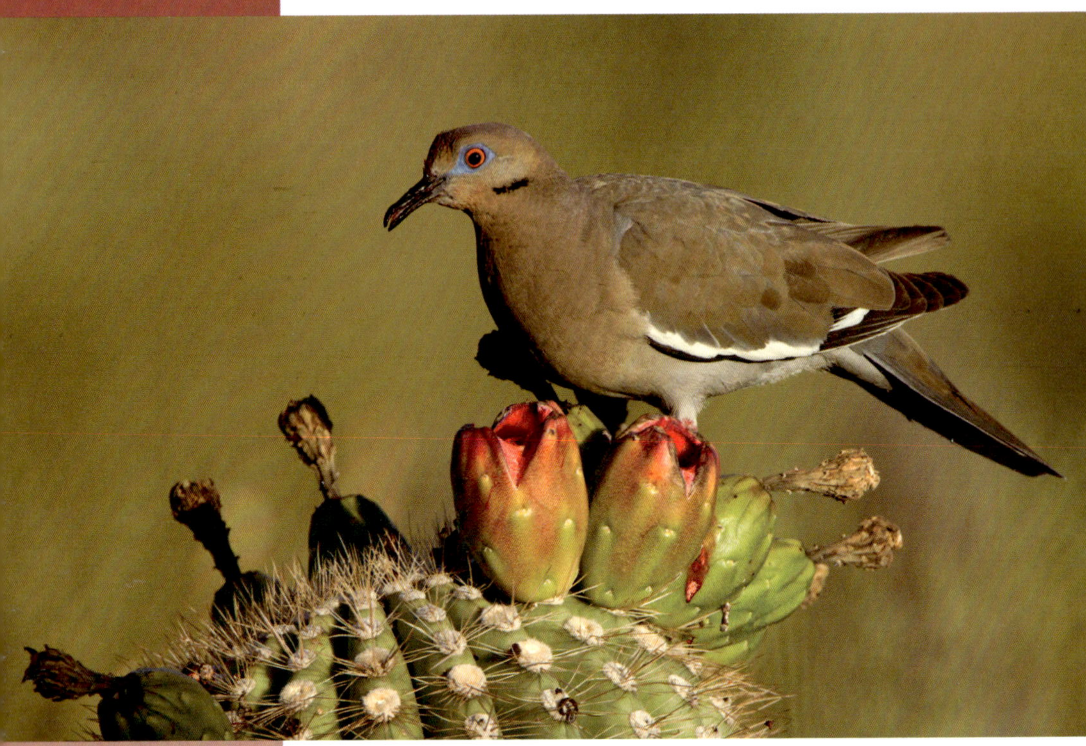

White-winged dove feasts on saguaro fruit. Canon 7D, 500mm, 1.4x teleconverter, ISO 200, f/8 @ 1/320 sec.

VIEW TIME
Year-round

IDEAL TIME OF DAY
Early morning to late afternoon

VEHICLE
Any

HIKE
Easy to moderate

Situated in the middle of the Papago Mountains next to the Phoenix Zoo, the Desert Botanical Garden not only maintains a world-class collection of desert plants, but also presents the photographer with one of the best places in Arizona to photograph desert wildlife. Stroll along any of the many walkways (some paved, some dirt) through flowering saguaro and prickly pear cacti, yucca gardens, and impenetrable cholla groves where birds, lizards, insects, butterflies, bullfrogs (which are non-native to the southwest and a threat to native frog species), and small mammals live.

Arrive early morning in late April, when the giant saguaro cacti begin to bloom. The saguaro bud begins to open about one hour after dark each day and stays open until the afternoon on the following day, when the unrelenting desert sun forces them to close. Although the blooms only last one day, each cactus produces flowers for a month or so. Birds, insects, and small mammals seek out the cactus blooms for their rich pollen and nectar. In the early morning hours, white-winged doves, Gila woodpeckers, Gambel's quail, gilded flickers, cactus wrens, house finches, European starlings (an exotic species), and many species of bees feed on the flowers to consume their nutrients.

The paths on the south end of the park feature the densest population of saguaro cacti. Walk slowly on the paths with your camera and **telephoto lens** (preferably on a **tripod**) ready. When you observe a bird feeding from a flower, move slowly and quietly to set your camera in position. Adjust your ISO speed to 400, and set the aperture to f/8 or less to maintain a shutter speed of 1/250 of a second or faster. Turn your camera drive mode to high-speed continuous shooting mode to take multiple frames when the action starts. Use continuous autofocus mode to track the birds as they reposition themselves on the cactus. After the flowering season, the saguaro cactus produces large, red fruit that are as sought after as the flowers.

During the same season (and likely initiated by the flowering cacti), many of the desert birds begin raising their families either on or in the saguaro. Gilded flickers, Gila woodpeckers, cactus wrens, starlings, and western screech owls nest in cactus holes. Mourning doves build flimsy nests balanced between the saguaro arms and its main trunk. When feeding their young, these birds franticly gather food and give it to their fast-growing and very vocal offspring. Once you find a nest hole or nest, set up your tripod with a telephoto lens. Set the camera to a relatively high shutter speed (e.g., 1/1000 of a second or faster) and continuous high-speed shooting mode. Wait for the parent birds to fly into the hole to feed their young. Pre-focus your lens a few inches in front of the hole, and when they approach, fire away! If you are lucky, you might get a great image of a fast moving bird with its wings spread or a rare image of a gilded flicker feeding its

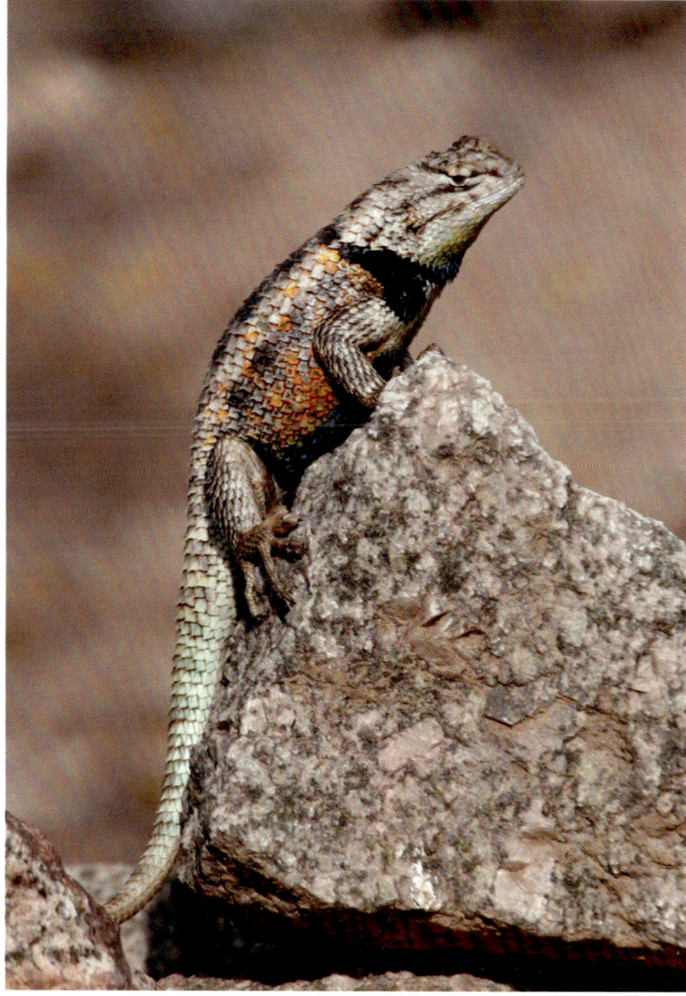

Colorful desert spiny lizard watches over his territory. Canon 20D, 100-400mm at 400mm, ISO 400, f/11 @ 1/1000 sec.

Desert Botanical Garden

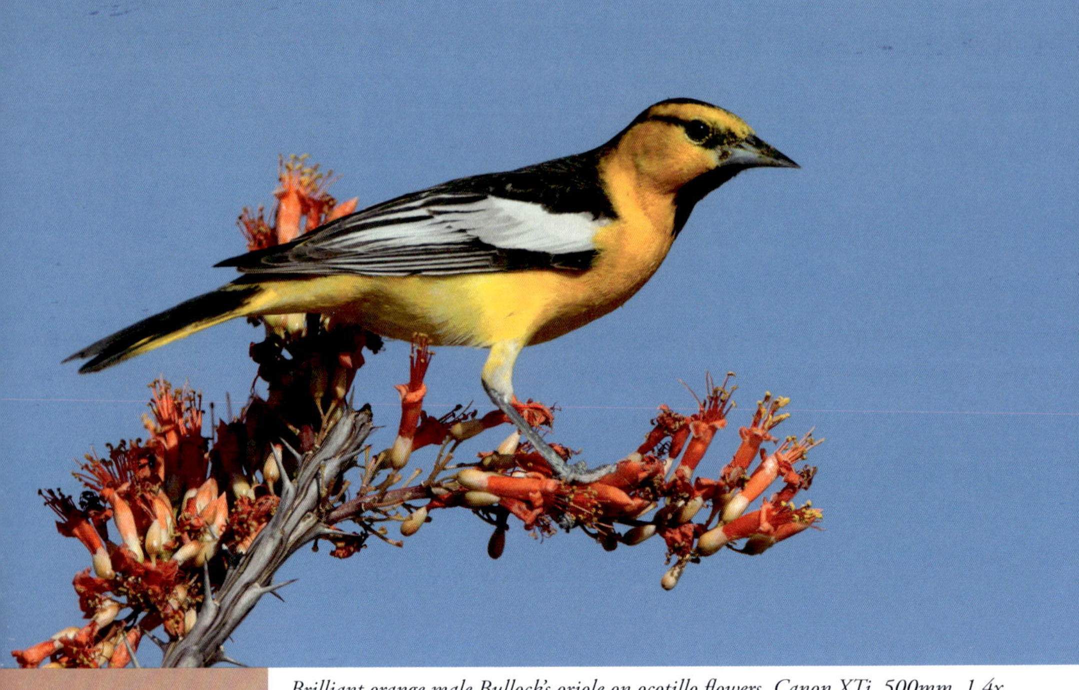

Brilliant orange male Bullock's oriole on ocotillo flowers. Canon XTi, 500mm, 1.4x teleconverter, ISO 200, f/8 @ 1/200 sec.

babies.

At any time of the year, visit the snack bar area to photograph the friendly Gambel's quail. During the warmer days, desert spiny lizards do push-ups to defend their territories and attract females for mating, as young birds learn to fend for themselves. Bullfrogs croak loudly in the pond, while hummingbirds are feeding on flowers at the aloe garden.

From March through May (check **www.dbg.org** for exact dates), the garden opens its butterfly pavilion for viewing. Giant swallowtails, zebra swallowtails, queens, painted ladies, malachites, and other species hover within arm's length. During October and November, as the temperatures begin to cool and monarch butterflies breed on desert milkweed, the garden displays a fascinating exhibit of these wondrous and delightful butterflies. Use a **100 or 200mm macro lens** (or a **fixed focal length of 50 to 300mm** with a **teleconverter** or **extension tubes**) and **fill flash**. Diffused lighting in the pavilion offers enough light to hand hold your camera, but use shutter speeds in excess of 1/200 of a second to ensure sharp images.

As summer turns to fall, many migrant birds such as Bullock's orioles and several species of flycatchers, sapsuckers, Cooper's hawks, and ruby-crowned kinglets visit the garden to live off the rich plants and insect life during the colder months.

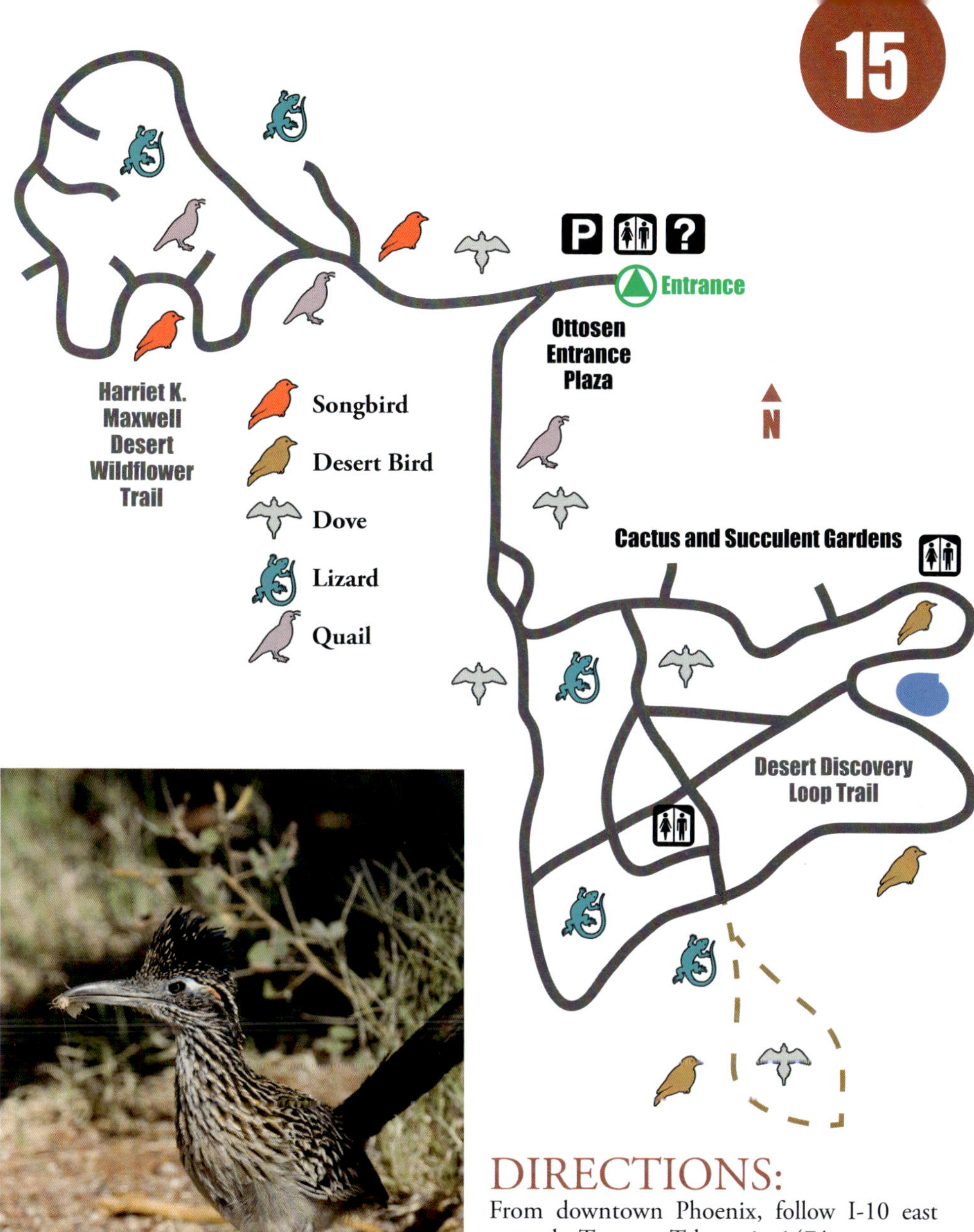

Greater roadrunner male tries to attract a mate with a grasshopper present. Canon 20D, 500mm, ISO 200, f/8 @ 1/640 sec.

DIRECTIONS:

From downtown Phoenix, follow I-10 east towards Tucson. Take exit 147A to merge onto AZ 202. Drive 4 miles (6.4 km), and then take Exit 4 for Van Buren Street. Turn left at Van Buren Street and drive just short of 1 mile (1.6 km) before turning left onto Galvin Parkway. Travel a little more than 0.5 miles (0.8 km) to a traffic roundabout. Take the first exit right at the roundabout into the Desert Botanical Garden and follow the signs to the parking lot.

The park charges an entrance fee. Learn more at **www.dbg.org**.

CENTRAL ARIZONA

Ponds at Papago Park

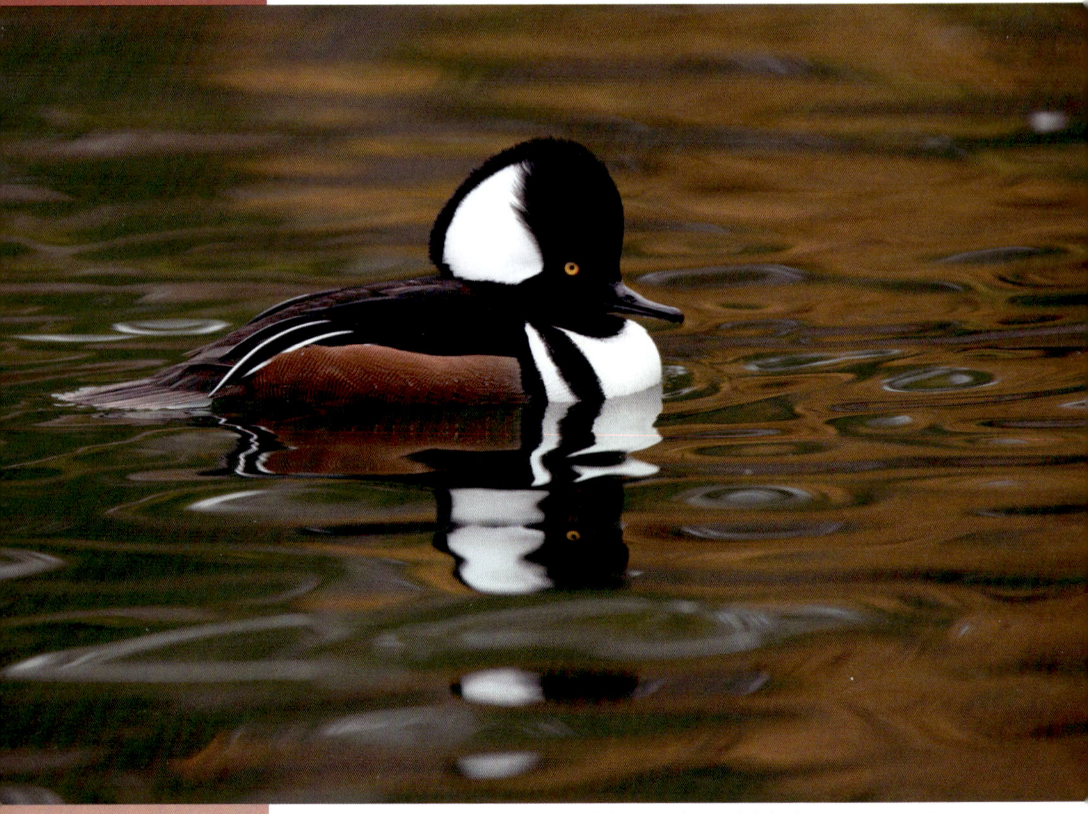

ABOVE: Hooded merganser, one of the most beautiful ducks in Arizona. Canon 5DMII, 400mm, ISO 800, f/6.3 @ 1/400 sec.
OPPOSITE: Green heron watches for its fish dinner. Canon 1DMIV, 500mm, 1.4x teleconverter, ISO 400, f/8 @ 1/400 sec.

VIEW TIME
October to February

IDEAL TIME OF DAY
Sunrise; late afternoon to sunset

VEHICLE
Any

HIKE
Easy

Urban pond photography, at least in the Phoenix metropolitan area, can yield rewarding opportunities. The ponds at Papago Park are great examples. Located between the Desert Botanical Garden and the Phoenix Zoo, the ponds (numbered 1, 2, and 3) are surrounded by a limited amount of vegetation and adjacent to parking lots, restrooms, and picnic areas. However, this metropolitan vibe does not stop a variety of waterfowl including canvasback, redhead, ring-necked duck, pintail duck, scaup, American coot, Canada geese, wigeon, gadwall, little green heron, and others who come here to lounge in the warmth of the desert southwest during the winter months.

Morning or evening, the cattails surrounding the first pond make green or golden reflections on the water. The second and third ponds are much larger and allow for better flight photography opportunities.

In order to make use of the best light and see the most wildlife, arrive at the ponds before sunrise or about two hours before sunset and bring an ample supply of bread or popcorn. Take a **telephoto lens**, **tripod**, and

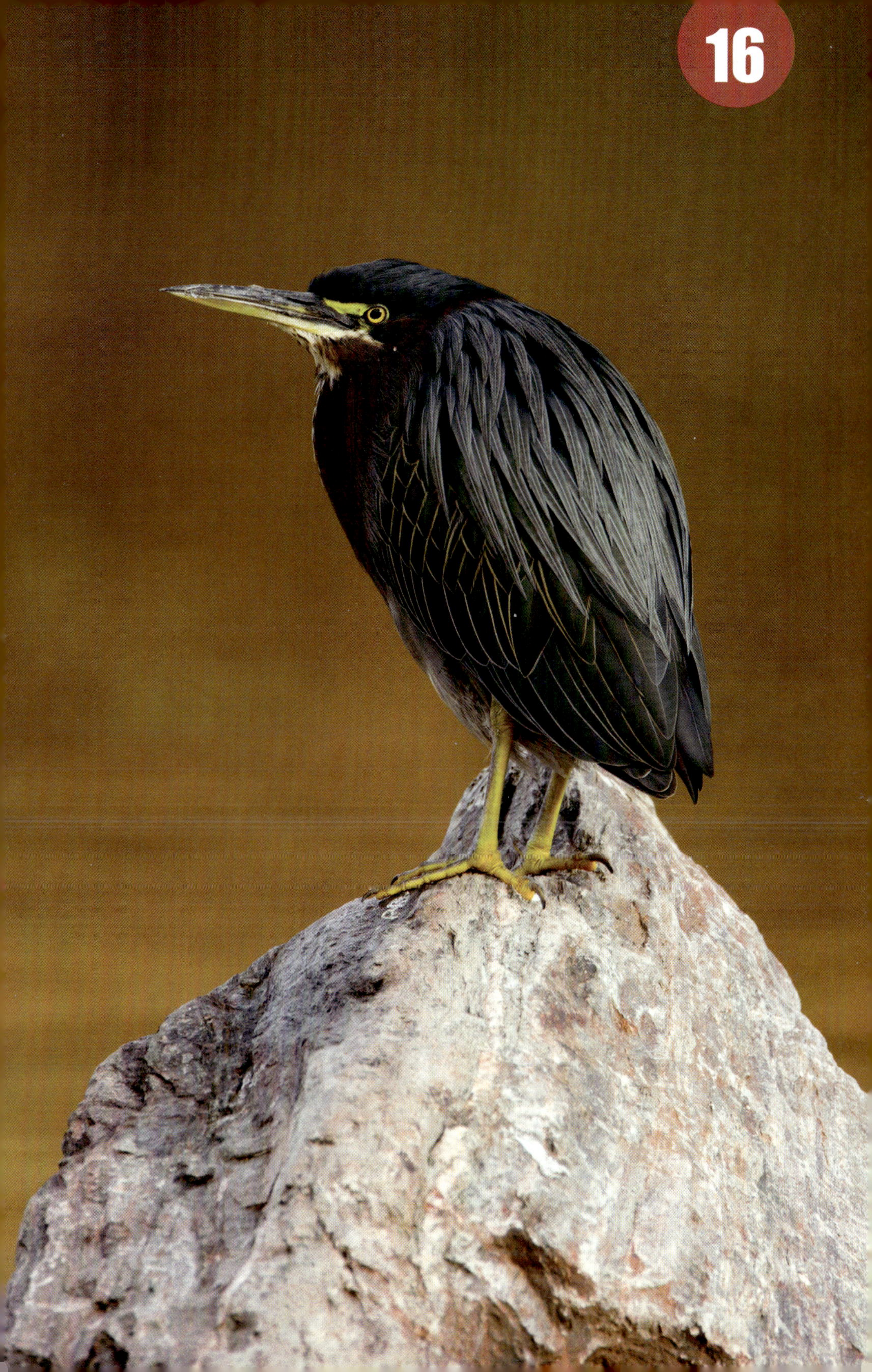

16

CENTRAL ARIZONA

Ponds at Papago Park

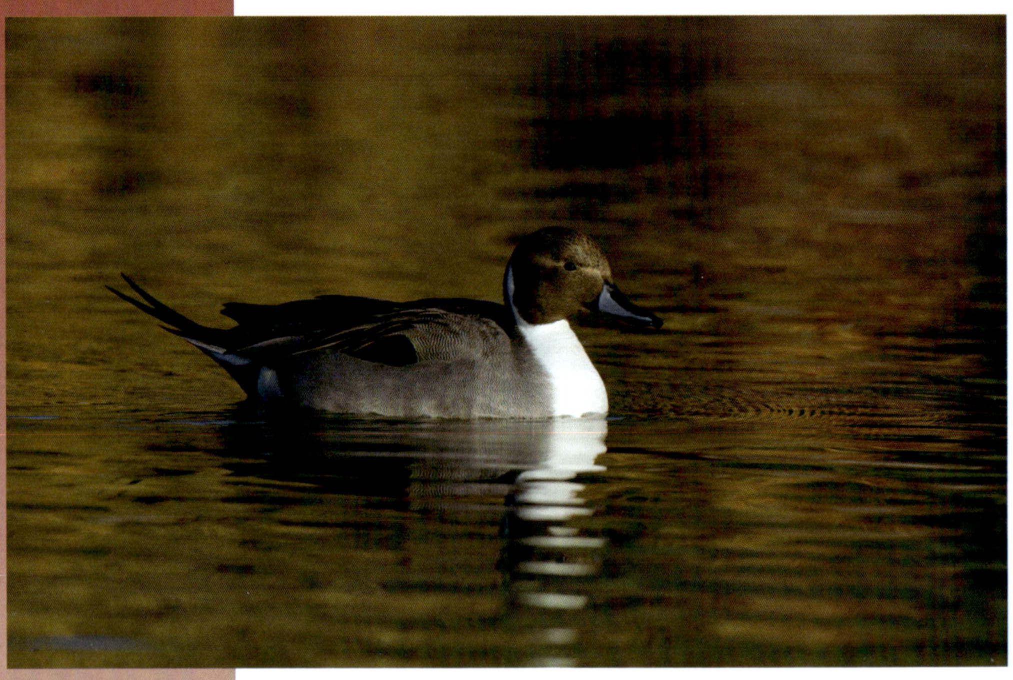

Notice the forked "pin" on the tail of the pintail duck. Canon 1DMIV, 500mm, ISO 400, f/8 @ 1/1600 sec.

a ground cloth on which to lie or sit. Position yourself on the side of the ponds where the sun falls on your back and your shadow is pointing towards the swimming waterfowl. Get your lens as close to the level of the water as possible, and look for reflections from the surrounding vegetation (mostly cattails and palm trees) turning the water a bright orange, yellow, or green. If possible, visit the ponds on sunny days when the reflections are bright and colorful (and the birds appear better lit). On overcast days, the water will appear dull gray and not very photogenic.

To photograph birds in focus while blurring the background, use an aperture around f/8 and a fast shutter speed (in excess of 1/500 of a second). Set the shutter on continuous high-speed shooting mode and the camera on continuous autofocus. Focus on the eyes of the swimming birds. Then, shoot in bursts of three to seven images in order to obtain at least one or two that are tack sharp as you track their movement.

When the ducks appear on the other side of the pond, throw some popcorn into the pond and watch the birds fly or swim to your location. Keep the camera on continuous focus and select the center collection (nine or so) focus points. When you notice the ducks flying to the food, lock focus as soon as you can and keep the shutter button half-depressed until the bird moves into your frame. Then, fire away using high-speed continuous shooting. Keep the shutter depressed until the camera's buffer fills or the bird moves out of the desired framing. Photographing flying birds takes lots of practice and patience!

Crowds are abundant any time, but mostly on the weekends. Arrive early in the day to secure a parking spot.

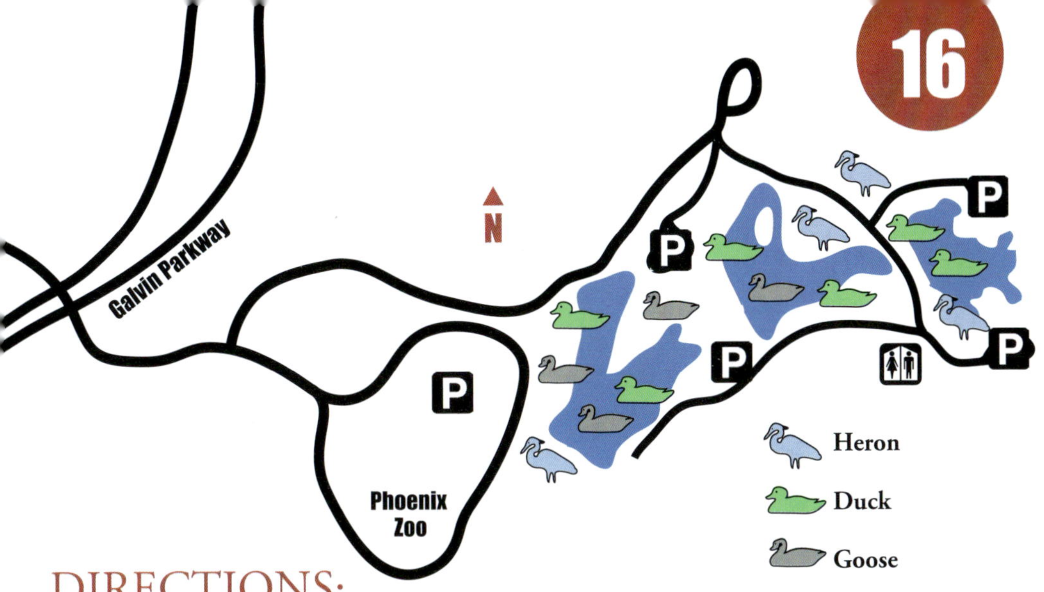

DIRECTIONS:

From downtown Phoenix, drive eastbound on I-10 for approximately 2 miles (3.2 km). Take Exit 147A for AZ 202 Loop. Continue driving an additional 3.8 miles (6.1 km) and take Exit 4 for 52nd Street/Van Buren Street. After exiting, turn right on North 52nd Street and drive 0.3 miles (0.5 km). Turn left onto East Van Buren Street. Travel 0.7 miles (1.1 km) to North Galvin Parkway. Turn left and proceed another 0.3 miles (0.5 km). Turn right into the entrance for the Phoenix Zoo. Take the first left, and follow the one-way road to the ponds.

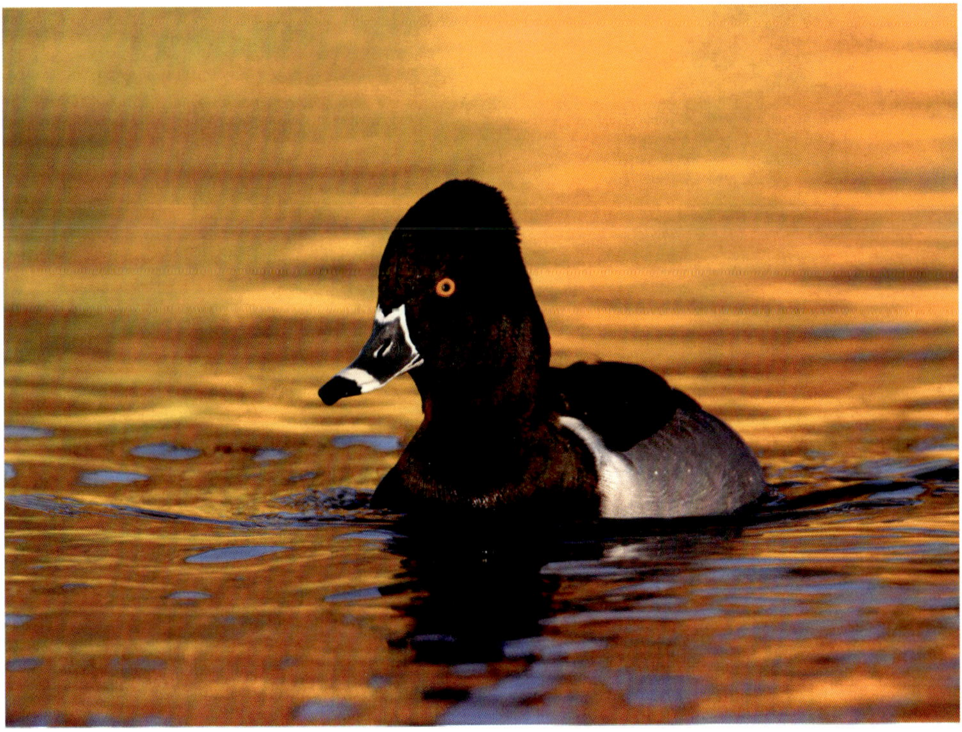

Ring-necked duck swims in the reflective waters in one of the ponds at Papago Park. Canon 50D, 500mm, 1.4x teleconverter, ISO 400, f/7.1 @ 1/1600 sec.

PHOTO TIP 7

The Eye: Get Down On It

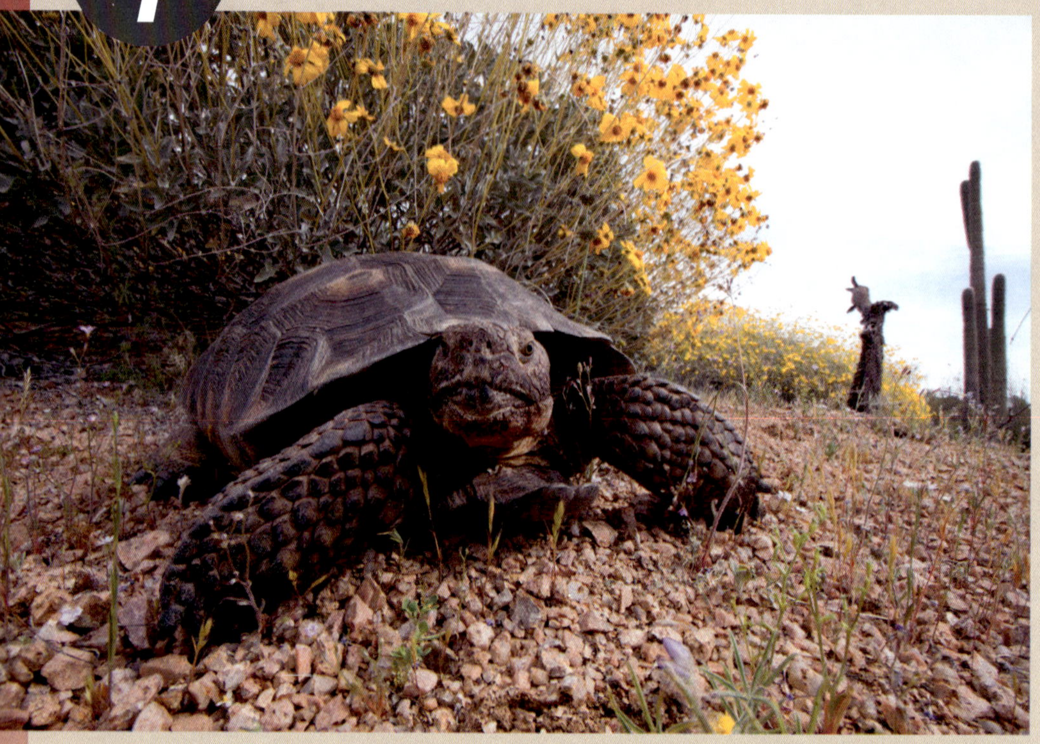

To get this eye-level image of a desert tortoise, I had to lie on my stomach and place the camera lens on the ground. Canon 50D, 10-22mm at 10mm, ISO 250, f/9 @ 1/250 sec., on-camera flash.

To capture an image that fully expresses the beauty of our subjects, photographers should photograph animals at their eye level (not ours). How do we make this happen in the ever-changing natural world? What separates a snapshot from a work of art? The photographer must get down low!

Hypothetically, let us say you are walking along the edge of a pond and spot a photogenic pair of mallard ducks. The male displays striking breeding plumage and swims in bright yellow reflective water. Resist the urge to take the photograph from a standing or kneeling position! By looking down at the duck, the image results in a view of the mallard's back in plain, non-reflective water without a direct path to the subject's eye. To capture a more effective image, sit or lie down on the ground. Place the lens as close as possible to the eye level of the duck. The beautiful reflection from the yellow-leafed cottonwood tree at the far end of the pond will surround the ducks. The subject appears silhouetted against an out-of-focus backdrop, and the foreground looks soft and unobtrusive. The duck's full face, including the eye, materialize.

However, this is sometimes easier said than done. Many reptiles and amphibians spend their entire lives within an inch or two (2.5 to 5.1 cm) of the ground. They crawl, slither, breed, feed, and die in an environment unfamiliar to humans. To take eye-level images of these subjects, one must kneel, sit, or lie down. For more comfort, wear knee pads or lie on a tarp. Place camera and lens close to—or on—the ground. Use autofocus and high-speed continuous shooting mode to get photos in a hurry.

On the other hand, maintaining the ability to snap an image at eye level of flying or perching birds offers a different challenge. One must elevate the lens to the eye level of the bird. Tall tripods usually only extend to seven feet (2.1 m) or so. The photographer can build scaffolding, use tall hunting platforms, tie their camera in a tree, or use all sorts of strange contraptions in attempts to raise the level of the camera to that of your feathered subjects. By far the best and most often used solution, though, is to use as long of a telephoto lens as possible to gain the proper perspective of the bird.

For example, say you have a 100mm lens on your camera and want to take an image of a red-tailed hawk that has perched itself in a tree 20 feet (6.1 m) off the ground. You also wish the hawk to fill about one half of your frame. A 100mm lens gives about 2x magnification on a full-frame sensor camera, which means you must position yourself within about 15 feet (4.6 m) of the bird to obtain the image size you want. If you are that close, and by some miracle the hawk stays on its perch, you are looking directly up at the underside of your subject. Not good!

Instead, a 400mm lens on your camera will render about 8x magnification. This allows you to stand at a further distance from the hawk (and to no longer look up at the bird's butt!). A longer telephoto lens coupled with teleconverters and/or a cropped frame camera enable you to keep your distance and provide a flattering perspective relative to your subject's eye level.

As you set up to photograph flying birds, position yourself in locations where you are likely to see birds flying close to the ground. Scout your subjects before shooting them to get an idea of the birds' behavior. If you can find their roosts, get close with a long lens. When the birds take off in the morning, snap eye-level photos immediately as they lift off the ground. Then, find their feeding areas and wait for them to land (preferably while wearing camouflage clothing). When they get close to the ground and start to land, take images of their wings in the forward position as they slow down.

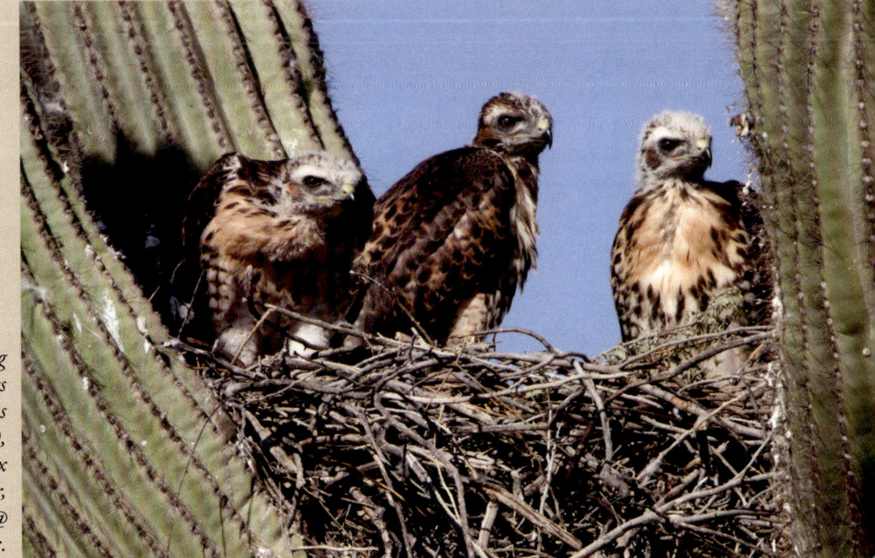

Three young red-tailed hawks in saguaro cactus nest. Canon 50D, 500mm, 1.4x teleconverter, ISO 250, f/22 @ 1/125 sec.

Scottsdale Lakes

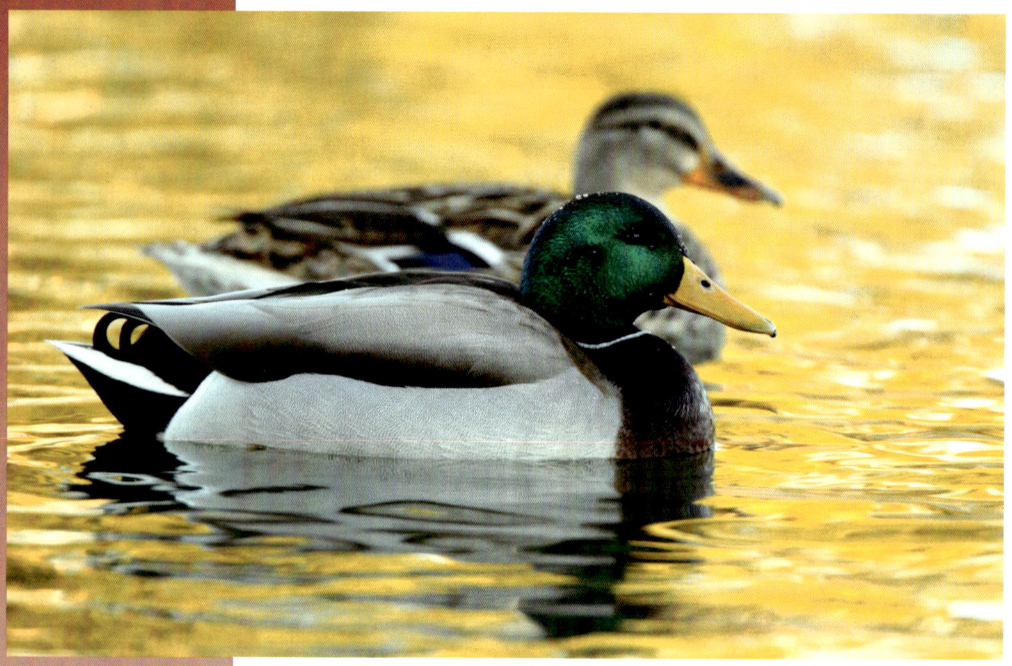

Male and female mallard duck in early morning light. Canon 1DMIV, 500mm, 1.4x teleconverter, ISO 400, f/6.5 @ 1/80 sec.

VIEW TIME
October to March

IDEAL TIME OF DAY
Sunrise; late afternoon to sunset

VEHICLE
Any

HIKE
Easy

The city of Scottsdale developed a series of recreational and flood control lakes to the east and west of North Hayden Road from East McDonald Drive on the north end to East McKellips Road on the south. During the winter months, thousands of ducks, geese, pelicans, and shore birds flock to these nutrient-rich lakes to escape the cold weather in the northern United States.

Arrive at daybreak or a couple of hours before sunset with your **longest telephoto lens** and a **tripod** with a **ball or gimbal head** along with plenty of memory cards and batteries. To photograph the paddling waterfowl, get low to the ground (consider taking a ground cloth to lie on). On manual camera mode, use continuous autofocus and high-speed continuous shooting mode. Blast away, shooting in bursts of five or more frames at a time. Keep your camera focused on the birds as they mill around. When you see one dipping its body in the water to bathe, set a focus on it. More often than not, the bird will prepare to lift itself from the water to shake its feathers—a great decisive moment to record with your camera.

To photograph flying ducks, set your camera on manual mode to keep from over- or underexposing the ducks as they fly from one background to the next. After finding an appropriate exposure in manual mode, the settings will not change as the background lighting does. However, if you set the camera on aperture priority, the camera will modify the settings as the background changes, resulting in an over- or underexposed bird.

Stand up while hand-holding the camera or place the camera and lens on a tripod. As the bird flies towards you, focus on the bird as soon as you are able and continue to follow it. Continuously focus as the bird moves.

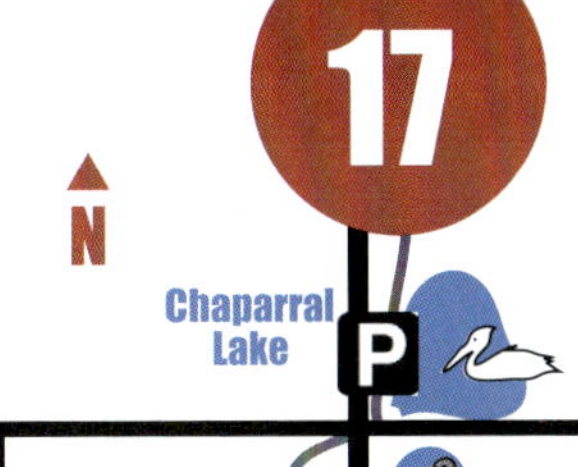

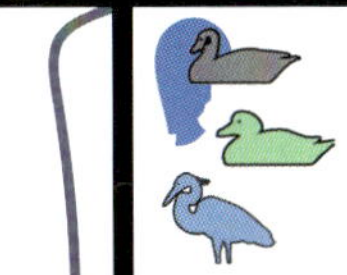

DIRECTIONS:

Chaparral Lake: From the intersection of Hayden Road and Chaparral Road in Scottsdale, travel eastward 0.1 miles (0.2 km) to the entrance to Chaparral Lake.

El Dorado Park Pond 1: From the intersection of Chaparral Road and Hayden Road, travel west 0.5 miles (0.8 km) to Miller Road and turn south. Travel 2.5 miles (4 km) to East Murray Lane and turn east. Travel 0.1 miles (0.2 km) to the parking lot.

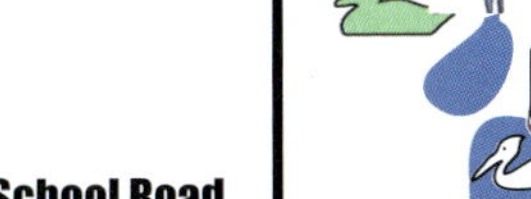

No-name El Dorado Park Pond: From the intersection of Miller Road and Murray Lane, travel 0.3 miles (0.5 km) south to the entrance of the south El Dorado Park Lake.

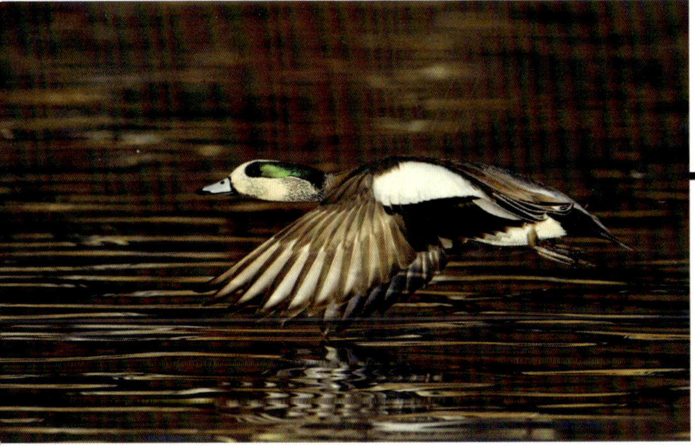

Flying American wigeon duck. Canon 1DMIV, 500mm, ISO 400, f/8 @ 1/3200 sec.

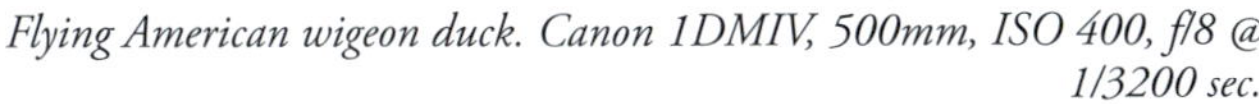

When the bird moves into a photogenic position, push the shutter button and take five or more images.

When shooting in "bursts," some photographs will appear sharper than others. Movement of the camera results when your finger first pushes the shutter button. The camera absorbs all of the inertia from your hand, causing potential camera movement. As you continue to depress the button that inertia disappears, resulting in less camera movement during the next few images.

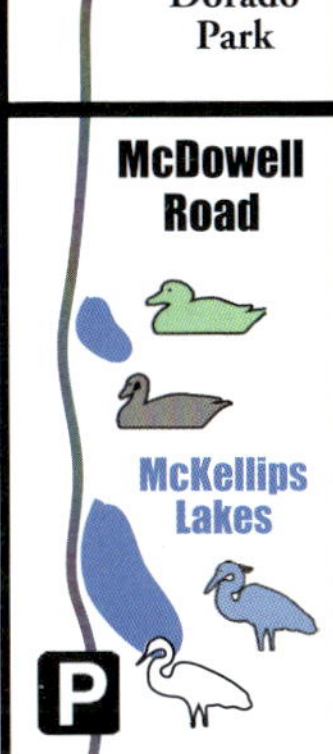

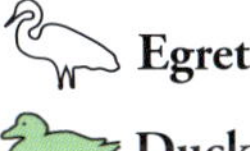 Egret

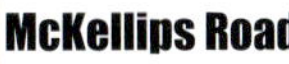 Pelican

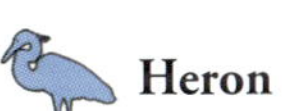 Heron

Duck

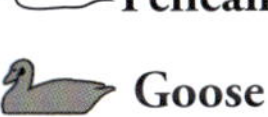 Goose

Making the Photo

3

Fish at Lunch

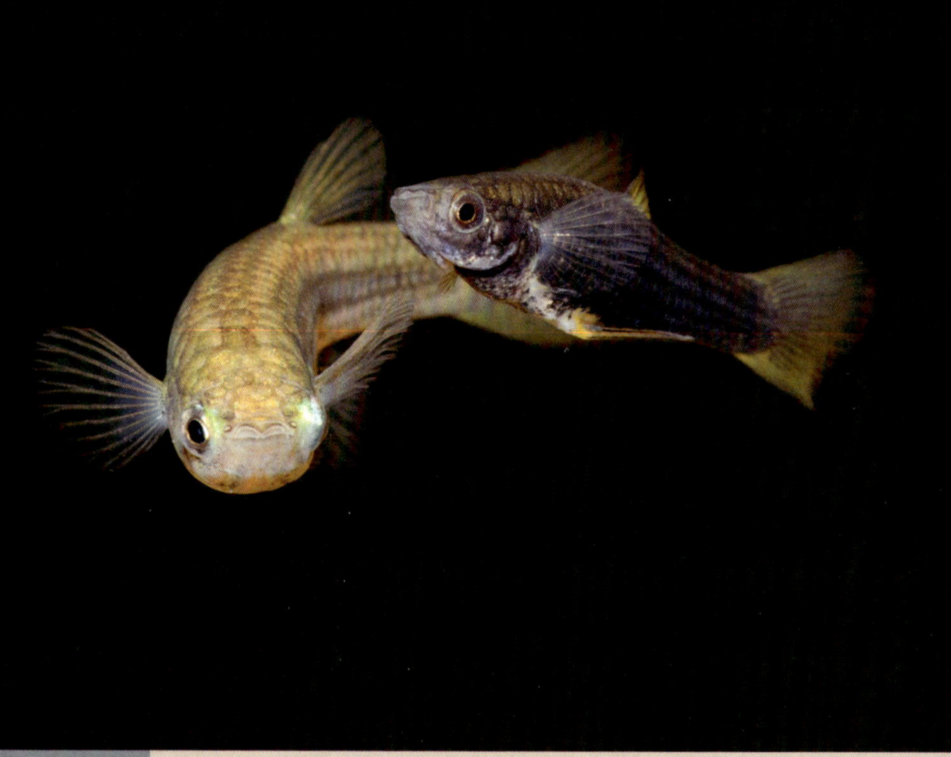

Despite the aquarium's small size (a 10-gallon tank), it took me several weeks to capture an image of this pair of courting Gila topminnows within the same frame. The darker male is attempting to woo the larger blond female. Canon 20D, 100mm macro, ISO 200, f/10 @ 1/200 sec., off-camera flash.

The question arose (as it does every day), "Where should I eat lunch?" I lived close enough to my office that I could go home, have a sandwich, take a little power nap, and get back to work within my allotted hour. Villa Deli was only a short drive away, or I could walk to Chino Bandido for Chinese-Mexican food. My final option—yuck—was to bring my lunch and eat it at work. I seldom selected the final option until I figured out that I could eat my lunch and have photography fun at the same time. Gila topminnow, along with most other native fish in Arizona, is a rare and protected species. I had an empty 10-gallon (38-liter) aquarium at home and decided to set it up in the lobby of my office so the public could enjoy it and, on occasion at lunch, I just might use it as a photo opportunity. I placed several male and female Gila topminnows in the tank and everyone enjoyed their antics. Soon they were breeding, giving birth to babies, and thriving. It was time to set up the camera.

I was a novice at aquarium photography. My horrible first attempts helped me learn, though. Who would have known that aquarium glass reflects and distorts the light (especially the light from the lobby window)?

I searched the Internet where, as usual, I found gobs of good and bad advice. To eliminate the reflection, I needed to reduce the room light as much as possible! Easier said than done, though, as our office was a full service agency and was required to have a human presence at all times during business hours. But my office sat at the far end of the building and my secretary tightly controlled access, therefore I could arrange an occasional period of "lights out." Problem solved.

I also learned to use two flashes, with one on each side of the camera at a 45-degree angle from the glass to eliminate the bright spots I experienced when using a single flash on the camera.

I continued having my lunch at work for the next several weeks. I set up my photo gear, turned off the lights, and began my quest for the perfect aquarium photograph of Gila topminnows. Most of the images I took appeared either out of focus or with the fish in a weird position. This was not as easy as I thought. I could see that if I was going to make a great photo, I was going to spend a lot of time eating lunch at the office.

After a while, I began to get the drift of aquarium photography. I successfully eliminated the reflected light and improved my ability to focus on moving targets.

Then one day, it happened. I recorded both the male and female in the same frame, in focus, and the male was courting the female. I was ecstatic! Not only had I created an image I was proud of, but I could also finally go home for lunch or better yet, join my friends at the deli or Chino Bandido.

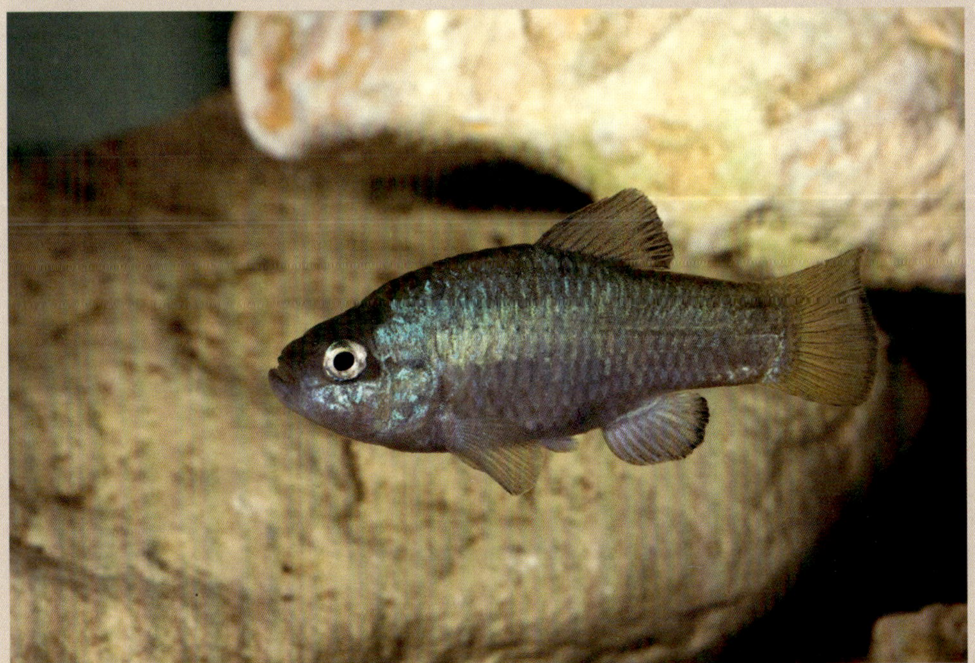

Desert pupfish enjoys life in an aquarium in my office. Canon 1DMII, 100mm macro, ISO 200, f/18 @ 1/250 sec., two off-camera flashes @ +1 FEC.

CENTRAL ARIZONA

Butterfly Wonderland

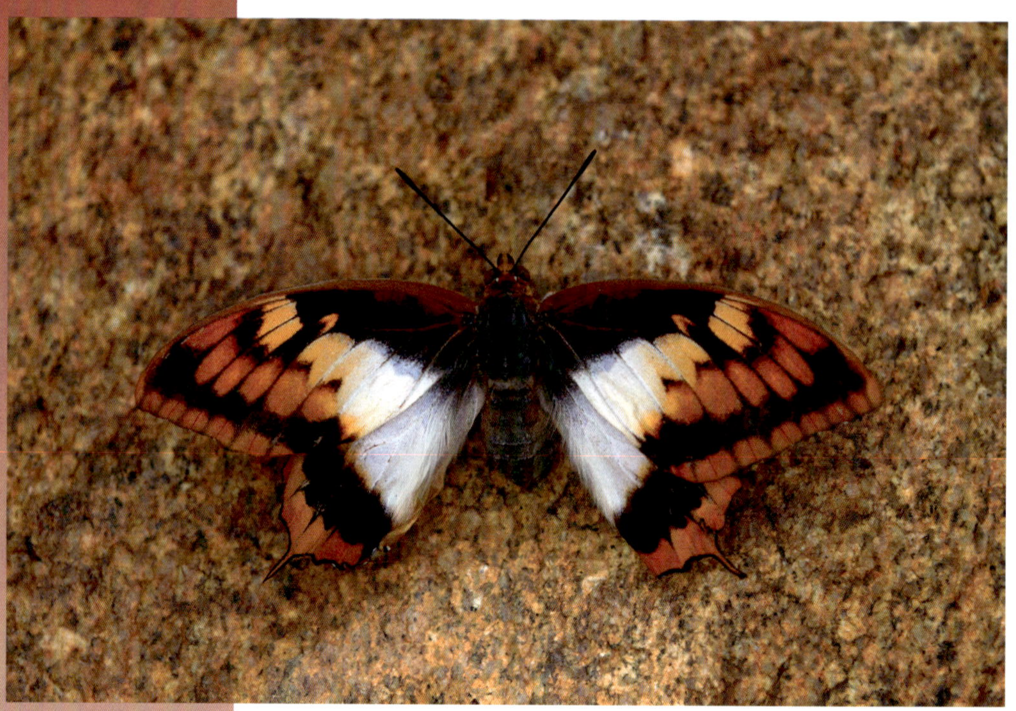

Female flame-bordered charaxes spreads its wings at Butterfly Wonderland. Canon 7D, 70-200mm at 126mm, 2x teleconverter, ISO 400, f/7.1/ @ 1/200 sec., on-camera flash.

VIEW TIME
Year-round

IDEAL TIME OF DAY
Early morning

VEHICLE
Any

HIKE
Easy

With their intricate patterns, bright colors, and fanciful flight, butterflies have long attracted not only photographers, but also the child in all of us. Touted as the largest butterfly pavilion in America, Butterfly Wonderland in Scottsdale provides a mecca for butterfly photography. Not only can a photographer take beautiful butterfly images but also have the space, time, and number of subjects to create hundreds of images in a morning. In the wild, this quality of shoot very seldom happens.

Start your outing by viewing the artful and educational three-dimensional monarch butterfly movie. After the film, and before you enter the atrium, linger a little at the butterfly emergence gallery. Although the area is brightly lit, behind glass, and painted white (and not ideal for photography), simply seeing the chrysalis and emerging butterflies bestows a visual treat.

As if designed specifically for photography, brightly-colored flowers fill the property. Seemingly no matter where you turn, hundreds of butterflies of numerous species flitter around and frequently land on the plants.

Tote a **macro lens** as you explore the conservatory. Set your camera (and lens, if required) on autofocus. Modify your aperture and ISO speed settings to allow for shutter speeds in the neighborhood of 1/250 of a second or faster.

Once a butterfly lands on a flower, approach slowly so as not to scare it away. Butterflies will often spook upon seeing shadows (memories of winged predators), so be aware of your position relative to the sun and avoid casting a shadow on them. Photograph them early in the morning when the

DIRECTIONS:

From the intersection of AZ 101 and AZ 202 in Tempe, drive 8.3 miles (13.4 km) north on AZ 101. Take Exit 43, for Via de Ventura. Turn right, and travel 0.3 miles (0.5 km) to the entrance.

Butterfly Wonderland charges an entrance fee unless you hold an annual membership pass (which gains you free access). For more information, visit **www.butterflywonderland.com**.

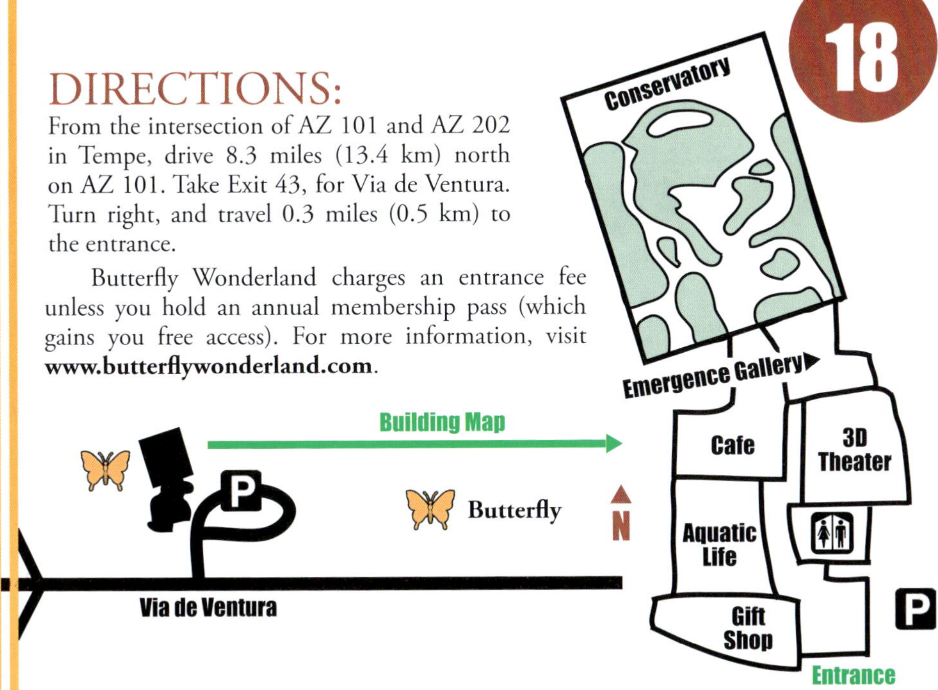

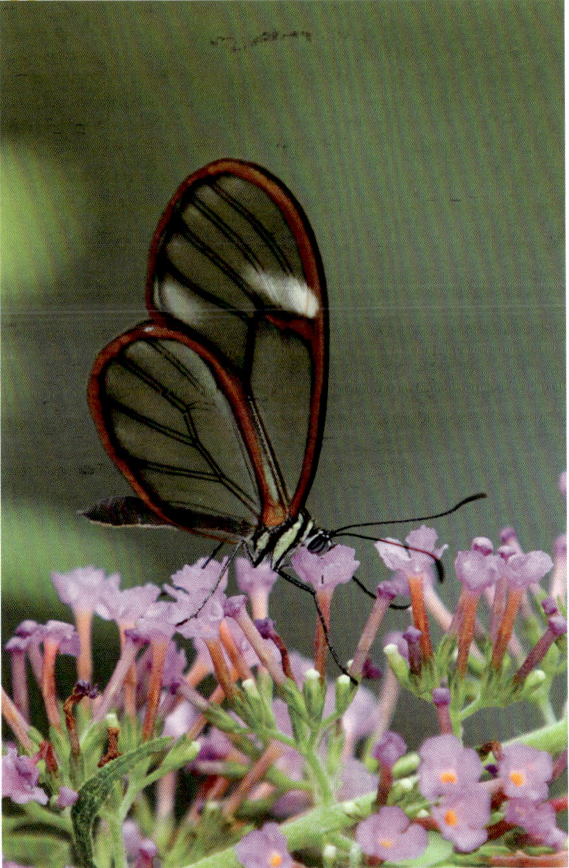

butterflies are relatively cool and not so flighty.

Tripods are not permitted in the conservatory (however, you can bring a monopod). While hand-holding your camera, focus on the butterfly's head. Keep your camera steady. Exhale and hold your breath before pushing the shutter button. Then, press and hold the shutter button down to take a few images at a time using continuous shoot drive mode.

If necessary, use **fill flash**, but turn the flash power down by about one stop and place the camera setting on single-shot shooting mode. Nothing looks worse than a butterfly image that looks obviously flashed. The soft natural lighting in the pavilion only needs a small amount of fill flash to accent the butterfly's colors. Too much flash will render the image harsh and flat.

Clearwing butterfly. Canon 7D, 70-200mm at 165mm, 2x teleconverter, ISO 400, f/7.1 @ 1/125 sec., on-camera flash @ -1 FEC.

Making the Photo

4

The Queen

My goal was to photograph the entire life cycle of the queen butterfly—from egg to metamorphosis to butterfly. Lucky for me, the queen butterfly's host plant, the desert milkweed, is widely used across the Phoenix metropolitan area as a landscape plant.

I knew that in late October and November, queen butterflies actively lay eggs on the milkweed. While searching milkweed plants at the shopping mall near Interstate 17 and the Carefree Highway, I noticed an adult queen butterfly had laid an egg near one of the plant's flowers. I quickly picked the portion of the plant with the fresh egg.

At home, I placed the plant stem in a glass of water to keep it turgid and alive. I photographed the egg using my Canon MP-E 65mm 1-5x macro lens at a magnification of about 4x. I positioned a diffused flash several inches from the egg. After hatching, the caterpillar eats the egg for its first meal and then begins the process of going through several molts.

Back at the strip mall, I photographed a series of wild caterpillars feeding on desert milkweed. I knelt at eye level with the plant to ensure that the background was sufficiently far from the caterpillar to soften it. I kept the long axis of the caterpillar's body parallel to the camera body to maintain sharp focus across the whole body. To keep the perspective I wanted, I used my favorite butterfly gear: a Canon 5D Mark III camera (which allows me to increase the ISO to achieve faster shutter speeds without having a disturbing amount of noise when shooting), 70–200mm lens, 2x teleconverter, and a little fill flash. Fortunately, with ample day light, I could use a fast enough shutter speed for me to hand-hold the camera.

Queen butterfly caterpillars take a few weeks to change into a chrysalis, and I was unable to spend days and days watching caterpillars in the field until they metamorphosed. In order to have an opportunity to make the best photos, I took some caterpillars home. I placed them in an aquarium with a lot of milkweed to eat. Every day, I went out to collect fresh milkweed to feed the caterpillars. When the caterpillars started getting large, I stayed home every day watching them so I would not miss the metamorphosis event. And I thought wildlife photography was a glamorous profession!

Finally, one of the caterpillars attached itself upside down from a milkweed branch in the aquarium—a position caterpillars take a day or so before they change into a chrysalis. As soon as I noticed the caterpillar in this position, I set up my "indoor photography studio" in my sunroom. The background I used was a complementary colored matte photograph of an out-of-focus scene. The print created contrast between the subject and the background. I hung the backdrop from a light stand using a double-sided clip. I removed the caterpillar from the aquarium and clamped the stick that it was on to another light stand and placed it between my camera and the backdrop. I set up my Canon 1D Mark IV camera with the 180mm macro lens on my Gitzo tripod and Really Right Stuff ball head. I lit the scene using two soft boxes (one on each side of—and slightly in front of

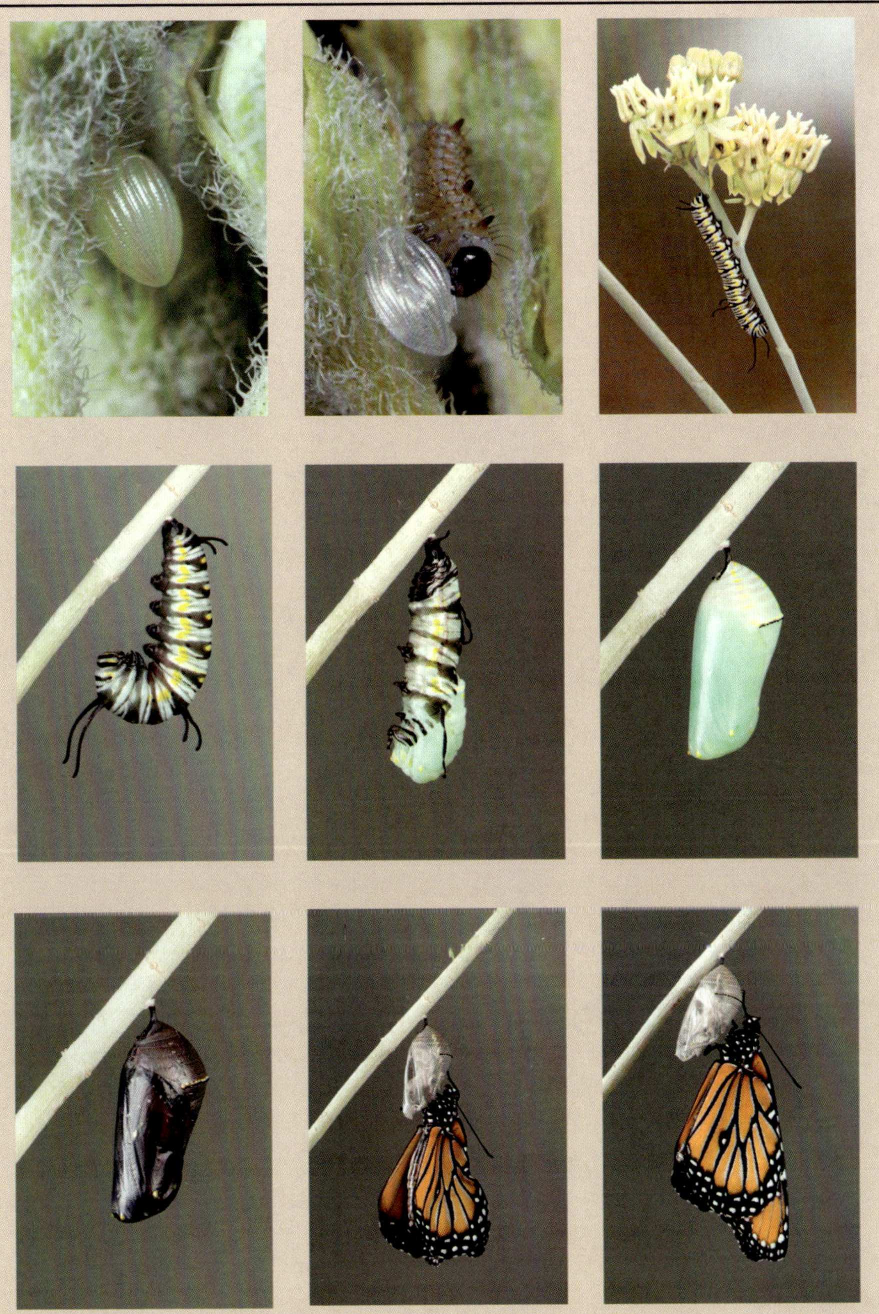

The magical changes that take place during a queen butterfly metamorphosis. Canon 5DMIII, 180mm macro, ISO 200, f/16 @ 1/200 sec., off-camera flash.

The Queen (continued)

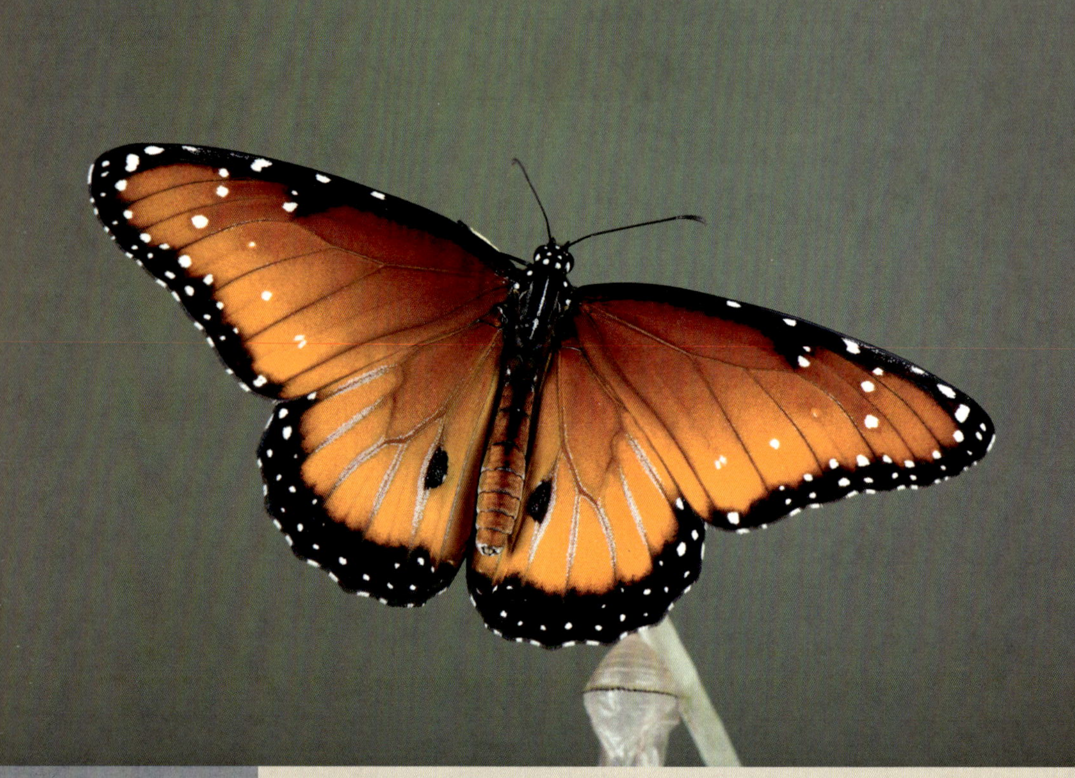

Adult queen butterfly spreads its wings just after hatching. Canon 5DMIII, 180mm macro, ISO 200, f/16 @ 1/200 sec., off-camera flash.

—the hanging caterpillar) that were slaved to my camera using the Canon ST-E2 infrared remote. The soft boxes act as a light modifier that diffuses and transforms harsh flash light into a softer, more even light.

Now the second waiting game began. I waited for the caterpillar to begin to make jerky movements (these movements are a prelude to it shedding its skin and developing into a beautiful jade-green chrysalis). For the next couple of days, I sat on a chair, at the ready, next to my camera. When the caterpillar began to shed its skin and develop into the chrysalis, I started taking photos every few seconds or minutes for the next 15 minutes. After 15 minutes, the caterpillar had completely shed its skin, and the chrysalis appeared. I waited five days for 15 minutes of photography!

Finally, I could take a few days off and relax. It would take about a week for the chrysalis to start the process of hatching. During this week or so, the very alien-looking being inside of the chrysalis transforms itself from a glob of cells to a beautiful butterfly complete with six legs, two wings, two antennae, and three body sections. This transformation, or metamorphosis, astounds me. How nature got to this level of sophistication is a wonder.

Fortunately, the queen butterfly, once again, gives me a clue as to when it might "hatch" out of the chrysalis. The shell of the chrysalis becomes more and more translucent, and a few hours before the butterfly emerges, you can see orange wings inside the chrysalis. At that point, I began another long wait in front of the television in a nearby room.

As long as I have been doing this type of photography, I have never been vigilant enough to catch the butterfly the moment it emerges from the chrysalis, so it must only take a matter of a few seconds. When the butterfly shakes loose from the constraints of the chrysalis, it looks like a fat bug with shriveled wings. The body contains a fluid that will eventually fill the wings so that the butterfly can fly. As soon as the butterfly hatches, it begins to pump fluid from its body into its wings until, a few minutes later, the wings are at their full size.

Each year, I continue to locate monarch and queen butterfly caterpillars so that they can transform into a chrysalis in my sunroom and, eventually, hatch into a beautiful winged creature. Most often I do not photograph the event but enjoy it for the wonder of it all and then release the butterflies into my backyard for them to begin another generation.

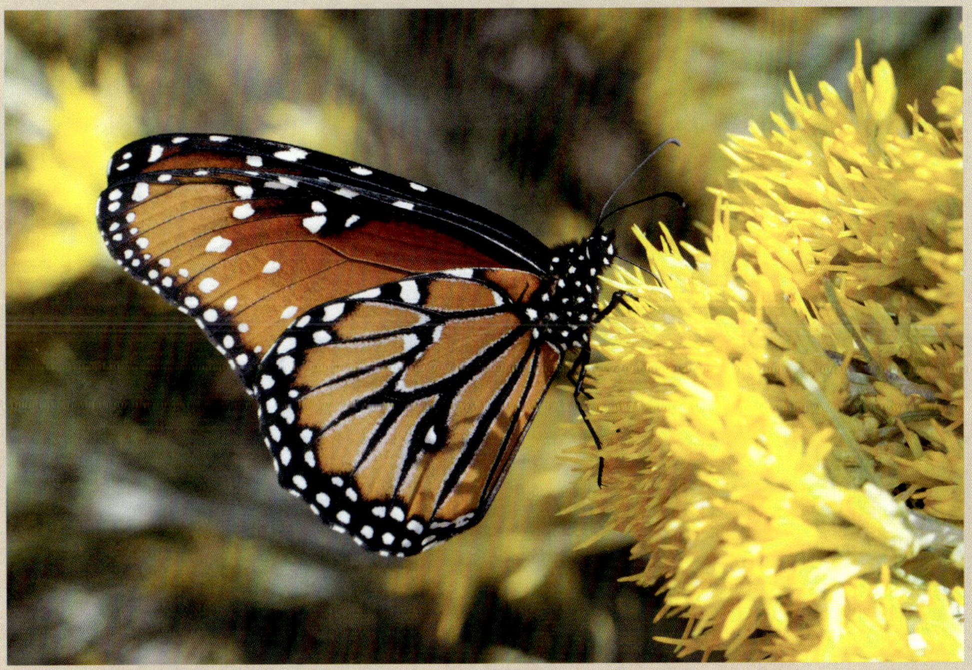

Queen butterfly feeds on a rabbitbrush bloom near Portal. Canon 20D, 100-400mm @ 105mm, ISO 200, f/11 @ 1/250 sec., on-camera flash @ -1 FEC.

CENTRAL ARIZONA

Veterans Oasis Park

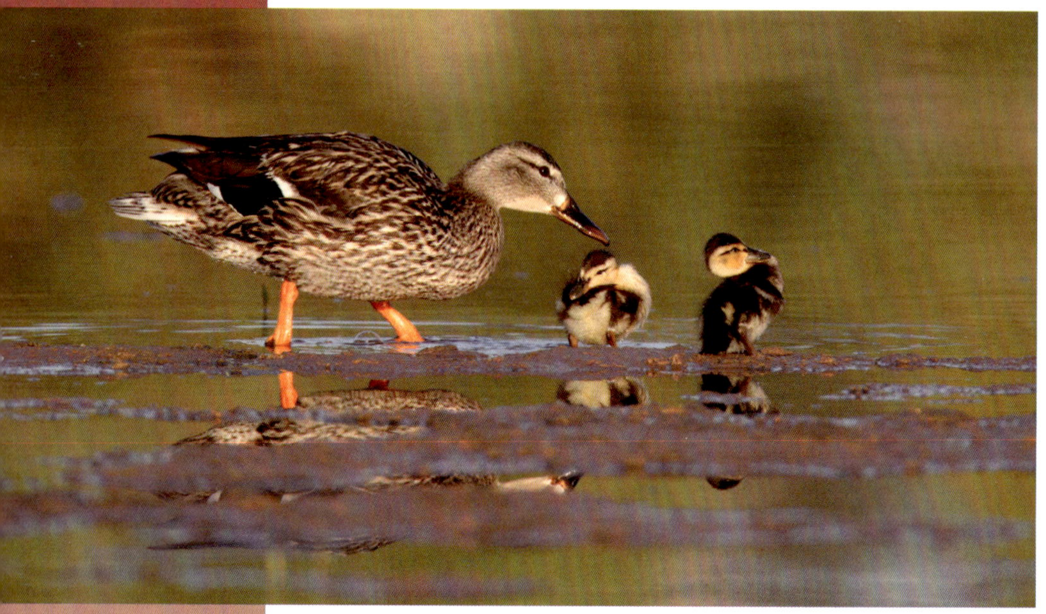

Mother mallard duck and ducklings. Canon 50D, 500mm, 1.4x teleconverter, ISO 400, f/6.3 @ 1/1000 sec.

VIEW TIME
October to March

IDEAL TIME OF DAY
Early morning and late afternoon

VEHICLE
Any

HIKE
Easy to moderate

Veterans Oasis Park in Chandler covers 113 acres (45 hectares) of aquatic habitats and desert uplands. To the east and northeast of the property lie five large water recharge basins that, during the winter months, serve as a home to thousands of water birds, ducks, and geese. The 4.5 miles (7.2 km) of trails crisscross throughout the property and allow for access to all of the best photography sites. Another one of the growing number of municipal water reclamation projects, Veterans Oasis Park is destined to become a fantastic series of ponds for both wildlife watcher and photographer.

With the exception of the ducks on the fishing lake, the wildlife at Oasis Park act more skittish and seem less easy to approach than at other urban parks. To make the most out of a trip to the park, take a portable blind and place it near the edge of one of the recharge ponds that has ample water bird activity. While you set up the blind, the wildlife will leave the immediate area for a while. Be patient and they should return within a half hour or so (after becoming accustomed to the blind).

Use a **moderate** or **long telephoto lens** situated on a **tripod** and sturdy **ball or gimbal head**. Set the camera on high-speed continuous shooting mode, an aperture around f/8 so the backdrop blurs, and an ISO speed that will allow the shutter speed to be at least 1/500 of a second. Remember to take a chair to use inside the blind as well as water and a snack.

The small hill in the center of the park hosts a large number of black-tailed jackrabbits (along with a few cottontail rabbits). Even though human traffic here is high, both species of rabbits are still wary and will run at your approach. Watch and locate an area where several rabbits congregate. Choose a vantage point with an unobstructed view, set your camera on a tripod with a long telephoto lens, and sit down and wait. Within a few minutes, the rabbits should begin to hop around and make imaging easy.

N

Guadalupe Road

DIRECTIONS:

From the intersection of US 60 and US 87 in Mesa, travel 5.9 miles (9.5 km) east. Take Exit 185 for Greenfield Road. Turn right onto Greenfield, and travel 1.6 miles (2.6 km) southbound to Guadalupe Road.

The reserve is located on the southeast corner of South Greenfield and Guadalupe roads. Parking is available near the library, just south of the library, and just east of the intersection.

For more information on park activities, go to **www.gilbertaz.gov/departments/parks-and-recreation/parks/riparian-preserve-at-water-ranch**.

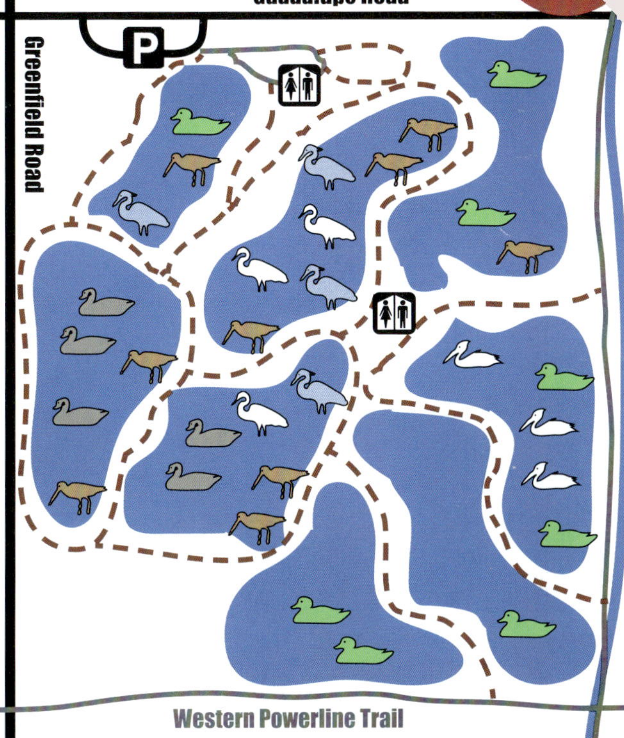

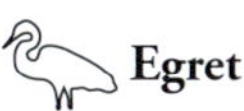

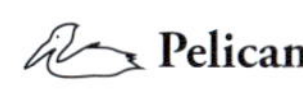

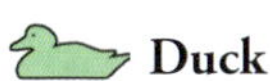

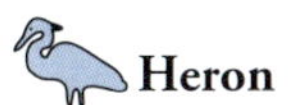

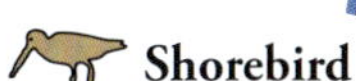

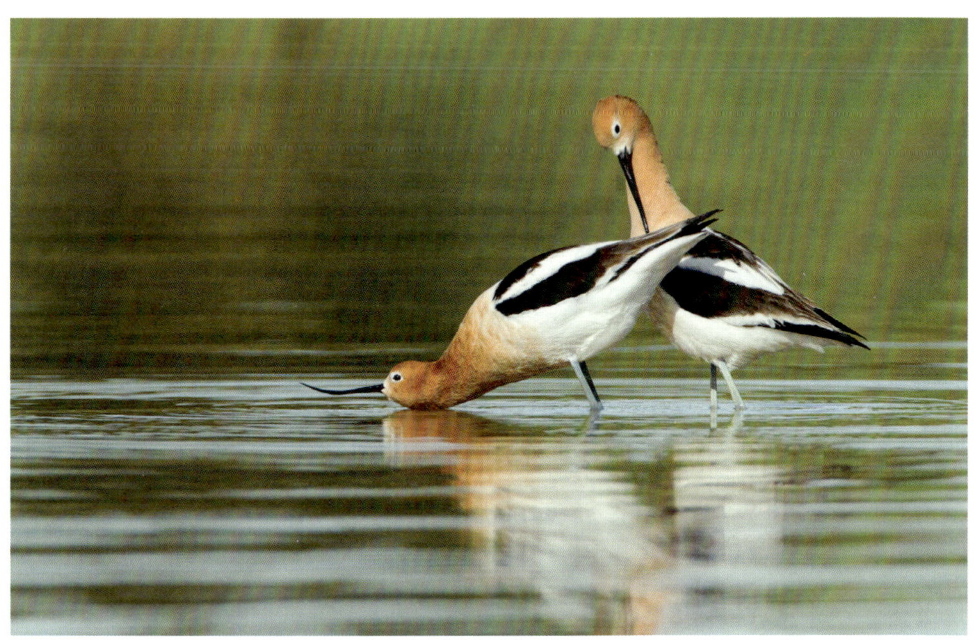

Going to the Dance

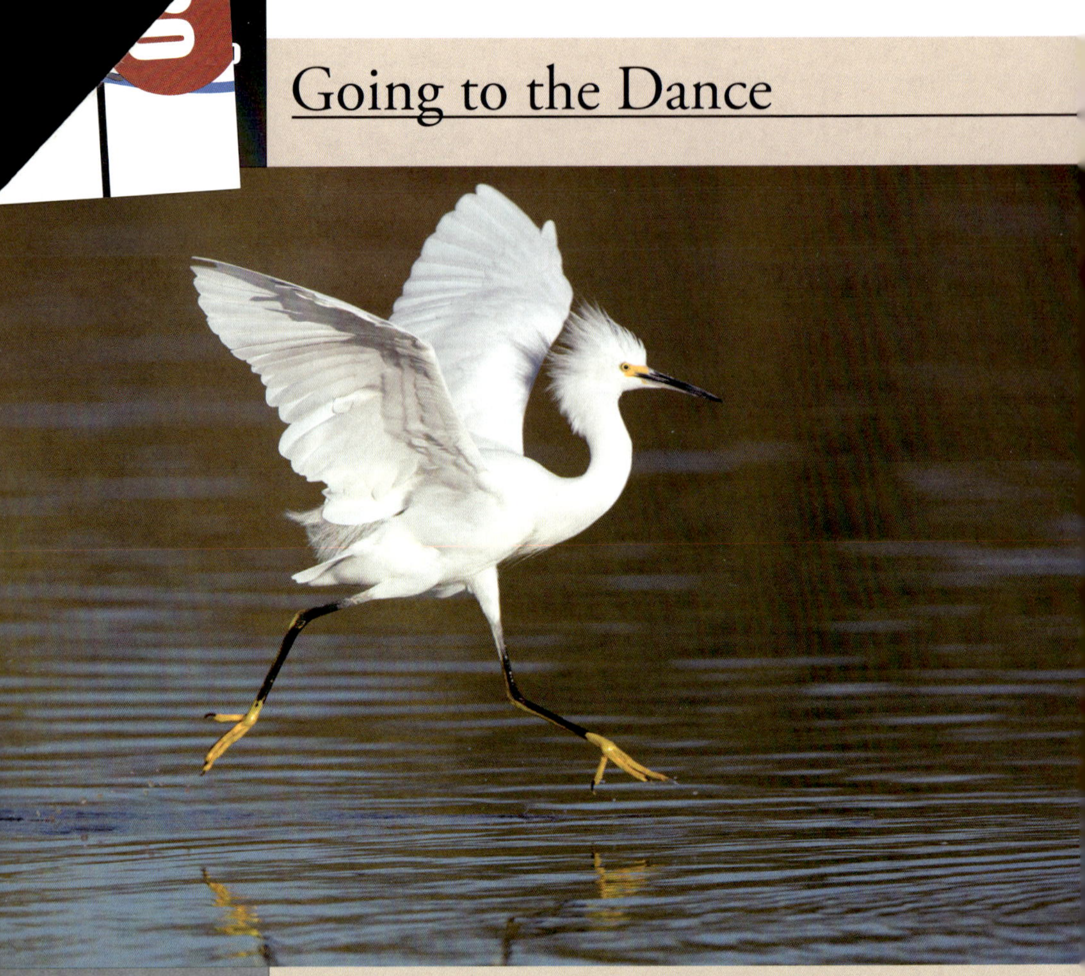

Snowy egret dances across the water while it defends its favorite fishing spot. Canon 1DMIV, 500mm, 1.4x teleconverter, ISO 400, f/8 @ 1/2000 sec.

In grade school, our teachers thought it was a good idea to have lunchtime dances once each week. We were all there—pimple-faced kids, too tall and skinny, trying to shed our prepubescent bodies. With girls on one side of the gym and boys on the other, we sized up the opposite sex while trying to muster enough courage to ask them to dance. We spent a lot of time on our separate sides of the gym, but once in a while, for some strange reason, one boy and one girl met in the middle and nervously swayed back and forth in sync to the music. The waiting was uncomfortable but the dance spectacular.

So it is with my photography at Gilbert's Riparian Preserve at Water Ranch. I put on my best dress (dark clothing and plenty of camera gear), drive to the Ranch (one hour from my house), and wait for the dance to begin. I am on one side of the pond waiting for the birds to travel closer so that I can record their antics.

So it was on October 10, 2013. I arrived at the park not knowing what to expect. It was a beautiful day, and there were a few snowy egrets feeding about 50 yards (45.7 m) from my location. Calm water and the blue sky welcomed me. Late autumn trees reflected off the water. I sat between two desert broom plants with a large tree to my back breaking up my silhouette. I placed my 500mm f/4 lens with the 1.4x teleconverter and Canon camera on my Gitzo tripod, Really Right Stuff ball head, and Wimberley Sidekick.

More snowy egrets began to arrive on the far side of the pond from my location. I could feel the tension in the air, but was it coming from me as I waited or from the egrets as they nervously approached each other? And then the dance began.

Snowy egrets were taking turns flying slowly across the pond while they periodically dappled their feet in the water. I had never seen this behavior before and was at a loss for what they were doing until I noticed that, every once in a while, the bird would stretch its neck out and grab a small fish from the water's surface. They used the slow flight and splashing feet to scare fish to the surface so they could eat them. I remembered seeing a similar type of feeding style while watching tri-colored herons fishing in a small pond in Florida. I kept my camera on high-speed continuous shooting mode, AI Servo, and manual camera mode (to ensure appropriate exposures as the birds moved). I shot hundreds of images of these beautiful white birds dancing across the water searching for tiny fish.

Just when I thought it could not get any better, two snowy egrets began to quarrel. They were just 60 feet (18.3 m) away from me and bathed in a beautiful light. The dominant bird kept chasing the subordinate bird away. Each time the dominant bird returned to feed, the other bird would try and horn in on its feeding site. This went on for 30 minutes or so, and I kept on shooting away. Seldom had I taken as many images in such a short period of time. Keep in mind that I do not measure success by the number of photographs I take, but this was special.

When I arrived home, I was anxious to look at my images on the "big screen." My time spent waiting for the dance was well rewarded. I had images of snowy egrets pirouetting across the water more like ballerinas than birds. I realized that this might be the shoot of a lifetime, at least when it comes to dancing snowy egrets, and that my patience and steadfast "waiting on my side of the gym" resulted in my participation in a beautiful dance.

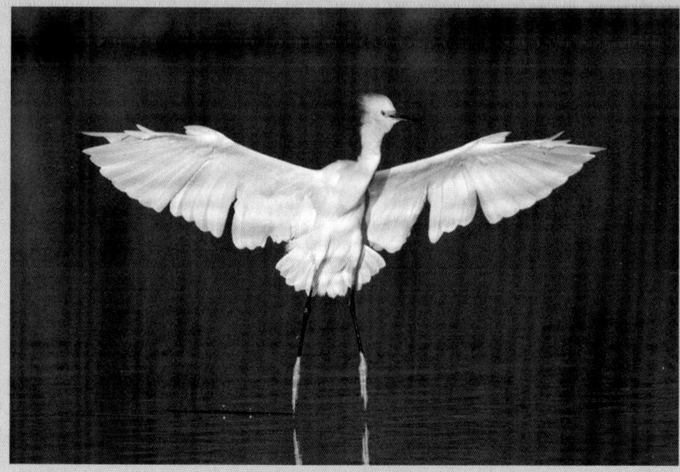

Snowy egret flutters across the early morning water at the riparian preserve. Canon 7D, 500mm, 1.4x teleconverter, ISO 400, f/5.6 @ 1/3200 sec.

LOWER Salt River Recreation Area

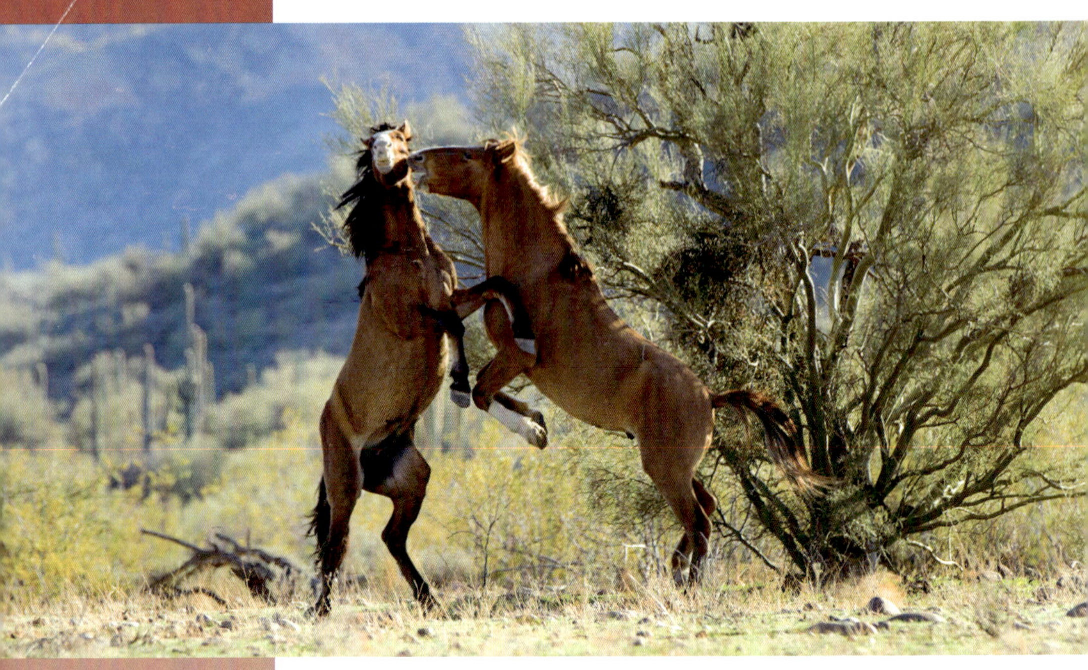

Wild horse stallions fight for supremacy. Canon 1DMIV, 70-200mm at 170mm, 2x teleconverter, ISO 500, f/6.3 @ 1/640 sec.

VIEW TIME
Year-round

IDEAL TIME OF DAY
Early morning and late afternoon

VEHICLE
Any

HIKE
Moderate

The ancestry of the horses along the Lower Salt River Recreation Area extends back to horses introduced in the late 1600s by the Spanish missionaries and settlers. Supplemented over the years through introductions by the US Calvary, ranchers, prospectors, pioneers, and many other sources, the horses are barely holding on in the face of increasing pressure from urbanization and recreational land use. Thus, finding and photographing wild horses among saguaro cacti and along the Salt River presents a unique and rare experience.

During the summer when temperatures increase to over 100 degrees F (38 degrees C), all of the horses keep near the water to stay cool and feed on the lush green aquatic vegetation. Quite likely, these horses would not survive without the sanctuary the riparian area and cooling water provides them. During the cooler months, they travel away from the water to feed on the desert vegetation.

Brown, white, palomino, spotted, and all shades of colors appear in these herds. For the most part, the Salt River horses stay in small herds as they roam. Herds consist of a dominant male, several females, and their young. These horses are easy to locate and approach for up-close images.

To locate a herd, drive along the Bush Highway and scan the surrounding desert using binoculars. Alternatively, visit the various camp and picnic areas, or hike along the river. Wild horses enjoy the company of a herd, so when you see one horse, others are nearby. In the heat of the summer, most of the horses stay near or in the river, living off the cooling water and submerged vegetation. At other times, they venture out into the desert to eat the leaves of the palo verde trees, ironwood trees, and other palatable vegetation.

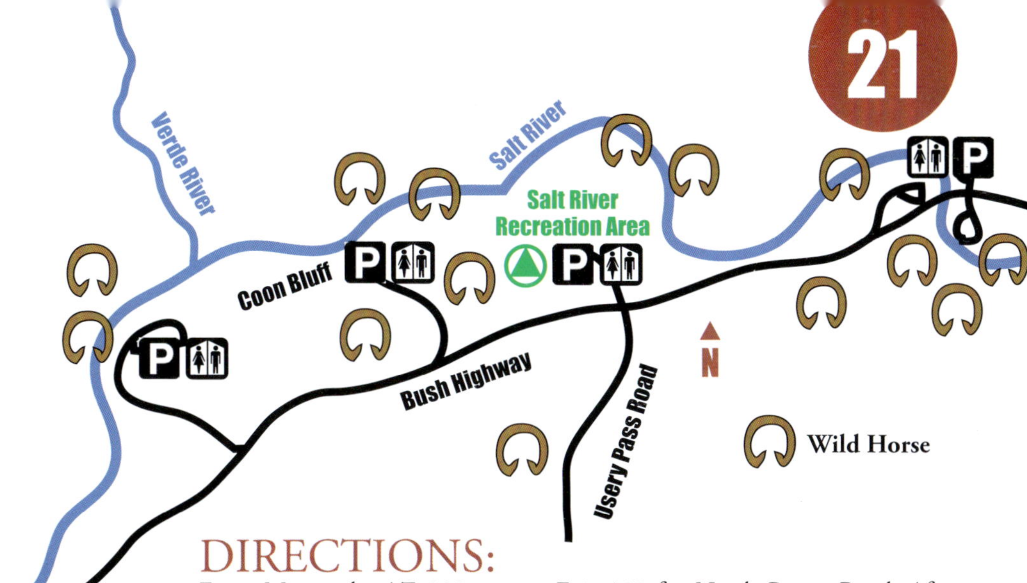

DIRECTIONS:

From Mesa, take AZ 202 east to Exit 188 for North Power Road. After exiting, turn left (north). Travel 2.4 miles (3.9 km). Power Road turns into the Bush Highway at the onset of the Lower Salt River Recreation Area.

The Forest Service requires a special Tonto Pass to park in this area. Learn more at **www.fs.usda.gov/main/tonto/passes-permits/recreation**.

When you locate a herd, walk towards them with a **moderate fixed or zoom telephoto lens (e.g., 300mm)** in hand. Position yourself with the sun at your back for good frontal illumination. If you follow the herd long enough, you may see foals feeding from the mares or stallions fighting. The horses appear unfazed by the presence of photographers but a safe distance of 50 feet (15.2 m) or more is suggested.

Watch your step in these areas as the loose desert rocks, large boulders, painful cactus spines, and occasional rattlesnake present hazards. Walking along the river makes for easier going, but venomous snakes may linger along the shore. Bring a hiking stick and/or wear leather snake chaps to help minimize the risk of getting bit.

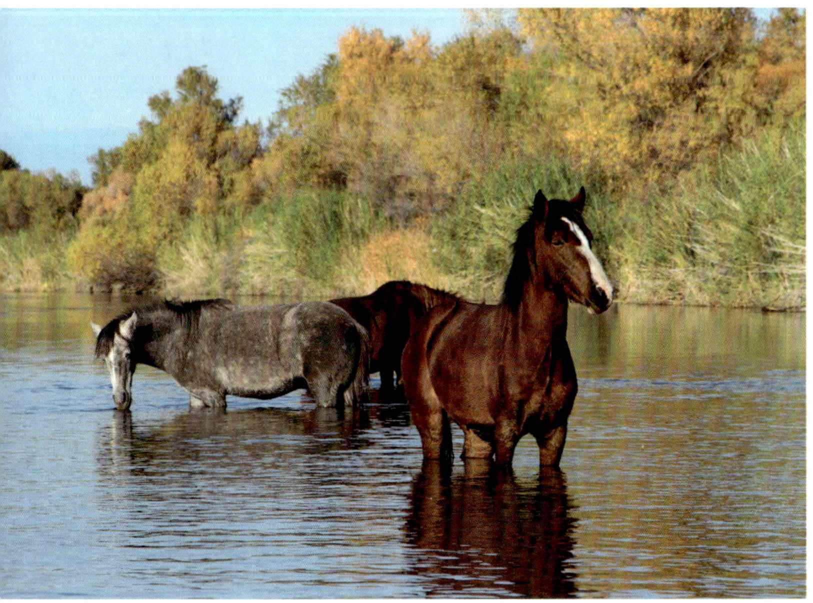

Horses at the Salt River Recreation Area use the cool waters to escape summer's heat, feed on the aquatic vegetation, and drink. Canon 7DMII, 24-105mm at 105mm, ISO 200, f/9 @ 1/500 sec.

CENTRAL ARIZONA

Canyon Lake

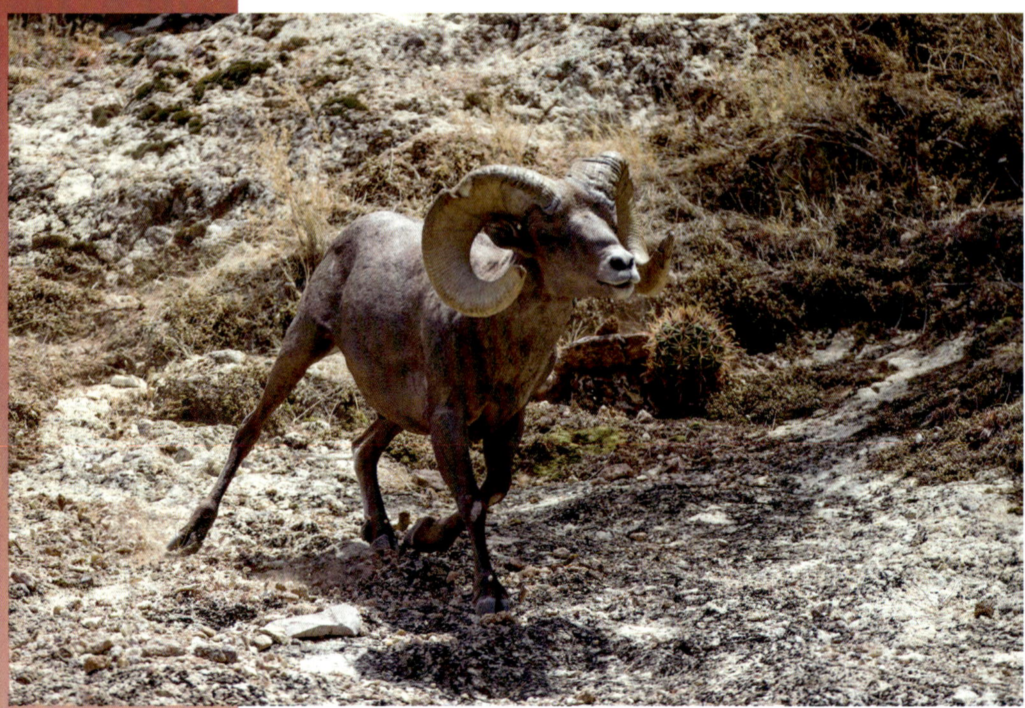

Ram desert bighorn sheep in attack mode. Canon 1DMIV, 500mm, ISO 640, f/4 @ 1/4000 sec.

VIEW TIME
May to July

IDEAL TIME OF DAY
Late morning to early afternoon

VEHICLE
Any

HIKE
None, boat required

The drive to Canyon Lake curves through lichen-covered rocks surrounded by saguaro cacti, ironwood and palo verde trees, canyon walls, and Sonoran Desert vistas. Canyon Lake seems out of place in this land of minimal rainfall and temperatures that top 110 degrees F (43.3 degrees C). The lake resulted from Mormon Flat Dam, one of four dams on the Salt River built to control water for agriculture, recreation, and flood control. The lake's cool water brings majestic desert bighorn sheep within close range of a well-prepared photographer.

Bring a boat (or rent one at the marina) for this photographic adventure, as well as plenty of drinking water, sunscreen, and snacks. Wet your clothes to cool your sweltering body. The hottest part of the summer does not faze the hardy sheep here.

Small by Arizona reservoir standards, Canyon Lake presents few navigational challenges (unless the wind is blowing). The main portion of the lake is circular, but the sheep seldom materialize in this area. Drive the boat to the opposite side of the lake from the boat ramps and travel towards the dam (get a map from the marina). As soon as the lake begins to narrow into more of a riverine habitat, begin scanning the shoreline and canyon walls on both sides of the reservoir for sheep.

Until 8:00 to 9:00 a.m., canyon walls shade the best sheep habitat, and the sheep do not travel to the water's edge for a refreshing drink until the air heats up later in the day. Large white rump patches help the viewer locate them among the desert's gray and brown environment. After spotting a group of sheep, drive the boat slowly to within 60 feet (18.3 m) of the

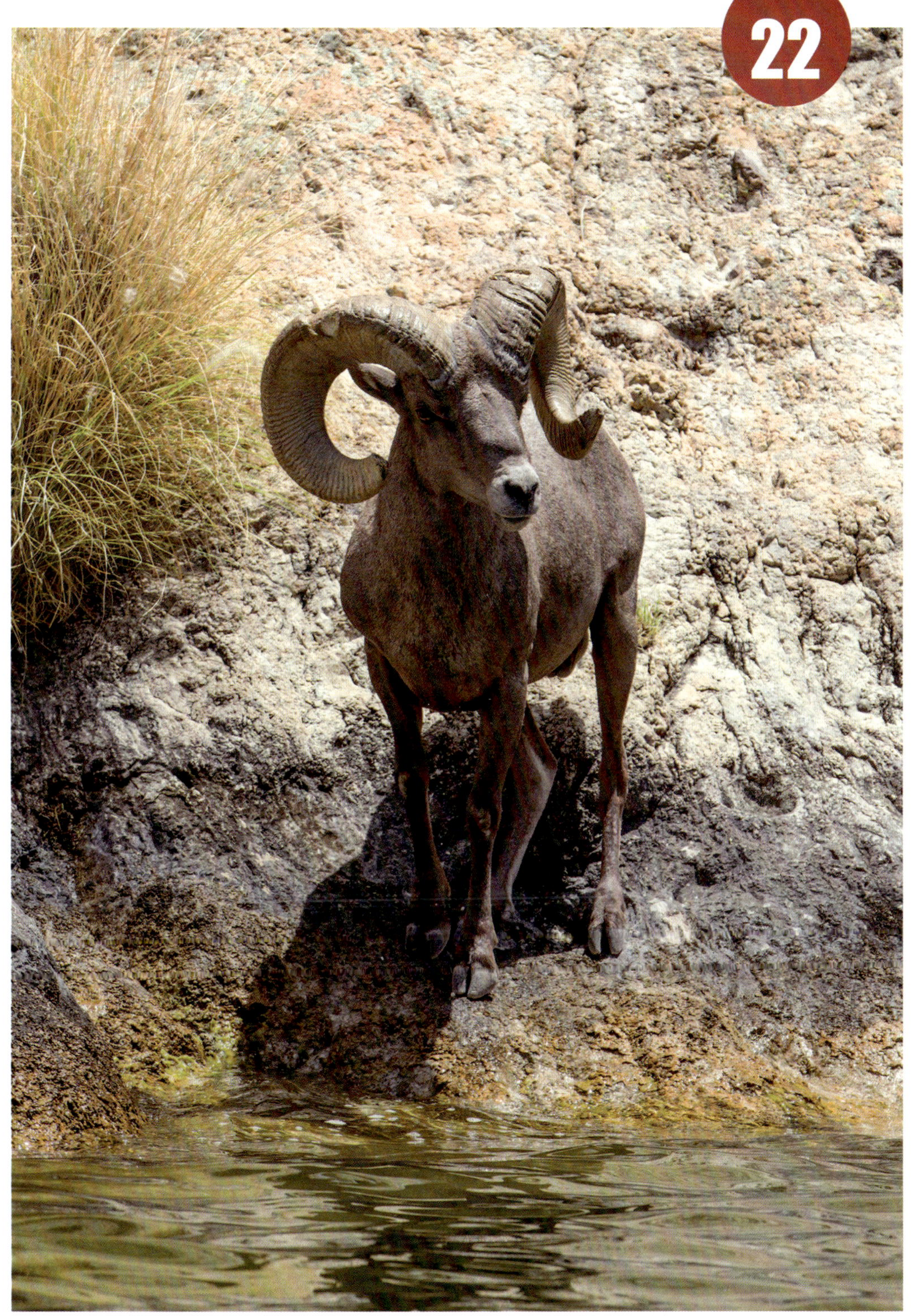

Mature ram desert bighorn sheep at water's edge. Canon 5DMIV, 70-200mm at 155mm, 2x teleconverter, ISO 800, f/9 @ 1/1600 sec.

Canyon Lake

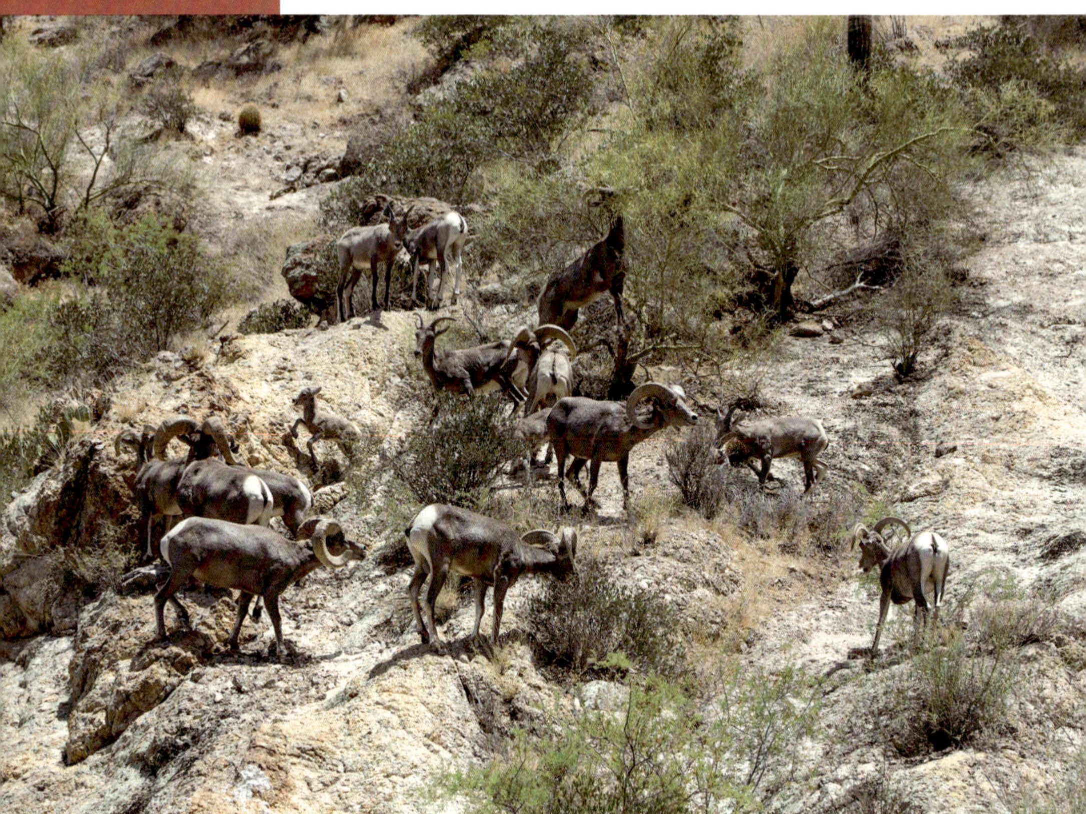

A herd of desert bighorn sheep socializes along the cooling waters of Canyon Lake. Canon 5DMIV, 70-200mm at 100mm, 2x teleconverter, ISO 800, f/8 @ 1/2000 sec.

shore. Stay there a while and photograph the sheep as they interact or drink water. If the sheep rest higher up on the canyon walls, drive the boat closer and capture what images you can. However, do not get too anxious or too close, and do not leave unless the sheep move out of camera range. Quite often, these sheep will appear as if they are just lounging and then a territorial dispute breaks out and two rams begin fighting.

Use a **moderate to large telephoto lens (e.g., a 100-400 mm telephoto zoom lens on a cropped frame camera)**. Tripods or monopods on the boat will only get in the way and will not help to stabilize the camera. Instead, to stabilize the lens and render sharp photographs in this situation, use a very fast shutter speed (1/2000 of a second or faster) combined with a fast ISO (e.g., ISO 400 or greater) and a wide aperture around f/5.6. Place the camera on continuous focus mode and use image stabilization (or vibration reduction) to keep your hand-held camera steady. In high-speed continuous shooting mode, take bursts of five or more shots simultaneously. While one or two sheep may attract your attention, occasionally zoom out to capture images of groups of sheep interacting.

DIRECTIONS:

From Phoenix, take US 60 east towards Apache Junction. Take Exit 196 for AZ 88 (Idaho Road). Turn left onto Idaho Road, and drive about 2.3 miles (3.7 km) before veering right onto the Apache Trail. Travel east on the Apache Trail (AZ 88) for 13.9 miles (22.4 km) to the Palo Verde Recreation Site 147 to reach a boat ramp.

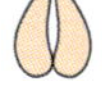
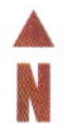
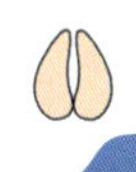
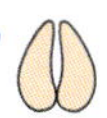
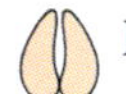
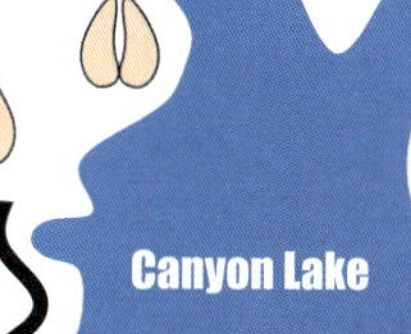
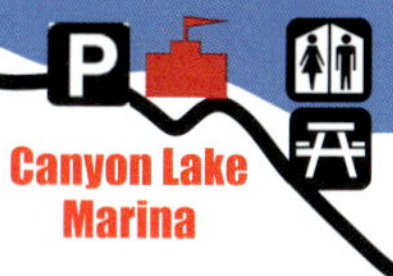

A Tonto Pass or one of the many other permits (Golden Age Pass, Life Time Pass, etc.) is required to visit this location. Learn more at **www.fs.usda.gov/main/tonto/passes-permits/recreation**.

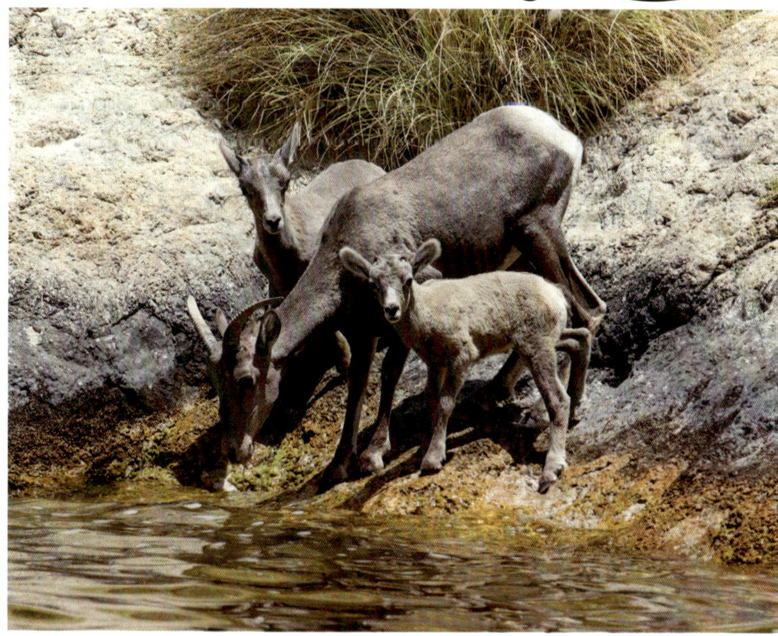

A ewe and two lambs drink the refreshing water of Canyon Lake. Canon 5DMIII, 70-200mm at 200mm, 2x teleconverter, ISO 800, f/10 @ 1/1250 sec.

Boyce Thompson Arboretum

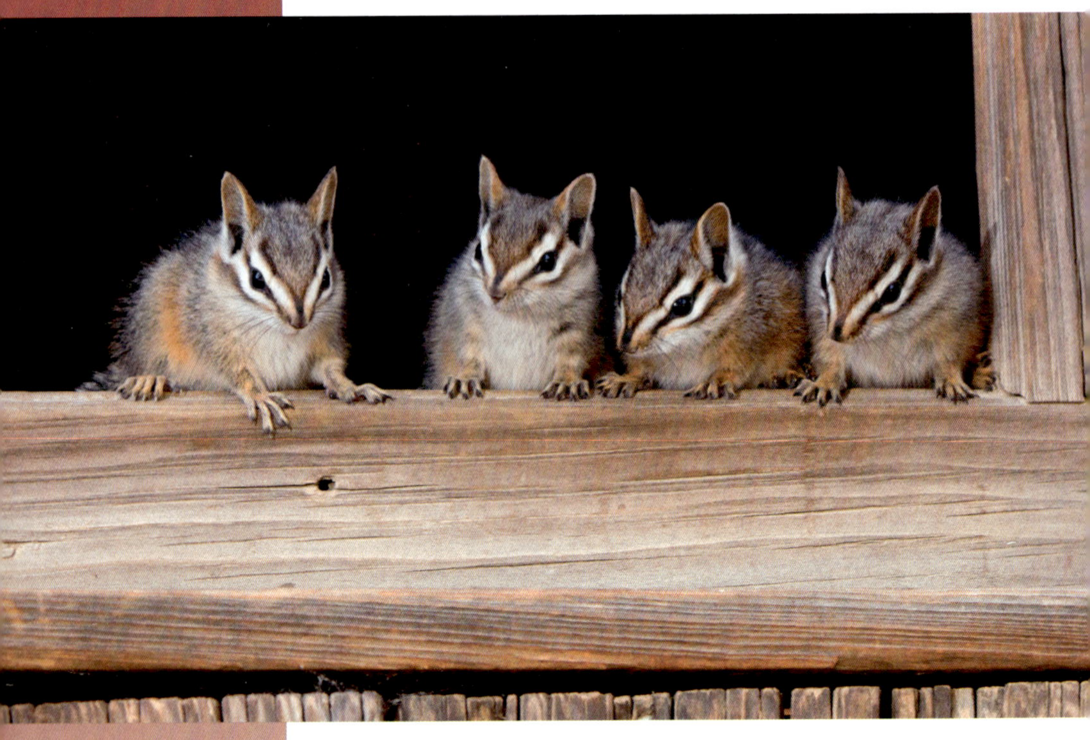

Four baby cliff chipmunks check out their new digs. Canon 70D, 70-200mm at 103mm, 2x teleconverter, ISO 640, f/8 @ 1/50 sec.

VIEW TIME
Year-round

IDEAL TIME OF DAY
Early morning and late afternoon

VEHICLE
Any

HIKE
Moderate

The name of Arizona's oldest and largest botanical garden comes from its founder, William Boyce Thompson. Within the 350-acre (141-hectare) arboretum, desert-adapted plants from around the world thrive. Ayer Lake and Queen Creek combine to create a variety of habitat types that provide homes to a large diversity of birds, mammals, insects, fish, reptiles, and amphibians. Time your visit for the desert spring (March through May), when the broad-tailed hummingbirds and cliff chipmunks breed and raise their young in the arboretum or during early summer for fun-filled lizard photography.

With a **moderate telephoto lens (e.g., 300mm)** perched on a sturdy **tripod**, photographers can create frame-filling images of female hummingbirds feeding their young. Since dense vegetation conceals the hummingbird nests, use fill flash to avoid harsh shadows and high contrast images. Use a relatively wide-open aperture (e.g., f/5.6 or f/8.0) to blur the background and emphasize your subject. Bring a comfortable chair and water to stay hydrated during your shoot.

Stay at least 15 feet (4.6 m) from any hummingbird nest, and do not photograph a nest until the young are large enough that you can see them above the rim of the nest. Remain quiet and still when photographing hummingbird nests. If the birds do not return to feed the young within 30 minutes, leave this location and look for another nest. It is better to abandon a photo opportunity than to have the hummingbirds abandon their young because they are afraid of your presence.

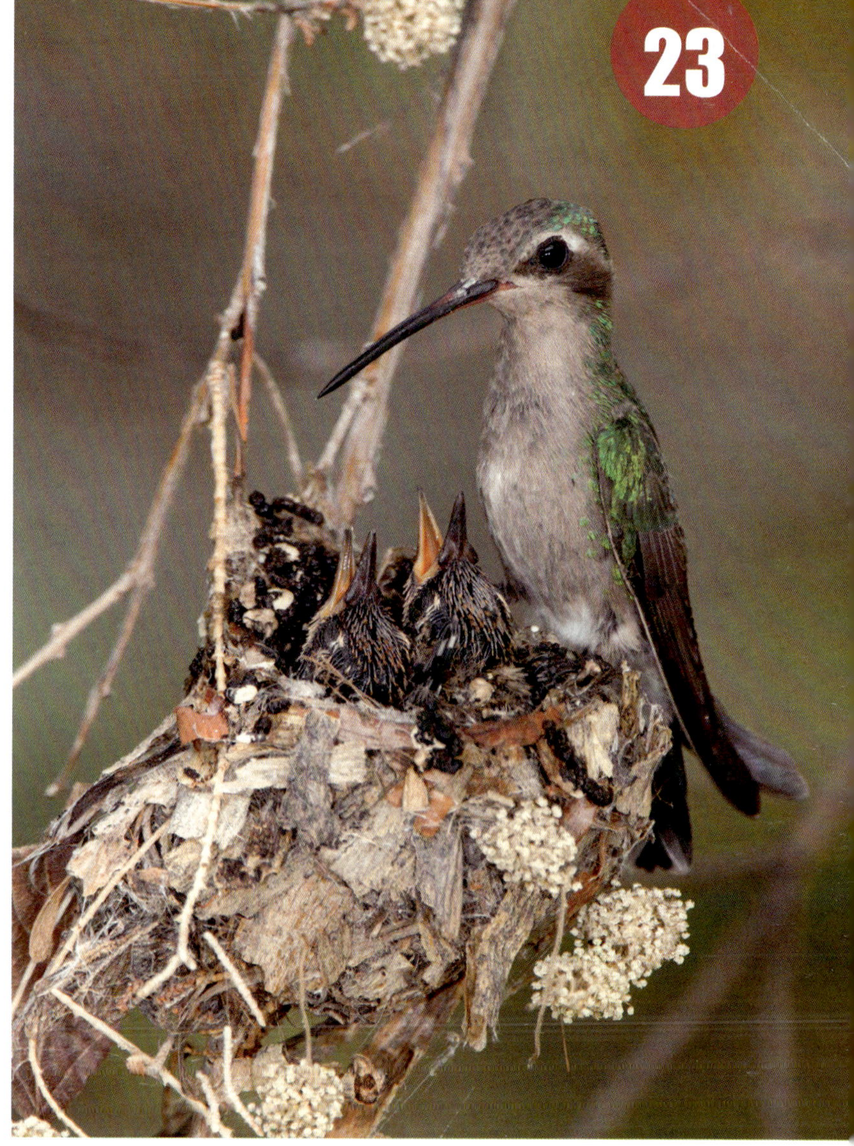

Broad-billed hummingbird female feeds its young at Boyce Thompson Arboretum. Canon 50D, 500mm, 1.4x teleconverter, ISO 500, f/10 @ 1/160 sec.

Cliff chipmunks scamper throughout Arizona, but nowhere do they appear as abundantly as at Boyce Thompson Arboretum. Walk the paths from the Smith Interpretive Center to the boojum trees, the area around Drover's Woodshed, Homesteader's Cabin, and the herb garden in the spring to see them feeding, chasing each other, and seemingly posing for photographs. Search old buildings where cliff chipmunks like to place their nests. When you sight one or more, stop, get into a comfortable position, and wait. Three or four of the young chipmunks will likely emerge and present you with an excellent opportunity to take a group shot. When you find chipmunks in shade, put your camera on a sturdy tripod and turn on your fill flash. If something scares the babies (such as the sudden movement of a photographer or hiker) and they hide, be patient. They will return.

Volunteer experts regularly lead "lizard walks" during the summer months. Guides show you the most likely spots to find collared, greater earless, desert spiny, common side-blotched, and even jittery tiger whiptail lizards. Most lizards show off their most colorful skin from late April through June. When you find a lizard, approach it very slowly. Get low to the ground (i.e., at the eye level of your subject; knee pads may prevent discomfort) with your telephoto lens

Boyce Thompson Arboretum

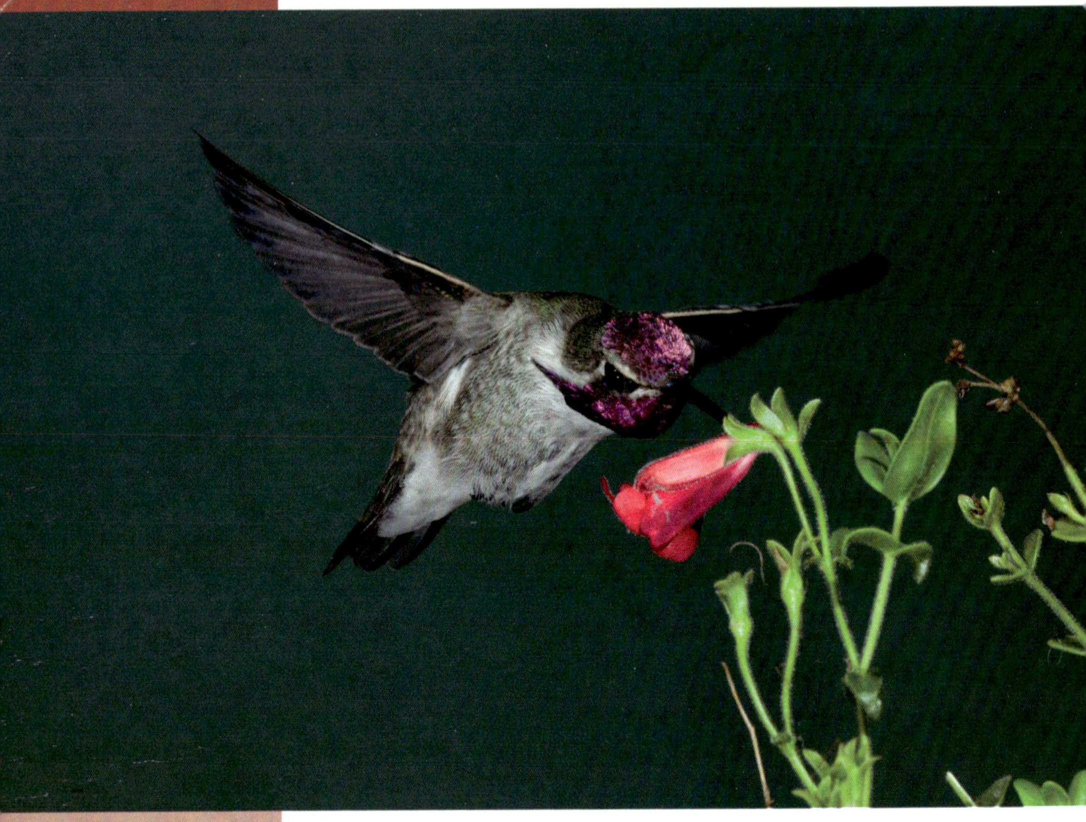

Costa's hummingbird. Canon 1DMIV, 70-200mm at 111mm, 2x teleconverter, ISO 400, f/18 @ 1/320 sec., high-speed hummingbird set-up.

mounted on a sturdy tripod and **ball or gimbal** head. If necessary, lay your lens on the ground to take a real lizard-eye-level image.

Consciously compose so that the backdrop offers a complementary color to the lizard. Set a wide aperture (such as f/4 or f/5.6) to blur the background. When the lizard is in the shadows or the sun is facing you, **fill flash** set to minus one stop compensation can make the difference between a lousy photo and a great one. Even when the sun falls directly on the lizard, use fill flash to open up the shadows on the belly of the subject.

For most of the year, resident black phoebes hunt for insects over Ayer Lake at the arboretum's east end. Black phoebes always hunt from the same perch. When you get to the pond, locate their favorite perches and set up within 20 feet (6.1 m) of one of them. Sit in the shade, both to be cool and unobtrusive, and wait. The bird will eventually return for its portrait—perhaps showing off an insect trapped in its beak! A telephoto lens propped on a sturdy tripod will enable sharp, frame-filling photos. Use fill flash to add light in the shadows during the harsh midday light.

Before you leave, look around the lake for other photo opportunities. Water in the desert is a strong attractant for many species of wildlife so always be on the lookout for the occasional coyote or coatimundi.

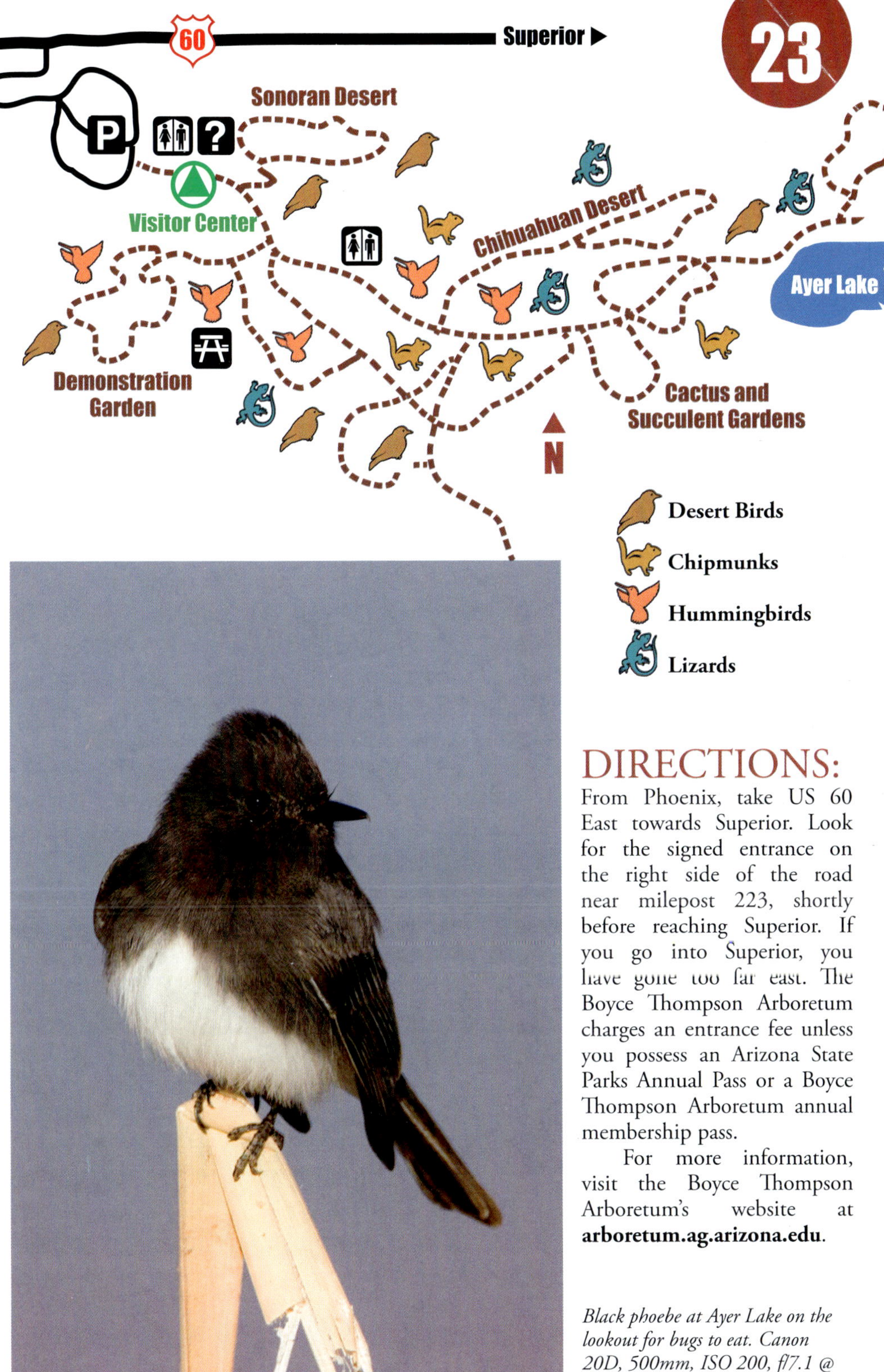

DIRECTIONS:

From Phoenix, take US 60 East towards Superior. Look for the signed entrance on the right side of the road near milepost 223, shortly before reaching Superior. If you go into Superior, you have gone too far east. The Boyce Thompson Arboretum charges an entrance fee unless you possess an Arizona State Parks Annual Pass or a Boyce Thompson Arboretum annual membership pass.

For more information, visit the Boyce Thompson Arboretum's website at **arboretum.ag.arizona.edu**.

Black phoebe at Ayer Lake on the lookout for bugs to eat. Canon 20D, 500mm, ISO 200, f/7.1 @ 1/1000 sec.

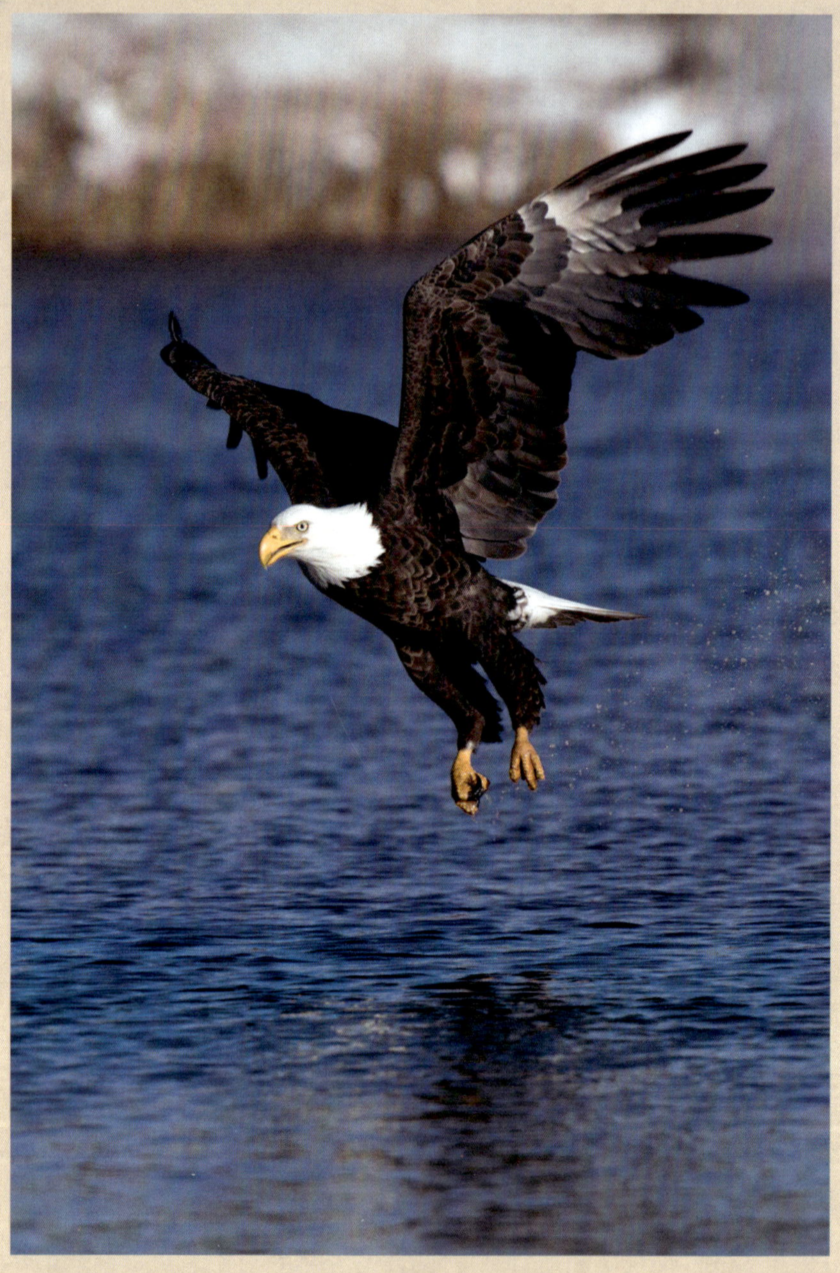

This bald eagle just missed catching a fish at Green Valley Lake in Payson. (No worry, its next attempt was successful!) Canon EOS 7DMII, 100-400 at 400mm, ISO 320, f/7.1 @ 1/2500 sec.

Wildlife of

EASTERN ARIZONA

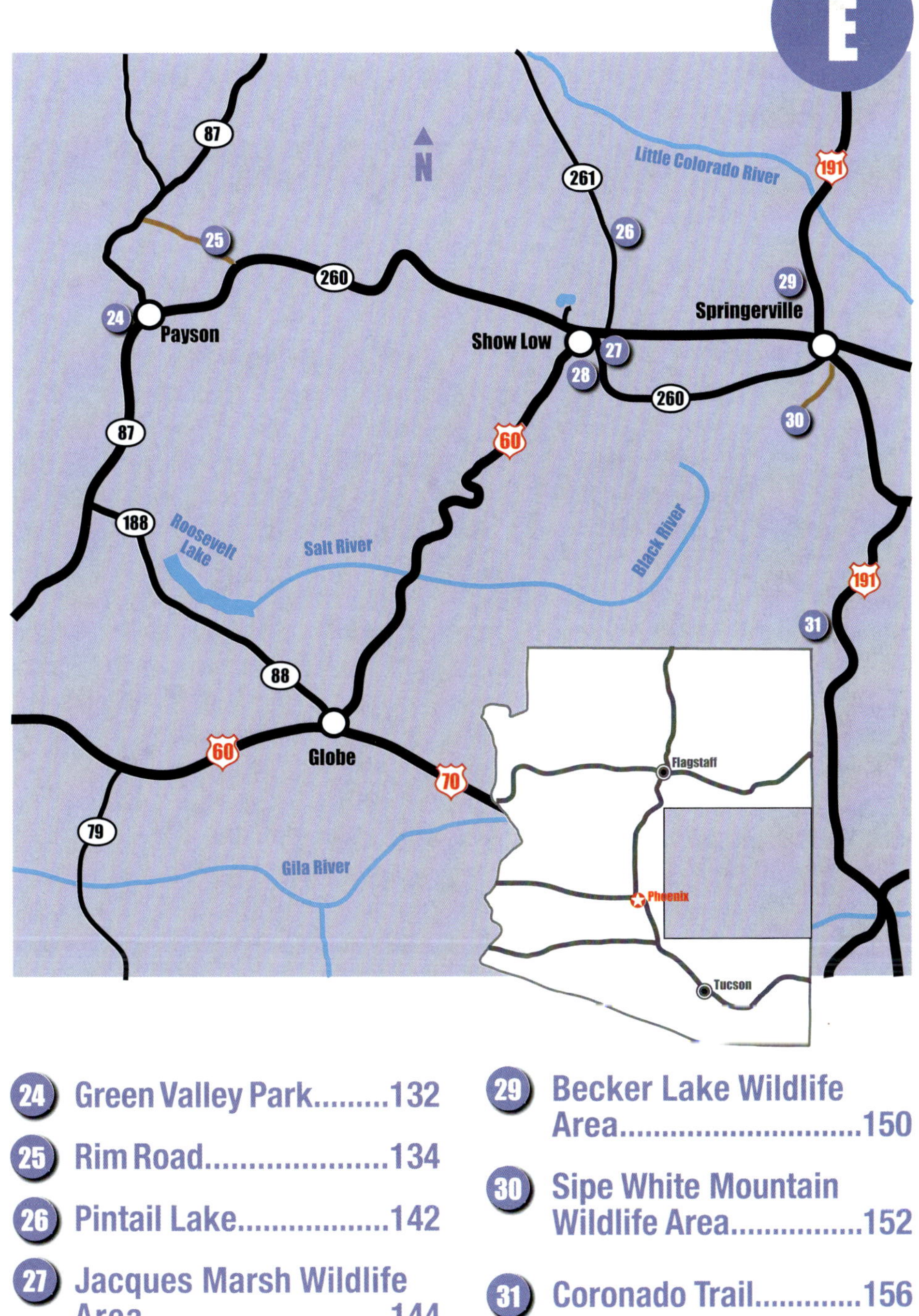
87
N
261
Little Colorado River
191
25
26
260
29
Springerville
24
Payson
Show Low
27
28
260
30
87
60
188
Roosevelt Lake
Salt River
Black River
191
31
88
Flagstaff
Globe
60
70
79
Gila River
Phoenix
Tucson

Green Valley Park

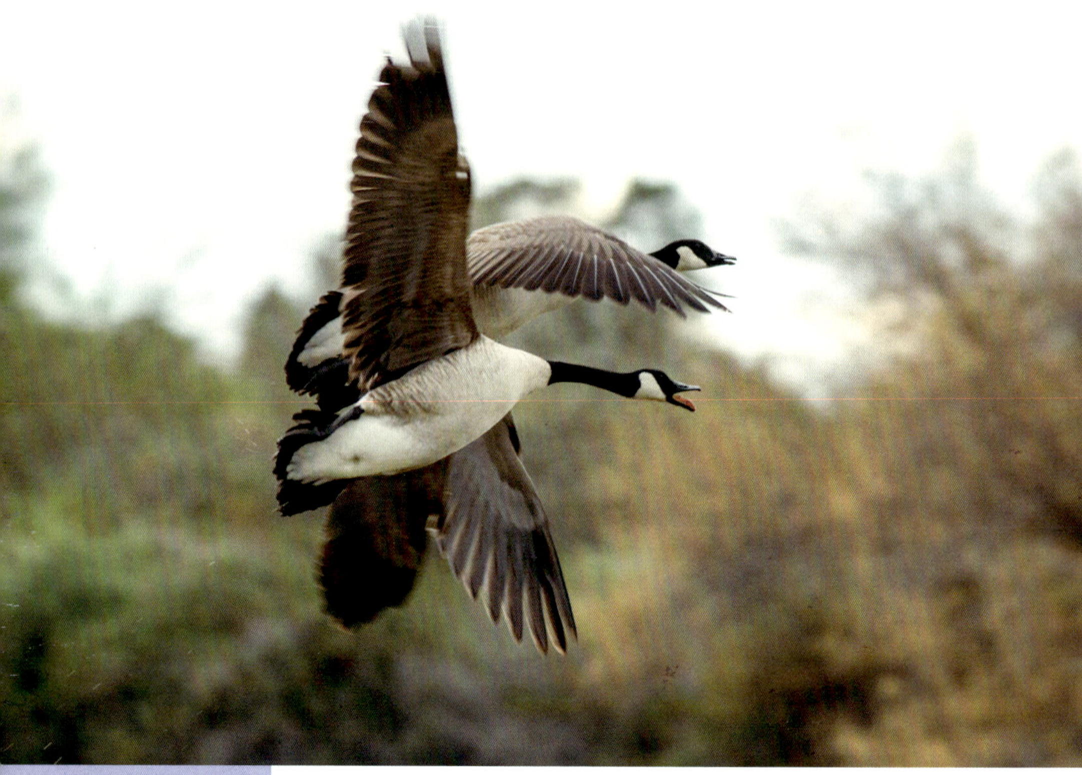

Canada geese cackle as they leave the pond at Green Valley Park. Canon 7D, 400mm, ISO 400, f/7.1 @ 1/250 sec.

VIEW TIME
October to April

IDEAL TIME OF DAY
Sunrise to early morning

VEHICLE
Any

HIKE
Easy

Considered "The Jewel of Payson" by the town, Green Valley Park was constructed in 1996 and features three lakes (comprising a total surface area of 13.1 acres (5 hectares)). The park entices outdoor enthusiasts with numerous recreational activities such as walking, hiking, picnics, and fishing. But the park's lakes are not just for recreational purposes. They also serve as an important water reclamation system. The golf course uses this water for irrigation, and through passive percolation, the surrounding water table can recharge. The value of these types of development projects cannot be overemphasized in the desert southwest. Not only do they assist in keeping communities viable, but they also provide incredible value to wildlife and their habitats.

During the winter months, the lake attracts hundreds of water birds including ducks, Canada geese, herons, and one or two bald eagles. The nearby golf course attracts wintering geese while the ponds and aquatic vegetation keep the ducks happy.

Arrive at the parking lot before sunrise to investigate the location of the largest congregation of ducks. Groups of wigeons commonly fly back and forth between the water and the grass to feed. Coots, mallards, and ring-neck ducks paddle calmly in small groups nearby. Situate yourself near the water's edge to obtain the best angle for both flight and swimming images.

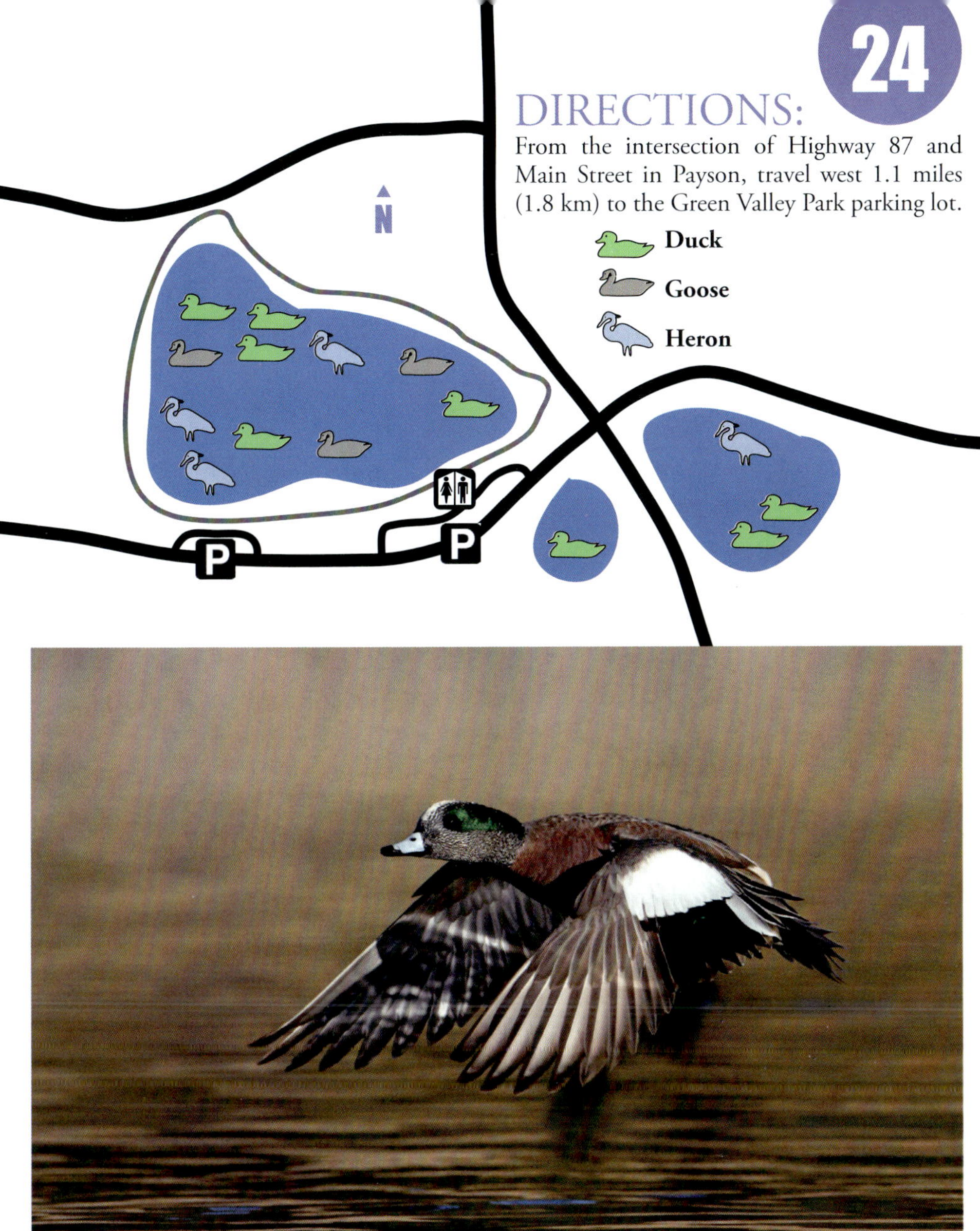

24

DIRECTIONS:

From the intersection of Highway 87 and Main Street in Payson, travel west 1.1 miles (1.8 km) to the Green Valley Park parking lot.

American wigeon flies across Green Valley Lake. Canon 40D, 500mm, ISO 400, f/5.6 @ 1/3200 sec.

Then, attach a **400mm or longer telephoto lens** to your camera. Whether you hand-hold or use a **tripod**, ensure a fast shutter speed (i.e., 1/500 of a second or faster) to freeze the waterfowl during the low lighting conditions of the morning.

When the first golfers get to the links, the geese will fly from the course to the ponds and offer great opportunities for flying shots. Do not worry about looking too hard for the geese. Just listen for their classic "honk" as they fly from land to water. On the water, Canada geese will often fight, chase each other, or conduct some sort of odd behavior worthy of a few frames.

EASTERN ARIZONA

Rim Road

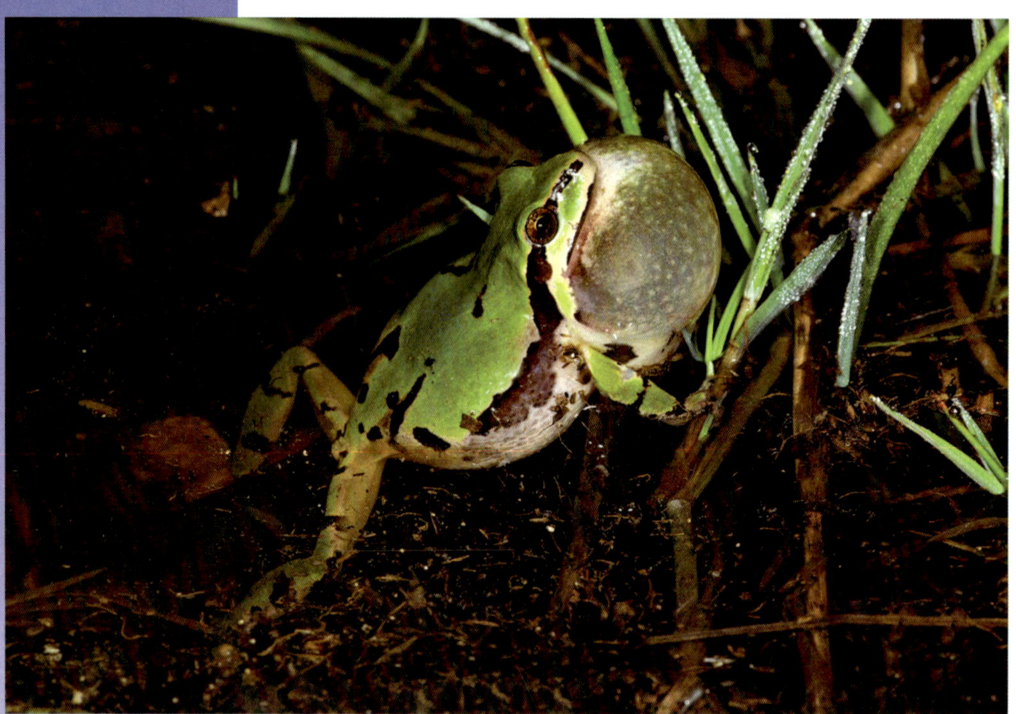

A male Arizona treefrog calls to attract a mate in the shallow, ephemeral water. Canon 5DMIII, 180mm macro, ISO 200, f/16 @ 1/200 sec., macro twin flash.

VIEW TIME
June and July

IDEAL TIME OF DAY
Night

VEHICLE
Any

HIKE
Easy

If you enjoy being alone listening to the primordial sounds of the monsoon as nightly storms move over the Mogollon Rim, then photographing Arizona tree frogs along the Rim Road (also referred to as Forest Service Road 300) to Woods Canyon Lake awaits you. Pine trees, cool weather, and wet meadows define the 7,500-foot (2,286-m) elevation drive. During late June and July, the hot, humid air flows over the rim—producing massive, localized rainstorms. Water from these storms fills low-lying meadows and creates the perfect habitat for frogs to breed.

Arizona tree frogs are possibly the most colorful of Arizona's frogs. Although some Arizona tree frogs appear dull brown in coloration, if you search hard enough, you will find others with striking green bodies. At two inches (5.1 cm) long, the best way to find this diminutive species is to listen for the males' high-pitched calls as they attempt to attract females. At dark, the male frogs (who have been hiding in the surrounding woods during the day) move to the ephemeral pools to compete for the attention of the females. From dark to just before dawn, listen for a constant chorus of frog calls. During this period, drive slowly along the road with your windows down and listen. When you hear a noisy group of male frogs, locate a safe place to leave your vehicle (you may have to drive further to find a pullout).

Bring either a **macro lens** or a lens coupled with **teleconverters**, **extension tubes**, or the like, so that you can obtain frame-filling images of these small frogs. Use a **flash** fitted with a **flash bracket** so that the flash fits directly over the top of the lens. Alternatively, use a specialized **macro flash** that fits on the front of the lens. Set the camera mode on manual and

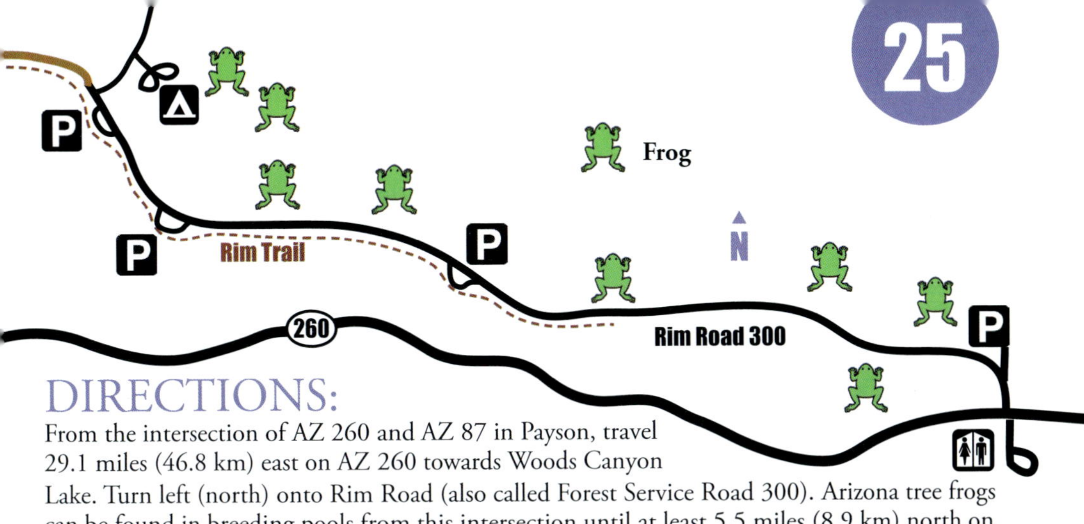

DIRECTIONS:

From the intersection of AZ 260 and AZ 87 in Payson, travel 29.1 miles (46.8 km) east on AZ 260 towards Woods Canyon Lake. Turn left (north) onto Rim Road (also called Forest Service Road 300). Arizona tree frogs can be found in breeding pools from this intersection until at least 5.5 miles (8.9 km) north on Rim Road.

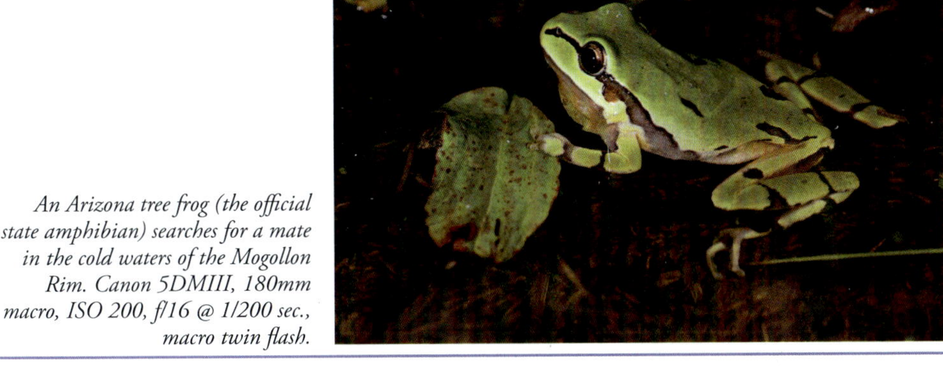

An Arizona tree frog (the official state amphibian) searches for a mate in the cold waters of the Mogollon Rim. Canon 5DMIII, 180mm macro, ISO 200, f/16 @ 1/200 sec., macro twin flash.

the flash on TTL. Set the camera shutter to the maximum sync speed for your flash (check your flash's instruction manual for this speed). Then, adjust to a small aperture (e.g., f/16) to ensure that your photograph has extensive depth of field.

At this elevation, nights get relatively cool, and the water always feels cold. Put on a pair of rubber boots, a warm coat, and a cap to keep from getting wet and chilly. Wear a headlamp not only to light your way, but also to illuminate the frogs for focusing. Wade into the large rain-produced puddles (only a few inches deep) and move towards a calling male frog. Move slowly and quietly. Frogs can not only hear you, but they can also see you in the moonlight or feel vibrations in the water as you move.

When you are close to a likely subject, kneel down and place your camera near the water's surface so that you can take your photographs close to their eye level. Frame the frog using the rule of thirds and take several images.

Check your histogram to ensure that your flash output is appropriate. As you modify positions or magnification, the histogram may change. If your highlight alert blinks, turn the flash output down by one-half stop at a time. Likewise, if the histogram shows a concentration to the left, add some flash power.

Take photographs of frogs sitting, calling, in amplexis (the male holds tight to the female while she sheds her eggs), and every other position you might encounter. This activity only takes place once each year, and if you are lucky enough to have the pleasure of experiencing it, blast away! You never know when another magical night will happen!

PHOTO TIP 9

Night Photography

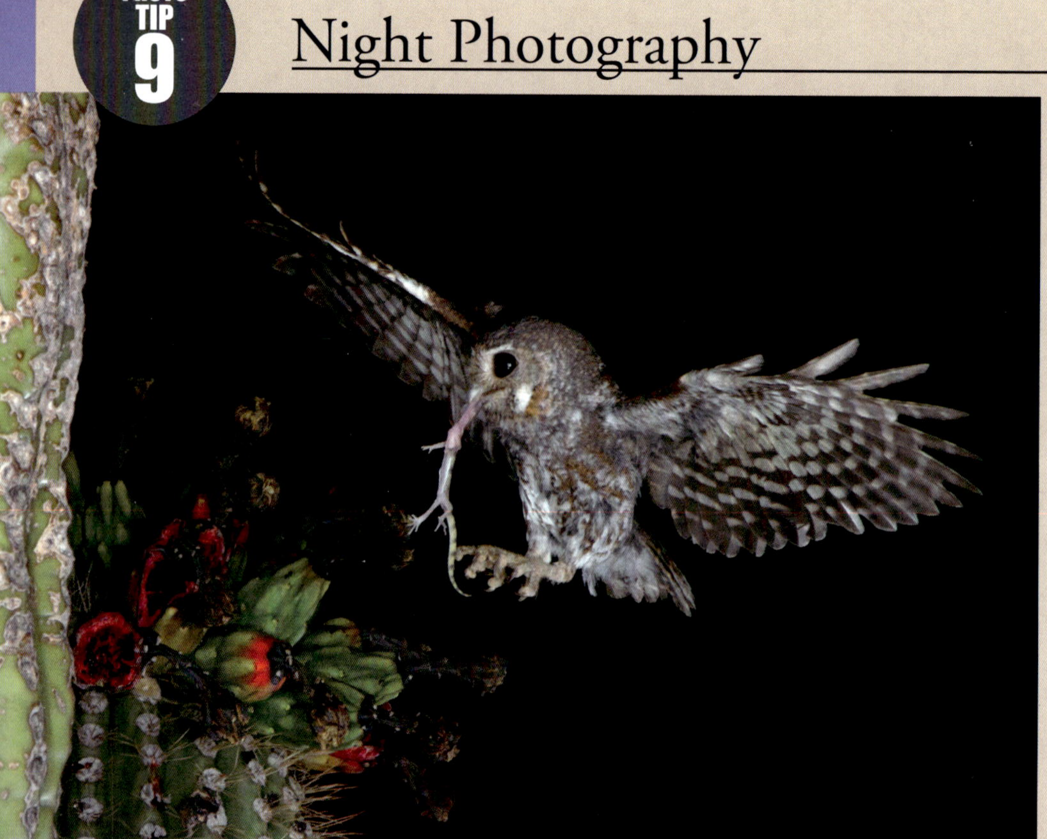

To photograph this elf owl carrying a banded gecko into its nest to feed its babies, I mounted my camera and flashes on a stand 16 feet (4.9 m) above the ground. Canon 7D, 24-105mm at 105mm, ISO 400, f/16 @ 1 sec., multiple off-camera flashes, Phototrap infrared trigger.

As the sun cracks over the horizon at the break of dawn, the resulting warm tones add an unmistakable and beautiful quality to one's images. As dawn turns to midday the nature of the light changes from golden oranges and reds to cold blue with deep, harsh shadows. Towards late afternoon, the light again begins to add warm colors to the scene, allowing the photographer to capture nature at its best. Shortly after sunset, most of the color disappears from the environment, and finding sufficient natural light to photograph nocturnal wildlife becomes a chore. Artificial lights to the rescue!

Flashes emit small, powerful beams of light with enough power to light up almost any critter in the dark. However, if you place the flash in your camera's hot shoe, the light will appear harsh, non-directional, and flat. Direct front light on a bird's feathers or the fur of a mammal will not show their delicate structure. Animal eyes will reflect the light from the flash and produce an annoying red or steel blue color. Objects directly behind your subject will show harsh shadows.

To resolve these problems, move the flash off-camera and away from the lens. Place the camera on a flash stand. Since this may necessitate moving the position of the camera (and yourself) several feet from the flash, use wireless remotes to set off the flash. Alternatively, use a flash cord and some sort of flash bracket to attach the flash to the camera hot shoe. Flash brackets either attach to the tripod or camera and allow the flash to be positioned several inches from the lens. By moving the flash only a few inches away from the camera, the light quality in your images will improve dramatically. Fur and feathers will have some detail, red (or steel blue) eye will go away and harsh shadows are reduced or disappear altogether.

For small subjects such as frogs, scorpions, and snakes, a flash on a tall flash stand will not lower to eye-level. Instead, when photographing ground-loving critters, use a macro bracket or specialized macro flashes which keep the flash close to your subject but away from your lens. An alternative is to use a remote system to operate the flash and place the flash on either a very short stand or on the ground a few inches from the subject. The key is to keep the flash from becoming an obvious artificial light source.

Normally you can illuminate small subjects using a single flash, but you may need several flashes to provide the appropriate amount of light for large critters such as ringtails or kit foxes. In these cases, set two or more flashes on light stands in a location near where you expect the animal to visit (along a trail, next to water, or at an animal's den). Place the first flash a few feet to the left of the location, and the second a few feet to the right. A third flash set behind the action will generate backlighting. The goal is to provide light from multiple directions and illuminate the subject in an appealing manner while rendering a strong enough light to make an exposure in the dark.

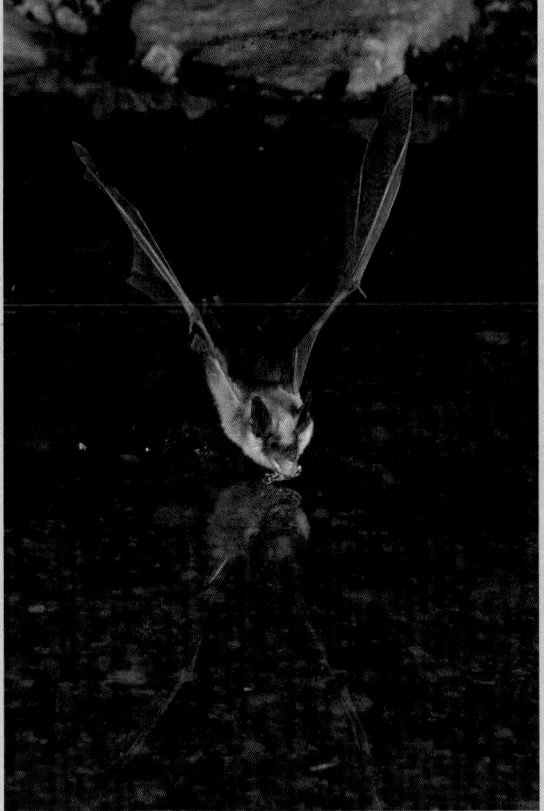

To focus on your animal, use a flashlight. You do not need to turn the flashlight off before taking the photo, as the light from the camera's flash will override the light originating from the flashlight. Place the camera on manual mode, and then set the shutter speed to the flash sync speed. At night, the light meter will advise using several second exposures if the camera is in aperture priority, shutter speed priority, or program modes. Since the flashes are providing the only light source, slow shutter speeds are unnecessary. In addition, use autofocus and single-shot drive mode. Set your flash to TTL.

Arizona is home to 27 species of bats including this Myotis spp. Bats obtain drinking water by flying over water sources and opening their mouths as they touch the water's surface. Canon EOS Rebel XSi, 24-105mm at 84mm, ISO 400, f/14 @ 10 sec., multiple off-camera-flashes, Phototrap infrared trigger.

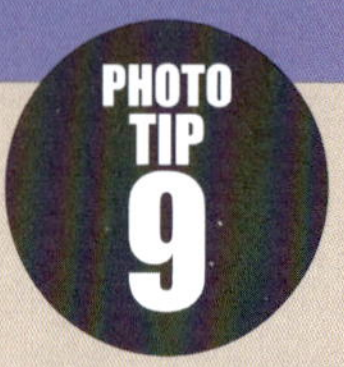

Night Photography (continued)

After taking the image, check your histogram to ensure a correct exposure. Making a judgment about exposure using the LCD image alone can be misleading. Using the LCD screen is similar to using a computer screen. At night, the image appears brighter than normal and during the day, the image looks too dark. Use the LCD for very general exposure determinations and the histogram for the finer aspects.

In the absence of flash units, you can employ flashlights to provide directional light on smaller animals like lizards, amphibians, and insects. These smaller, but potentially powerful, lights move around easily, provide enough light for nighttime photography, and, when used appropriately, can add a dramatic look to images. When using flashlights as the main light source, place the camera on a tripod, pre-focus on the subject, and then place the lens on manual focus. Place the camera on Bulb mode so that you can take several images at different exposures without going to the camera and changing the shutter speed. Point one or more flashlights at the subject and trigger the shutter using a remote cable release. Hold the shutter open for one second while illuminating the subject. Try another exposure while holding the shutter open for two seconds and then maybe another at three seconds. Review your image on the camera's LCD screen and histogram to evaluate the light quality and quantity.

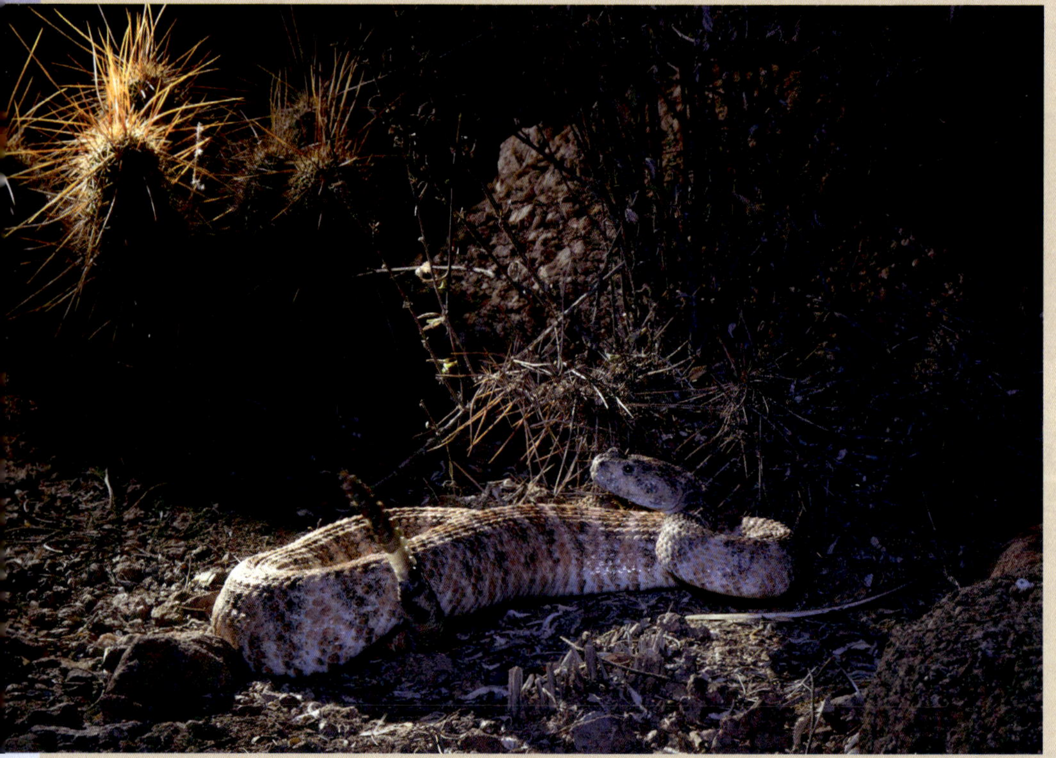

A coiled speckled rattlesnake waits for its prey to get close. Canon 50D, 24-105mm at 70mm, ISO 200, f/22 @ 8 sec., illuminated by flashlights.

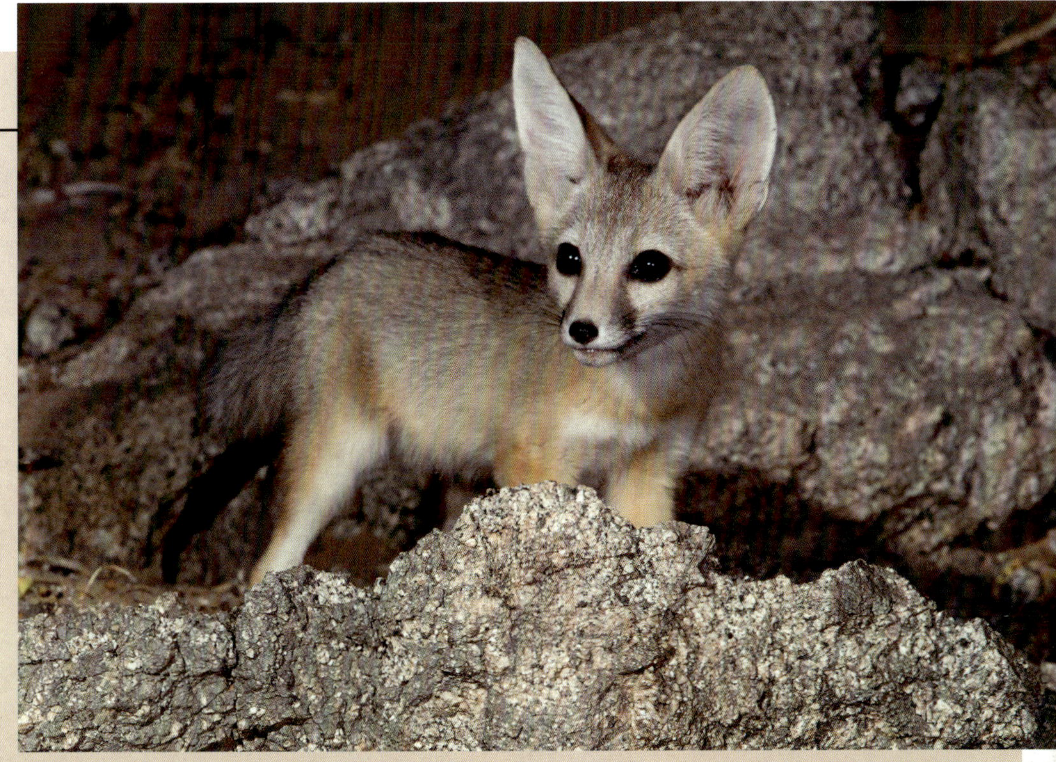

Kit fox kit. Canon 1DMIV, 70-200mm at 200mm, 2x teleconverter, ISO 400, f/22 @ 1/250 sec., off-camera flash.

Experiment with changing the flashlight's direction and the amount of time you open the shutter. Move the flashlights around during the exposure, and selectively light portions of the scene until you achieve the results you seek.

Because many large and cautious animals such as coyotes, fox, and deer will not approach if they sense your presence, use a blind while taking nighttime photos of these species. Arrive on location early in the afternoon to not only set up the blind, but also flashes, remotes, light stands, and any other gear necessary to take a successful nocturnal image.

For night photography, a security camera can help you track your subject from a blind. These portable and inexpensive cameras use 12 volts for power and an infrared light to see subjects after dark. Place the security camera where you expect the subject to appear, get into your blind, and turn on a small television or computer to watch the action. When the critter moves into your frame, remotely release the camera's shutter button. Obviously, this form of photography means that you must have considerable knowledge about your subject's behavior, location, and movement patterns (see page 180 for tips on using remote equipment to help you take a great nocturnal image).

Insects leave their daylight sanctuaries at night to feed, breed, and do whatever insects do. To attract insects for a photograph, use either a black light or a mercury vapor light. Adjacent to the lights, place a large white bed sheet perpendicular to the ground where the lights can reflect off the sheet's surface and enhance the perceived amount of light. Insects of all sorts will fly to the light and land on the sheet where you can photograph them in situ or collect and move them to a better, and more pleasing, backdrop.

Making the Photo 6

Nocturnal Moths

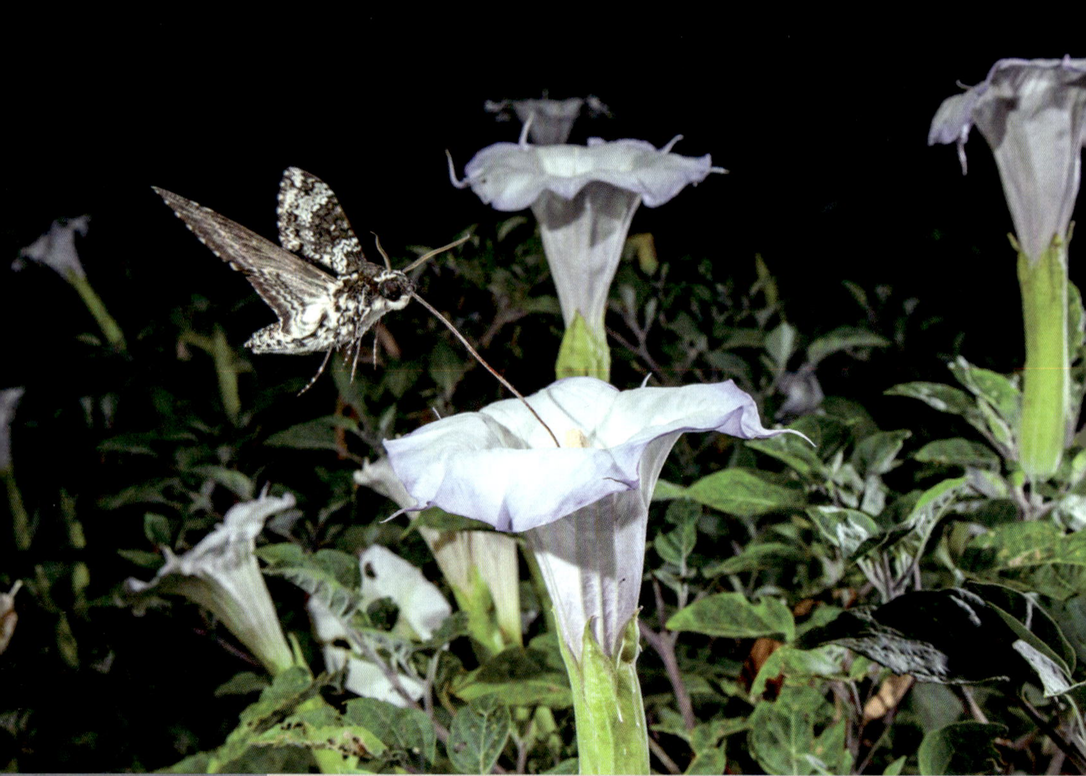

A sphinx moth feeds from sacred datura flower. Canon 5DMII, 24-105mm at 50mm, ISO 400, f/18 @ 1/250 sec., off-camera flash, Phototrap infrared trigger.

I love to photograph at night. The critters behave differently, I can create any light I wish, and I have access to many subjects that are not available during the daytime. Also, because not many photographers venture out in the dark, nighttime photos set my work apart from many others. Not only do I love to photograph in the dark (my friends say I am always in the dark anyway), I also prefer photographing small, moving subjects. My favorite nocturnal macro photography subject is the sphinx moth.

Sphinx moths feed on flowers that bloom day or night. For nocturnal photography, the beautiful nocturnal-blooming flower of the sacred datura fits the bill. The pleasantly fragrant sacred datura can display massive numbers of large, white blooms during the monsoons. Sphinx moths find the flowers using olfactory and optic cues (they smell and see the flowers). They select the sacred datura plant because it offers them nectar for food and serves as a nurse plant for their caterpillars. These vibrant blooms start opening up at dusk and are ready for both moths and photographers by about one half hour after sunset. As is the case with butterflies, all I have to do is find the flower and the moth will find me.

When I first started photographing sphinx moths feeding on sacred datura, I used a 100mm macro lens. I loved the results and thought I had found the best perspective possible. That is, until a friend of mine, Randy Babb, created an inspirational photograph of sphinx moths feeding on a sacred datura plant in his backyard using a wide-angle lens. Not only did he record a large moth feeding on a white flower, but he also included more flowers in the backdrop which allowed the photograph to tell a stronger story.

After seeing his image, I changed my technique. Now I use both a 100mm macro on a full-frame camera and a 17–35mm wide-angle lens on a cropped frame camera (which translates to the equivalent 35mm focal length of 27-56mm, considering the aspect ratio). By using the wide-angle lens on this smaller sensor, the resulting magnification allows me to make the moth large in the frame but still maintain a wide-angle view.

During one afternoon, I set my camera close to a flower that I expected to open that night (the bloom of the datura only lasts for a single night, so a photographer must plan ahead). To get set up, I placed my camera on a tripod and manually set my aperture to f/16 and shutter speed at the camera's flash sync speed.

Then, I positioned three flash units atop light stands about two feet (0.6 m) from the flower. I moved one flash to the right of the flower (off-camera right), one flash to the left, and one to the rear of the bloom to provide some backlighting. At night, my flashes are the only source of light and the speed of the flash is the final effective shutter speed. To keep the flash duration fast enough to stop the moth's wing movement, I dialed each flash to 1/16 power. I knew that by reducing the power of the flash, I was also reducing the flash duration. Because my entire light source is from the flashes I reduced the flash power to 1/16 of full and the resultant 1/15,000 of a second stops the moth's wing movement. I attached an Impact wireless remote receiver to each flash and placed a transmitter in my camera's hot shoe. In the past I would wire all the flashes together, but using wireless equipment eliminated lots of set-up time—and reduced the likelihood of me tripping over the wires and pulling the flashes and flash stands down.

Although I have tried many, many times, I have found it almost impossible to depress the shutter button manually at exactly the same time that the sphinx moth is feeding on the flower. Also, unless a bright moon lights up the night, just seeing the flying moth is a chore. When it comes to taking photographs of moving or flying subjects at night, an infrared trip mechanism comes to the rescue!

In the past, I have created my own infrared trips from electronic parts purchased from various sources and asked my electronically-talented friends (Charlie Cobeen and Vern West) to assemble them. Then I purchased Phototrap's high-speed camera infrared trigger. "The Trap" (as it is called for short) emits infrared beams of light and automatically trips any camera and/or flash units plugged into the system as soon as a subject crosses those beams. The system is so easy to use, one does not have to know anything about electronics.

With the trigger in place, I mounted the infrared light source so it projected the beam over the top of, and parallel to, the face of the flower bloom. I manually focused the camera in the middle of the flower opening. When the moth came to feed on the flower, it interrupted the beam. The camera and flashes instantly fired. I created a photograph of a sphinx moth with its long proboscis sticking into the flower to gather its sweet nectar. In addition, through the use of a wide-angle lens, I was able to show several of the datura blooms which added to the story telling capabilities of the image.

Pintail Lake

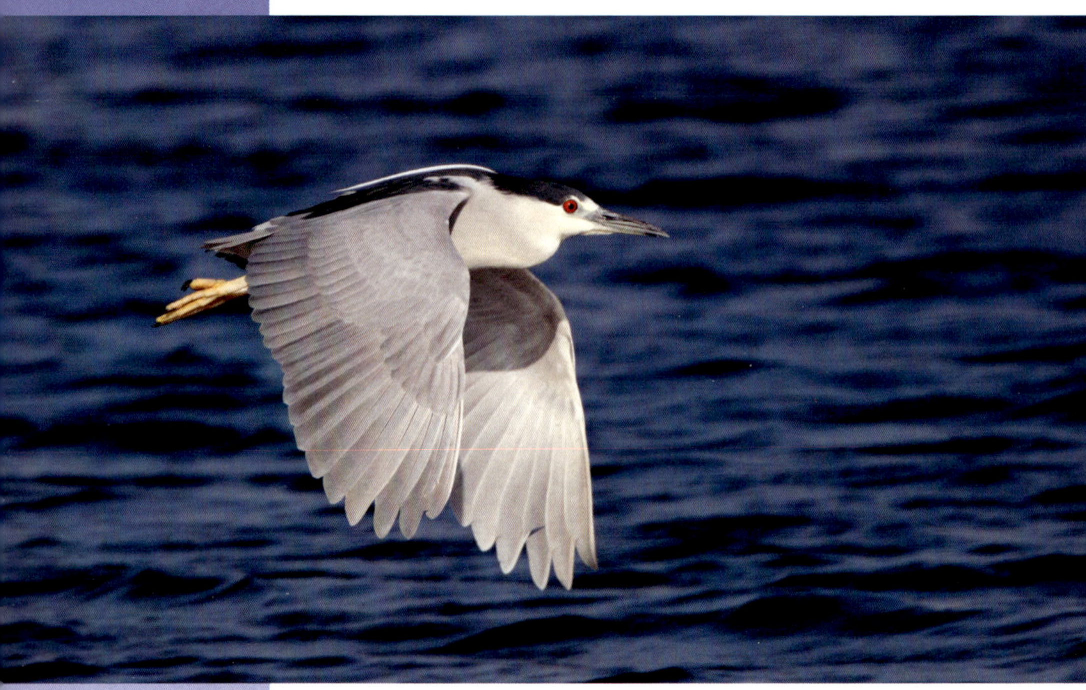

Black-crowned night heron. Canon 50D, 500mm, 1.4x teleconverter, ISO 400, f/5.6 @ 1/1600 sec.

VIEW TIME
Late May to September

IDEAL TIME OF DAY
Early morning and late afternoon

VEHICLE
Any

HIKE
Easy to moderate

North of Show Low, in the midst of high elevation grasslands, the water and habitat of Pintail Lake offers a unique photographic experience. One of only a few water reclamation projects developed in a natural setting, Pintail Lake concentrates wildlife for the photographer's shooting pleasure. Although Pintail Lake was developed with recreationists in mind, it has not been maintained as such. The birdwatching blinds are in need of repair as are the trails. Planted with native aquatic vegetation and developed with several islands, Pintail Lake provides a lush water sanctuary for ducks, avocets, bald eagles, blackbirds, herons, ibis, rails, stilts, and other water birds.

But Pintail Lake offers more than just birds. As you drive along Pintail Lake Road, scan the rocks along the roadway for an eastern collared lizard basking in the sun or waiting for an unlucky lizard (or insect) to crawl by and become its next meal. As you approach the lizard, roll the car window down and situate your **telephoto lens** on a bean bag flung over the window's ledge. Set your camera on an aperture of f/8, or wider, to selectively focus on the critter. To render the sharpest image possible, use a shutter speed faster than 1/250 of a second and enable high-speed continuous shooting mode as well as image stabilization (or vibration reduction).

Frame the collared lizard according to the Rule of Thirds and the Rule of Space, so that the lizard appears in the lower part of the image and has more room ahead of its direction of travel. If the lizard's skin has heated up in the sun, the lizard will show off a multi-hued color palette. Unfortunately, only the males make this vivid change; the females only display light cream or white with a dark collar around their necks.

Continue down the road and park in the cinder lot. Pass through the

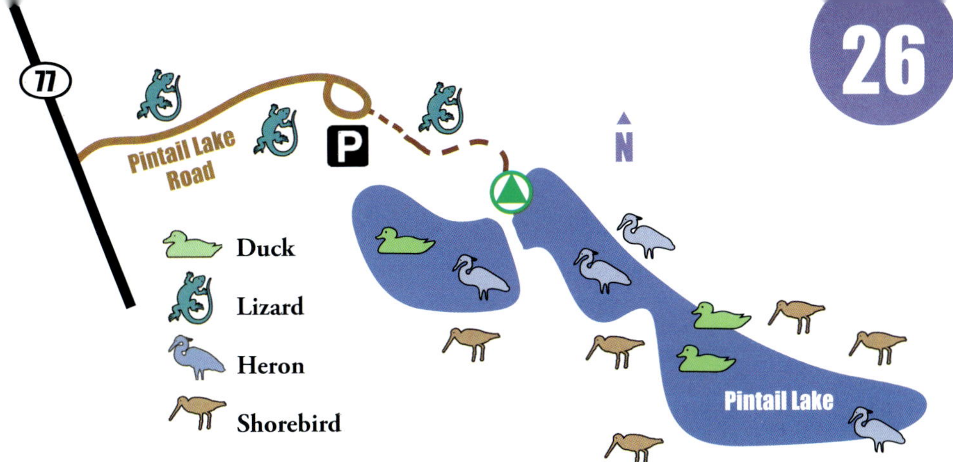

DIRECTIONS:

From the intersection of AZ 77 and AZ 60 in Show Low, drive north on AZ 77 for 3.9 miles (6.3 km). At the sign for Pintail Lake and Pintail Lake Road, turn right. Travel 0.4 miles (0.6 km) on the dirt road to the parking lot.

gate and walk the blacktop trail to the lake. Along the way, keep an eye out for more collared lizards and the occasional greater short-horned lizard.

Walk the various trails with your telephoto lens and tripod while searching for birds. Note where the birds are concentrated or have established breeding areas. Take the occasional image if it presents itself, but on your first foray around the lake, focus instead on scouting and finding the most photogenic spots.

Set up a blind in an area that has some bird activity—breeding, loafing, swimming, or feeding. Give special attention to the mud flats where avocets and stilts build their nests and feed. Open water areas supply prime locations for swimming waterfowl while cattail marshes often see blackbirds breeding and defending their territories. Watch the pinyon trees for hunting raptors and the larger trees for the occasional bald eagle.

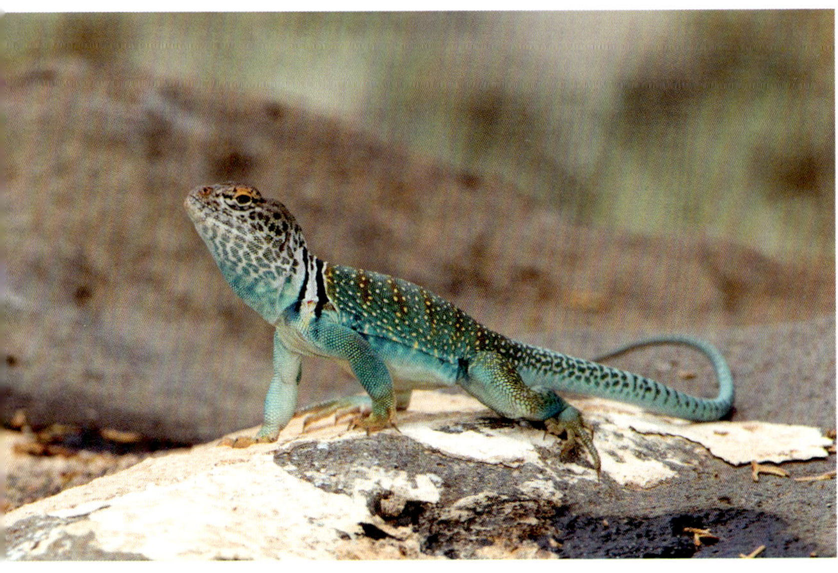

A brightly colored collared lizard sits high on a rock and watching for an unsuspecting meal. Canon 1DMIV, 500mm, 1.4x teleconverter, ISO 100, f/11 @ 1/125 sec., on-camera flash @ -1/3 FEC.

Jacques Marsh Wildlife Area

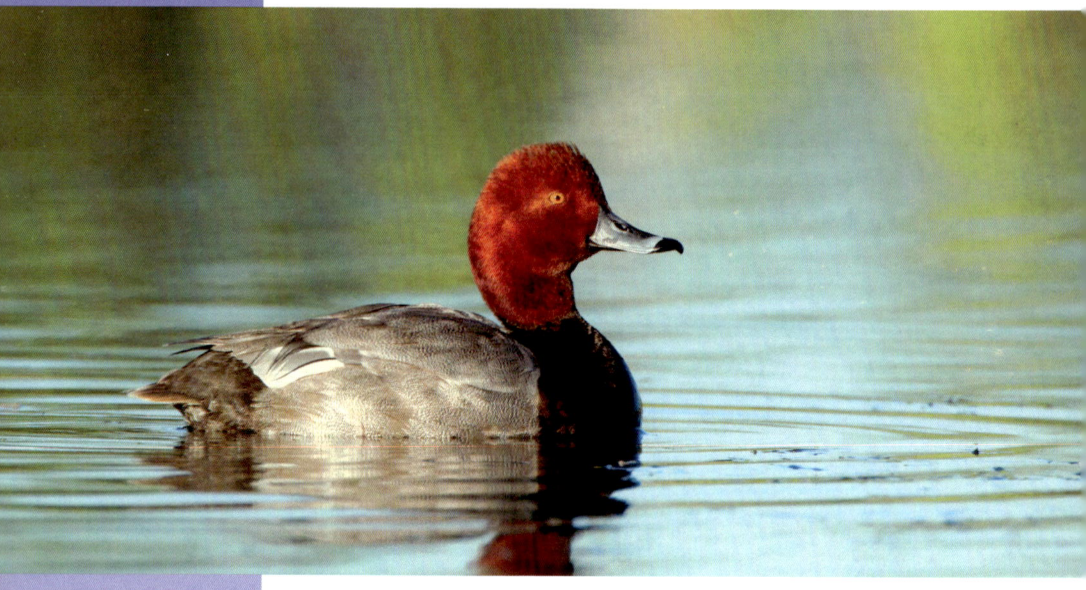

A redhead duck swims in the early morning at Jacques Marsh Wildlife Area. Canon 50D, 500mm, 1.4x teleconverter, ISO 400, f/7.1 @ 1/1250 sec.

VIEW TIME
Late April to July

IDEAL TIME OF DAY
Sunrise and sunset

VEHICLE
Any

HIKE
Moderate

Serving as Pinetop-Lakeside's water treatment plant, treated wastewater moves through seven ponds which cleanse nitrogenous wastes from the water until it is clean enough to be discharged. Do not let that ignoble beginning tarnish your desire to visit Jacques Marsh to photograph some of Arizona's unique species!

Planted with native aquatic vegetation in the early 1980s, Jacques Marsh teems with plants and animals. Along with many species of small birds, aquatic species such as yellow-headed blackbird, redhead, belted kingfisher, sora, and the incredible blue-billed ruddy duck make this marsh home. Next to the marsh, in the pinyon-juniper woodland, lurks the mighty Rocky Mountain elk.

The best time to visit Jacques Marsh for water birds is late April through July when water birds are setting up territories and breeding, and the elk come in at night and early morning to eat the verdant vegetation and drink water.

A series of berms impound the water and elevate the ponds above the surrounding ground. This arrangement makes it easy to walk along the lower edge of the berms and go undetected by the birds inhabiting the marsh. When you hear birds calling or making other noises, sneak up the berm to the edge of the water. Stay low so that the wildlife does not see you, and place your **long telephoto lens** on the top of the berm for stabilization. For this type of commando-style photography, you will not need a tripod or monopod, but remember to enable image stabilization (or vibration reduction) as you hand-hold your camera. Use continuous autofocus mode to help focus on the birds swimming in the water or rummaging around in the vegetation.

Ideal light occurs during the morning and evening. Keep the sun to your back and your shadow pointing at the subject to evenly expose your

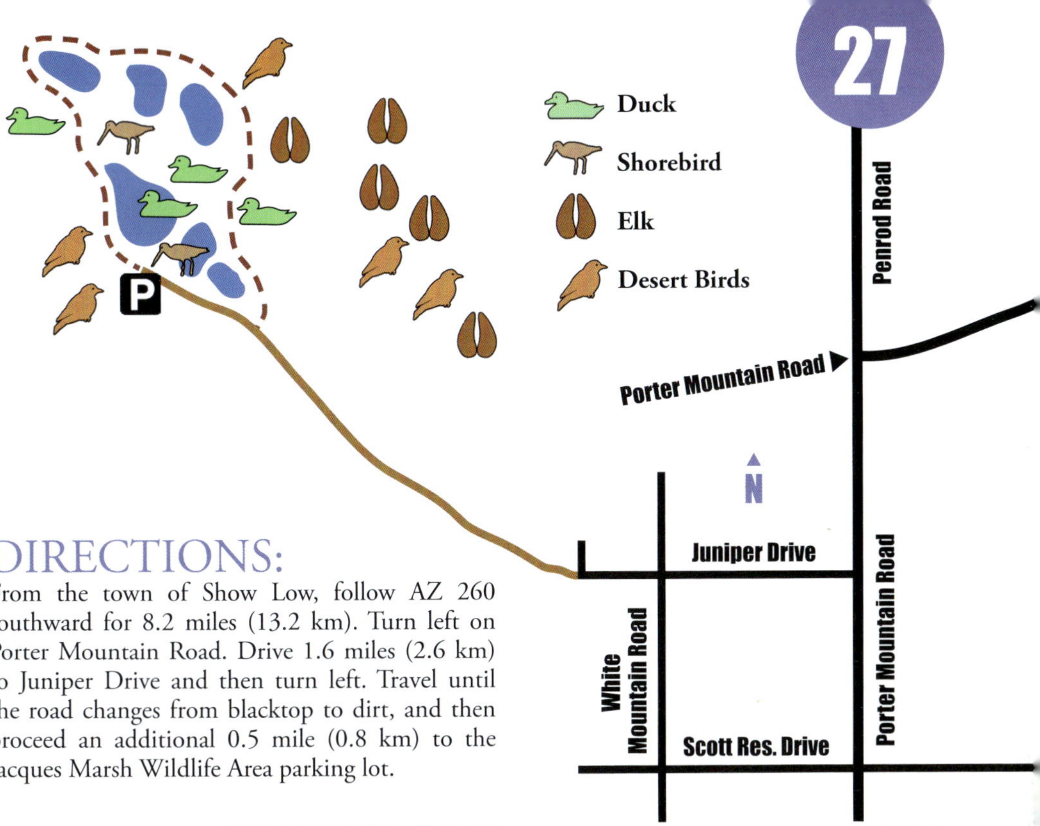

27

Duck
Shorebird
Elk
Desert Birds
P
Penrod Road
Porter Mountain Road
N
Juniper Drive
White Mountain Road
Scott Res. Drive
Porter Mountain Road

DIRECTIONS:

From the town of Show Low, follow AZ 260 southward for 8.2 miles (13.2 km). Turn left on Porter Mountain Road. Drive 1.6 miles (2.6 km) to Juniper Drive and then turn left. Travel until the road changes from blacktop to dirt, and then proceed an additional 0.5 mile (0.8 km) to the Jacques Marsh Wildlife Area parking lot.

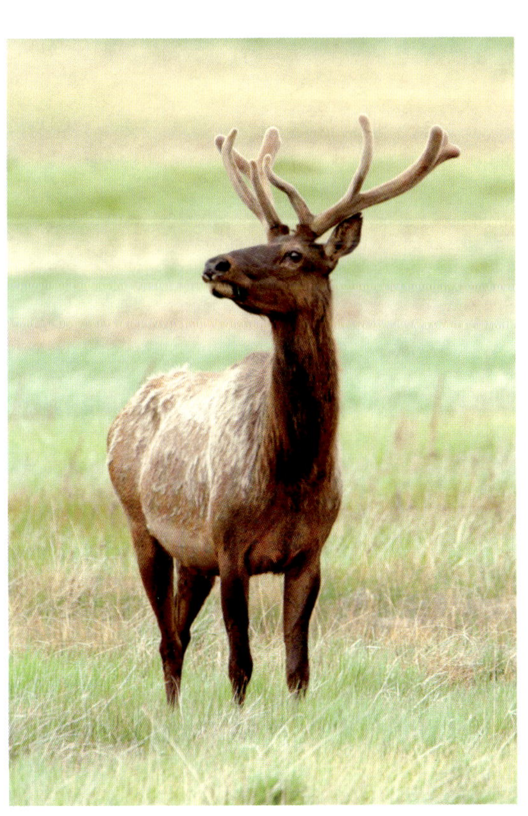

photograph.

Playing artificial bird calls will get the bird's attention. As the bird swims towards you, snap some images. After the waterfowl gets bored with you, play the call again. Hopefully, the duck will respond with a territorial display or a call of its own.

As you walk around Jacques Marsh, use binoculars to scout for elk. During June and July the bulls will be in velvet (a fuzzy look to the growing antlers). Once you spot this large mammal, think about how to sneak to within frame-filling range. Use trees or large bushes to hide your approach, and be as quiet as a church mouse. Use a solid **tripod** and the appropriate **ball or gimbal head** to keep your camera still while you shoot. Since most elk leave the marsh area early in the morning, use a fast ISO speed and a wide aperture setting to keep your shutter speed fast enough to prevent blurry images in the low light.

Rocky Mountain elk in velvet. Canon 1DMII, 500mm, 1.4x teleconverter, ISO 250, f/8 @ 1/250 sec.

Woodland Lake Park

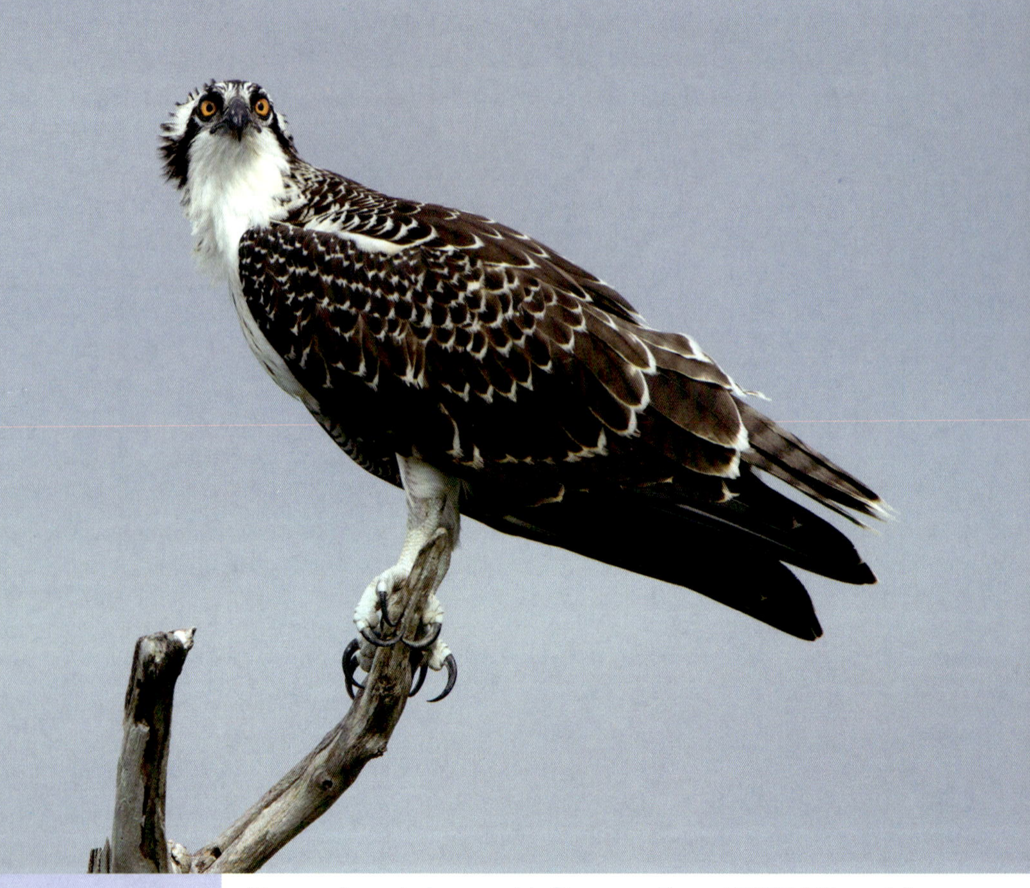

The stare of a comical cross-eyed-looking osprey. Canon 1DMIV, 500mm, 1.4x teleconverter, ISO 400, f/6.3 @ 1/2000 sec.

VIEW TIME
April to October

IDEAL TIME OF DAY
Early morning and late afternoon

VEHICLE
Any

HIKE
Moderate

Escape the low desert's scorching summer sun by heading to Pinetop-Lakeside to explore Woodland Lake Park high in the cool White Mountains. Surrounded by ponderosa pine, Gambel oak, and grassland, the lake combines open water, marsh, snags (dead trees for roosting and nesting birds), and plenty of live trees with nest cavities, making this watering hole the perfect habitat for a variety of birds. Because the lake is open to fishing and walking (and there is a public softball field nearby), the wildlife visiting this refreshing watering hole are habituated to human presence and do not spook easily.

From May through July, action shots abound as the birds set up and defend their territories while breeding, feeding, and raising their young. Yellow-headed blackbirds nest in the cattails that surround much of the lake. During May and June, if you are patient, you may observe adult blackbirds feeding their insatiable young. Fearless males call and set up their valued territory. When you spot a male singing in their unmistakable creaky-door, high-pitched voice, position yourself within 20 to 40 feet (6.1 to 12.2 m) of the bird with the sun to your back and your shadow pointing towards the subject.

Use a **telephoto lens** on a sturdy **tripod** and **ball or gimbal head**. In order to produce out-of-focus backdrops for your image, use a relatively wide aperture (e.g., f/4 or f/5.6). To obtain sharp images, dial in your shutter speed setting to 1/250 of a second or faster. Wait for the male birds to sit on top of the numerous cattails to sing in defense of their territories. Use continuous high-speed shooting mode and continuous focus.

Ospreys also make their presence known at Woodland Lake. This raptor looks more at home in wetter states like Florida but spins through the higher elevation lakes in Arizona in the summer when breeding. The osprey enjoys soaring above the lake until it spots a fish near the surface. Then, the aerodynamic predator will dive at high speeds, put out its talons, pull its wings back, and hit the water—hopefully, catching the intended meal.

Keep your telephoto lens ready as you keep your eye on the sky for this large, soaring raptor. Hand-holding the camera (instead of using a tripod) will allow you to respond more quickly to a diving or feeding bird. Set your camera on high-speed continuous shooting mode and continuous autofocus. Ensure your shutter speed exceeds 1/1200 of a second to freeze the osprey's mid-air motion. A relatively wide-open aperture (e.g., f/4 or f/5.6) and fast ISO speed provide enough light to enable even faster shutter speeds. When you have the osprey in focus, blast away and take as many images as your camera's buffer will allow.

For more exciting wildlife photography opportunities, scan the Gambel oak trees for woodpecker nesting cavities, specifically those of the Lewis's and hairy woodpeckers. While the females sit on eggs, the males will guard the territory and occasionally bring food to their mates. After the eggs hatch, both parents will feed the young and ensure an almost constant flow of food to the hungry and fast-growing chicks. When the young birds are a few days old, the adults may feed them as often as every 15 minutes.

When you have found the nesting cavity, get to a comfortable location and set up your telephoto lens on a solid tripod and head. Pre-focus on the nest cavity, sit down with a remote control shutter release, and wait. Although a blind is not necessary, be very quiet and still. Any movement will cause the feeding to be aborted and your opportunity lost. When the adults come in to feed the young, shoot as many photographs as you can.

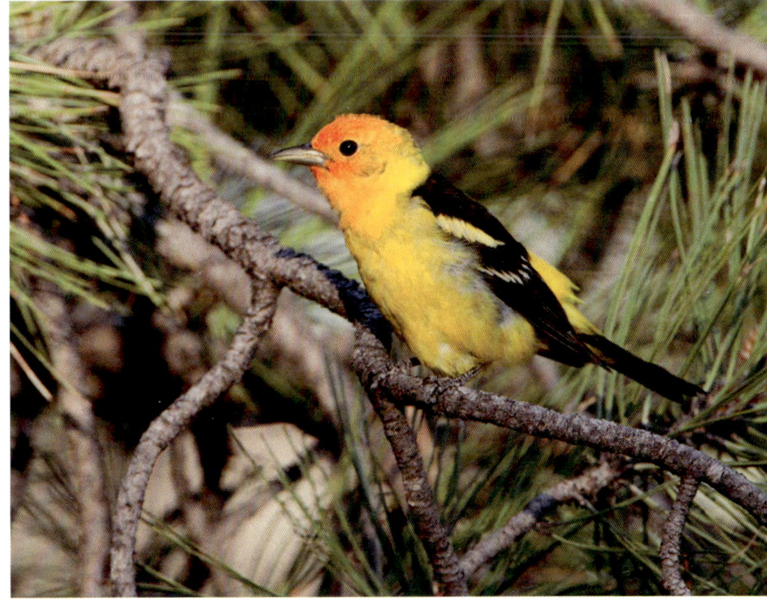

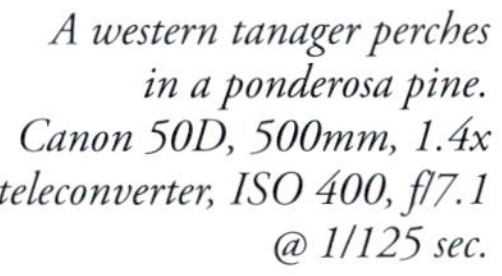

A western tanager perches in a ponderosa pine. Canon 50D, 500mm, 1.4x teleconverter, ISO 400, f/7.1 @ 1/125 sec.

Woodland Lake Park

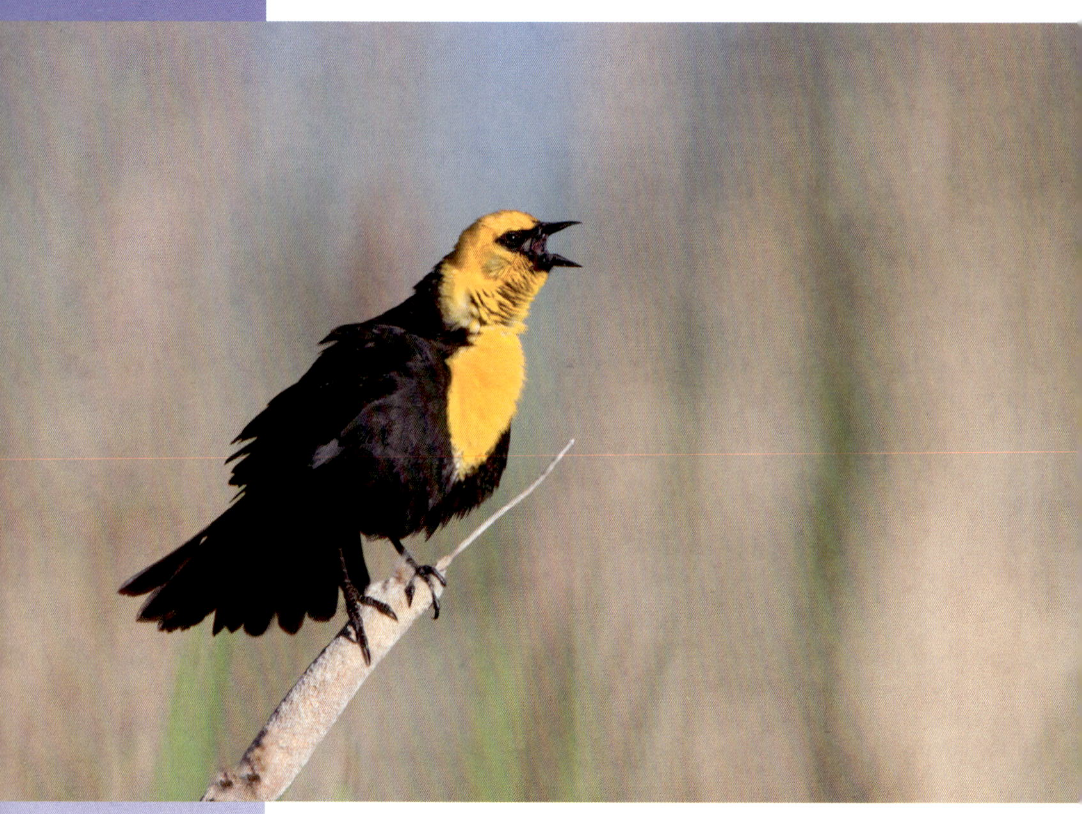

A yellow-headed blackbird calls to defend territory. Canon 50D, 500mm, 1.4x teleconverter, ISO 400, f/7.1 @ 1/1600 sec.

If necessary, use **fill flash** to open up shadowed areas on the subject and to make the birds' colors pop. Use high-speed continuous shooting mode and a fast shutter speed (e.g., 1/250 of a second or faster) to record this furious activity.

Several pairs of pied-billed grebes also make Woodland Lake their home. During the spring, these grebes make their nests out of floating debris and incubate their eggs literally over water. After the babies hatch, the parents carry them on their backs while the little ones learn to swim and fend for themselves. The parents take turns between tending the young and diving for insects, fish, and small clams to feed them.

To photograph grebes with their young (or other shy waterbirds) at water and eye level, use a floating tube (an inner tube with a harness that allows the photographer to float in the water and approach the birds at close range). Camouflage the floating tube with cattails or some other local vegetation in order to look like a floating mat. Use a telephoto zoom lens and paddle slowly to within range of the birds.

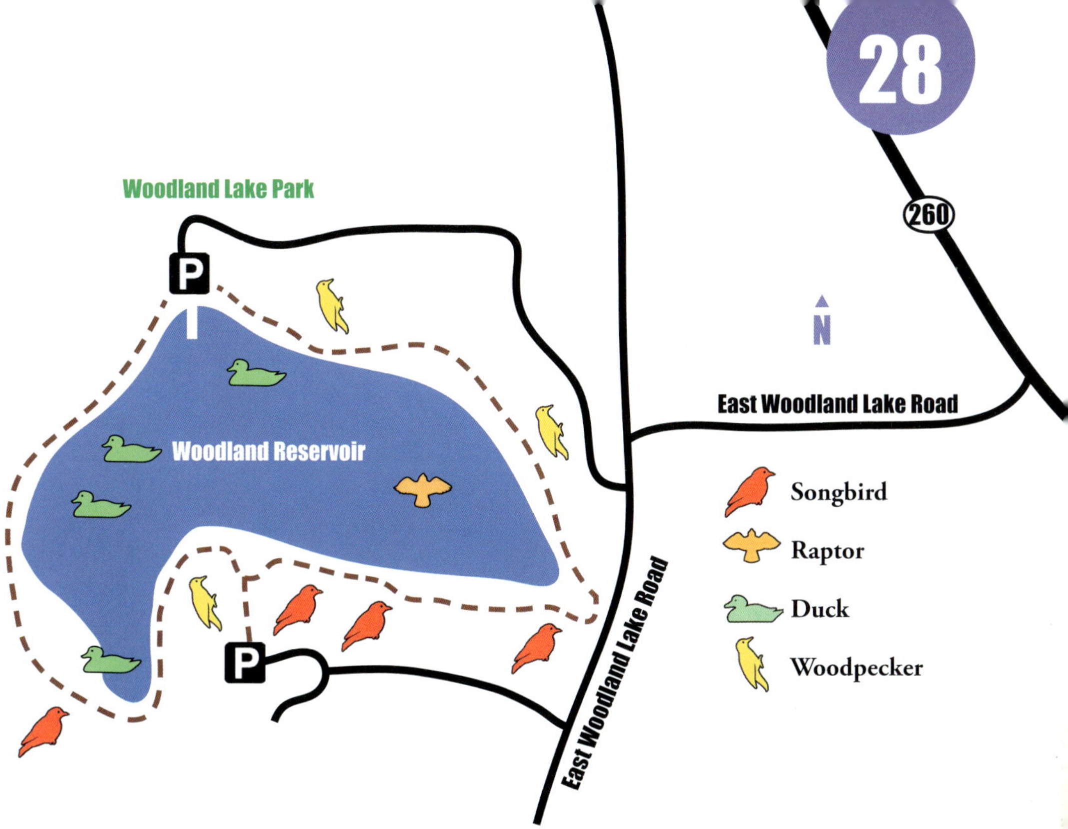

28

Woodland Lake Park

260

N

East Woodland Lake Road

Woodland Reservoir

East Woodland Lake Road

Songbird

Raptor

Duck

Woodpecker

DIRECTIONS:

From downtown Show Low, follow AZ 260 southbound for 11.3 miles (18.2 km). Turn right on Woodland Lake Road. The entrance to Woodland Lake Park appears after continuing another 0.4 miles (0.6 km) to the parking lot.

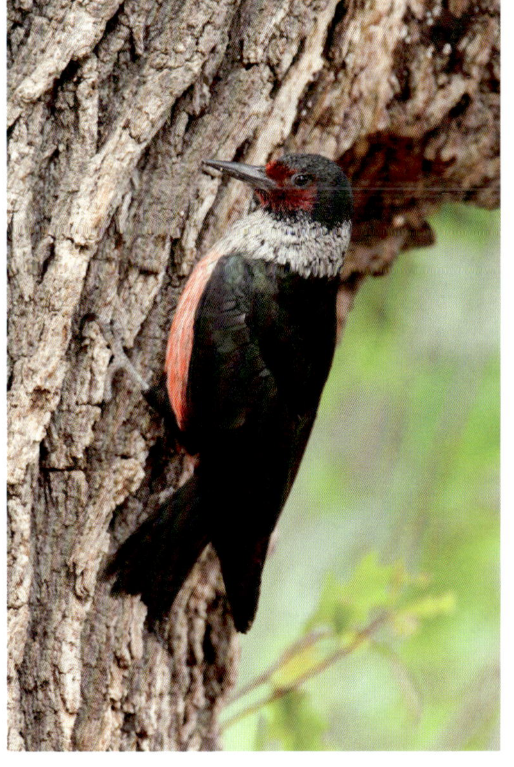

Lewis's woodpecker near its oak tree nest. Canon 50D, 500mm, 1.4x teleconverter, ISO 400, f/5.6 @ 1/30 sec.

EASTERN ARIZONA

Becker Lake Wildlife Area

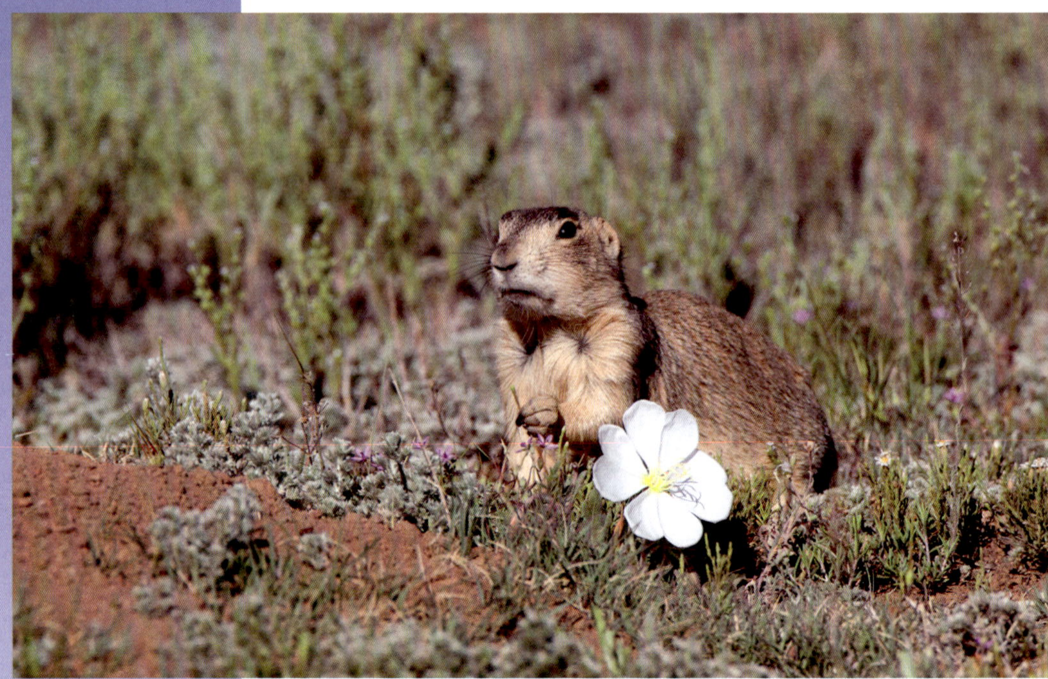

Gunnison's prairie dog. Canon 50D, 500mm, 1.4x teleconverter, ISO 200, f/8 @ 1/2000 sec.

VIEW TIME
Late May to September

IDEAL TIME OF DAY
Early morning

VEHICLE
Any

HIKE
Easy

Built in the 1880s as a water supply for the town of Springerville, outdoor enthusiasts currently enjoy Becker Lake for sport fishing, boating, and picnicking. Although the teardrop-shaped, cattail-lined, shallow lake offers limited possibilities for waterfowl photography, the land to the north and northwest features a large colony of playful—and photogenic—Gunnison's prairie dogs.

Historically, Gunnison's prairie dogs lived across Arizona, but today, only a few places present the chance for the photographer to observe the scurrying critters at close range. During the early spring and summer, these prairie dogs actively collect food to store in their burrows while raising their young. In early June, the mischievous and energetic young dogs begin to show themselves above ground. The inexperienced pups are not yet savvy enough to avoid predators and therefore are easier to approach now than at other times of the year.

Arrive early in the morning and set up a blind close to, but not in, the town (a prairie dog colony is referred to as a "town"). Place your **long telephoto lens** atop a **tripod**, and wait for the sun to begin warming the landscape. At 7,000 feet (2,134 m) elevation, cold nights cause the prairie dogs to come out of their burrows in the early morning to warm themselves.

As you photograph, be very quiet and move as little as possible. Prairie dogs are extremely aware of their surroundings and will retreat to the safety of their burrows at the slightest sign of anything unusual. When the young prairie dogs begin to come above ground for the first time, they will stay very close to the mound and each other, making for great group shots of cute, young dogs.

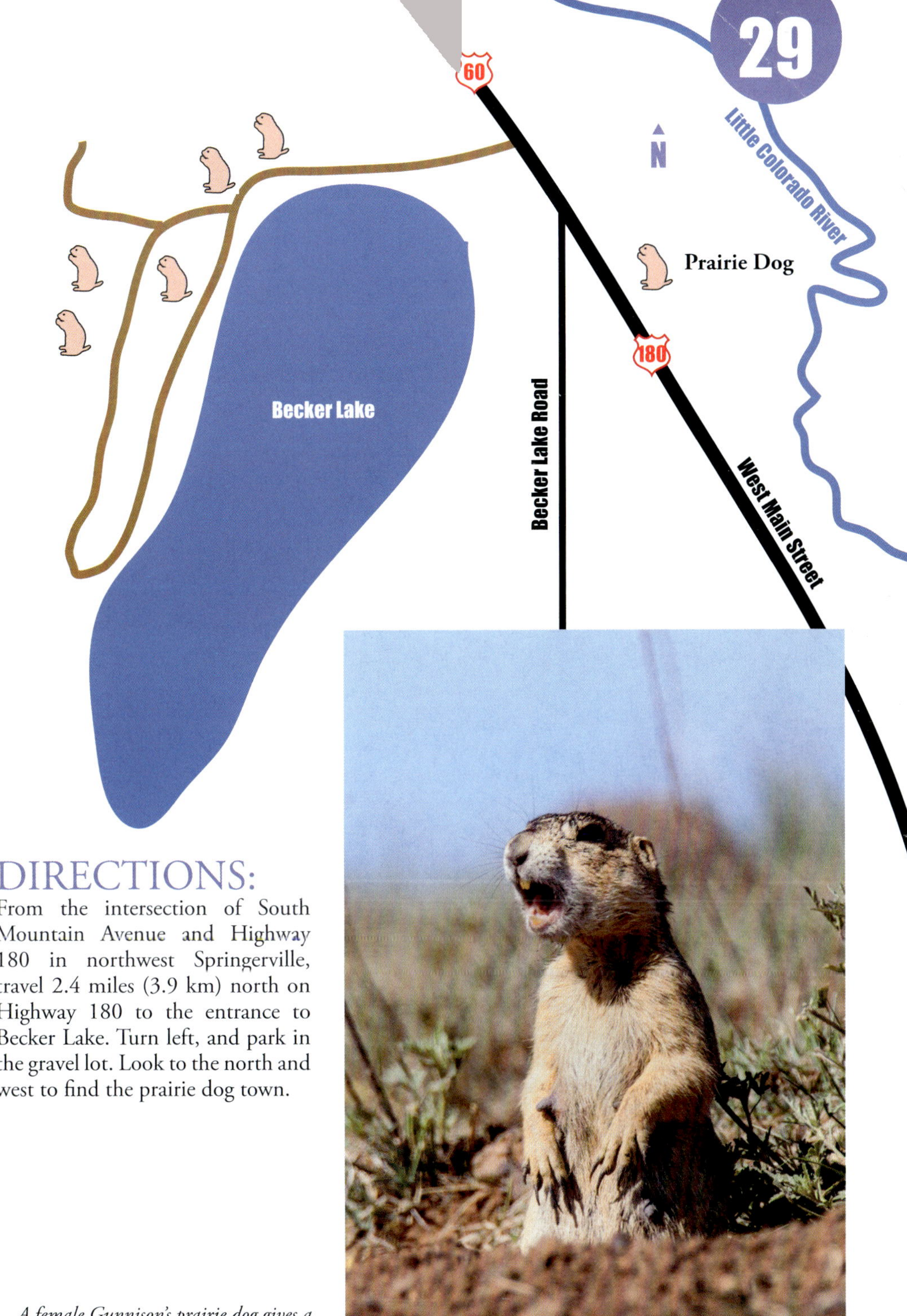

DIRECTIONS:

From the intersection of South Mountain Avenue and Highway 180 in northwest Springerville, travel 2.4 miles (3.9 km) north on Highway 180 to the entrance to Becker Lake. Turn left, and park in the gravel lot. Look to the north and west to find the prairie dog town.

A female Gunnison's prairie dog gives a danger call. Canon 50D, 500mm, 1.4x teleconverter, ISO 400, f/7.1 @ 1/1600 sec.

Sipe White Mtn. Wildlife Area

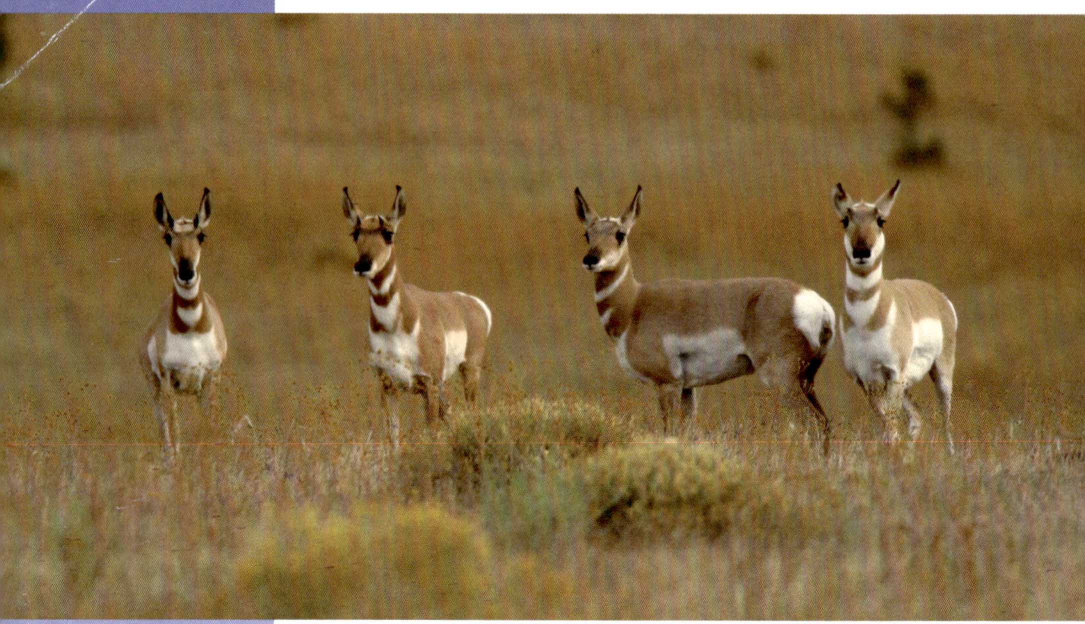

Four doe pronghorn near the road into Sipe White Mountain Wildlife Area. Canon 60D, 500mm, 1.4x teleconverter, ISO 200, f/6.3 @ 1/250 sec.

VIEW TIME
Year-round

IDEAL TIME OF DAY
Sunrise; late afternoon

VEHICLE
2WD high-clearance

HIKE
Easy to moderate

With over 13,000 acres (5,260 hectares), the Sipe White Mountain Wildlife Area features large expanses of meadow, pinyon pine, and juniper habitats, along with a small creek and several lakes. A drive into this wildlife area reveals pristine, rolling grassland hills intermixed with strings of trees and waterways—a setting difficult to replicate anywhere else in Arizona. The iconic Rocky Mountain elk, pronghorn, and mule deer roam this unique landscape while a variety of smaller species also attract wildlife photographers. However, not all animals appear in abundance year-round and, as such, it takes some time and effort to locate them.

Soon after the monsoons begin, migrant rufous, calliope, broad-tailed, and black-chinned hummingbirds arrive to feed on the plentiful high desert flowers and several hummingbird feeders at the visitor center. Numerous other hummingbirds sit on the small bushes around the visitor center as they await their turn at the feeders. Use a **telephoto lens** on a sturdy **tripod** with a small amount of **fill flash** to capture images of perched birds. A small apple orchard with several picnic tables provides the perfect locale to set up your high-speed hummingbird equipment (see "Hummingbird Photography" on page 154). Keep your high-speed hummingbird gear away from the visitor center so as not to disturb bird watchers. Hummingbird activity peaks from mid-July through mid-September.

In addition, starting in July, large herds of elk (including cows, calves, and young bulls) feed on the meadow grasses during the early morning and late afternoon. Be extremely quiet and shoot from your vehicle. These herds can get nervous and will leave the area if they sense a potential threat. To keep your camera still, use a telephoto lens situated on a bean bag resting on the car door window. Use high-speed continuous shooting mode, continuous

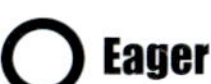

DIRECTIONS:

From the intersection of AZ 260 and US 180 in Eagar, travel south on US 180/191 for 2.4 miles (3.9 km) to the entrance sign for the Sipe White Mountain Wildlife Area. Turn right onto County Road 4131. Drive for 1.4 miles (2.3 km), and then bear left on Forest Route 57 (stay on the well-traveled road). Continue for 3.7 miles (6 km) to the ranch headquarters where ample free parking exists.

autofocus, and a relatively fast shutter speed (in excess of 1/250 of a second). By sunrise, the elk begin moving to the protection of the forest, so use a fast ISO speed (e.g., ISO 800) to maintain a fast enough shutter speed to freeze their motion. Elk photography is best from July through March.

Photographing pronghorn and mule deer often feels like a "catch-as-catch-can" experience. After you spot a group of pronghorn, they usually pose for merely a few shots before bounding away. The mule deer have a similar habit of giving the photographer a few seconds before disappearing into the surrounding trees. Have your camera and lens at the ready with the window down and bean bag in position. Drive slowly and look for a photo opportunity. Be sure and shut the engine off before attempting an image.

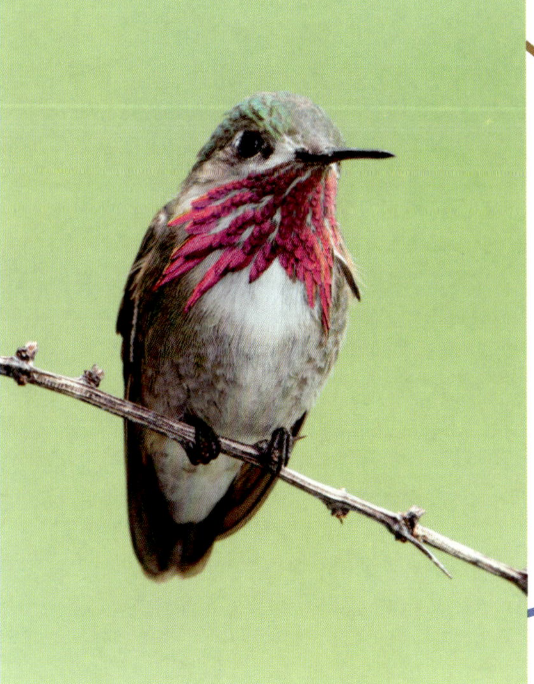

A male calliope hummingbird guards the porch at Sipe White Mountain Wildlife Area. Canon 70D, 70-200mm at 200mm, 2x teleconverter, ISO 400, f/9 @ 1/200 sec., off-camera flash @ +1 FEC.

PHOTO TIP 10

Hummingbird Photography

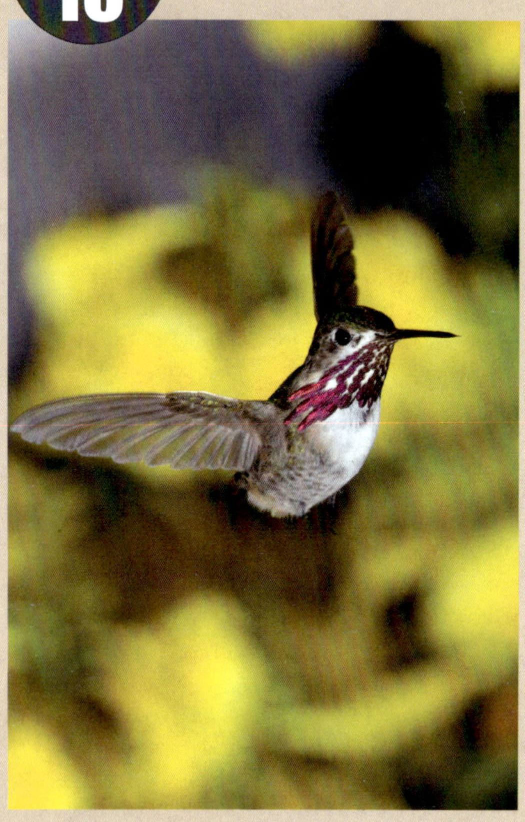

Calliope hummingbird. Canon 1DMIV, 70-200mm at 150mm, 2x teleconverter, ISO 400, f/16 @ 1/320 sec., high-speed hummingbird set-up.

How do you take a stop action photograph of a two-inch-long (5.1 cm), fast-flying hummingbird whose wings beat in excess of 60 times each second? And make sure that the background appears out-of-focus? And get the hummingbird to feed on a colorful flower for the photograph? You cannot just sit next to a beautiful flower in hummingbird habitat and wait for the perfect moment—you need to set up an "outdoor hummingbird photo laboratory."

For most flying birds, it only takes a shutter speed of about 1/1500 of a second to stop the wing motion and take sharp images. However, because hummingbird wings move faster than those of most birds, you will need a shutter speed in excess of 1/8000 of a second to freeze their motion in your frame. A 1/8000 of a second shutter speed is the maximum speed for many DSLR cameras and you will still need bright sunlight, a wide-open aperture, and a quick ISO speed to take sharp images. With an aperture of f/5.6, the depth of field for a close-up photo of a flying hummingbird is so narrow that it will be difficult to keep the bird's eye in focus, much less the entire body.

Manual mode on your flash can help resolve this predicament. The speed of most flashes varies from about 1/800 of a second to 1/1500 of a second. To speed up a flash (decrease the duration), place your flash on manual mode and turn the power down. Although many flashes allow you to reduce the power to 1/128 of full power, the best setting for hummingbirds is around 1/16. At 1/16 power, the flash speed will be approximately 1/15000 of a second—plenty fast to stop the movement of a hummingbird wing, right?

Not so fast! When you reduce the power of the flash, you reduce the light output. The flash will not put out enough light to illuminate the bird. However, by using multiple flashes simultaneously (generally three to six) you increase the light output overall and provide enough energy to take an image of a flying hummingbird with the flashes situated about 18 inches (45.7 cm) from the bird at an ISO speed of 400 and a small aperture of f/16.

When you take a properly exposed image with those flash speeds and aperture, the backdrop will look jet black. The flashes do not emit enough light to illuminate the trees, bushes, or other features. To solve this, bring in a brightly-colored artificial backdrop. To make the backdrop, find an out-of-focus photograph of a pleasing scene, and then print the image on 13"x19" matte printer paper. Place the backdrop about two to three feet (0.6 to 0.9 m) behind the hummingbird's anticipated position to keep the backdrop out of focus. The light falloff from the flashes will sufficiently illuminate the backdrop without drawing attention to it. To eliminate any risk of ghosting from ambient light, set up in a sheltered and shadowed area so that no sunlight hits either the bird or backdrop.

To tempt the hummingbird into your desired position, use a hummingbird feeder. Place three flashes on flash stands in front of the feeder (two above and one below the level of the feeder), and wait for the hummingbird to feed. When feeding, the bird will frequently back off and look at you—snap the photo when this occurs. Use one of the other flashes to light the backdrop and, if necessary, place more flashes for accent lighting.

If you prefer to photograph a hummer feeding off a flower, simply replace the feeder with a flower (in the exact same place), and wait for the hummingbird to return. Even though they are not used to the flower, they will often take some nectar from the flower just because it is there. When they leave after feeding from the flower, use an eyedropper to put some sugar water on the flower so that when they return, a reward awaits them.

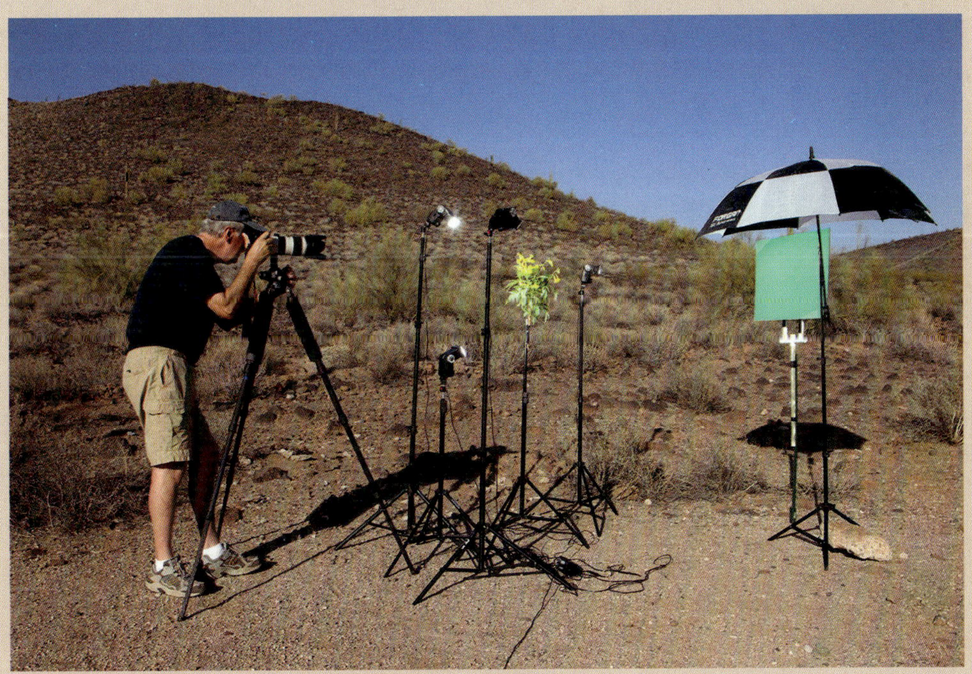

I check my exposure settings while I prepare to photograph hummingbirds with my hummingbird equipment set-up. Canon 1DMIV, 24-105mm at 28mm, ISO 400, f/11 @ 1/800 sec., off camera flash.

Coronado Trail

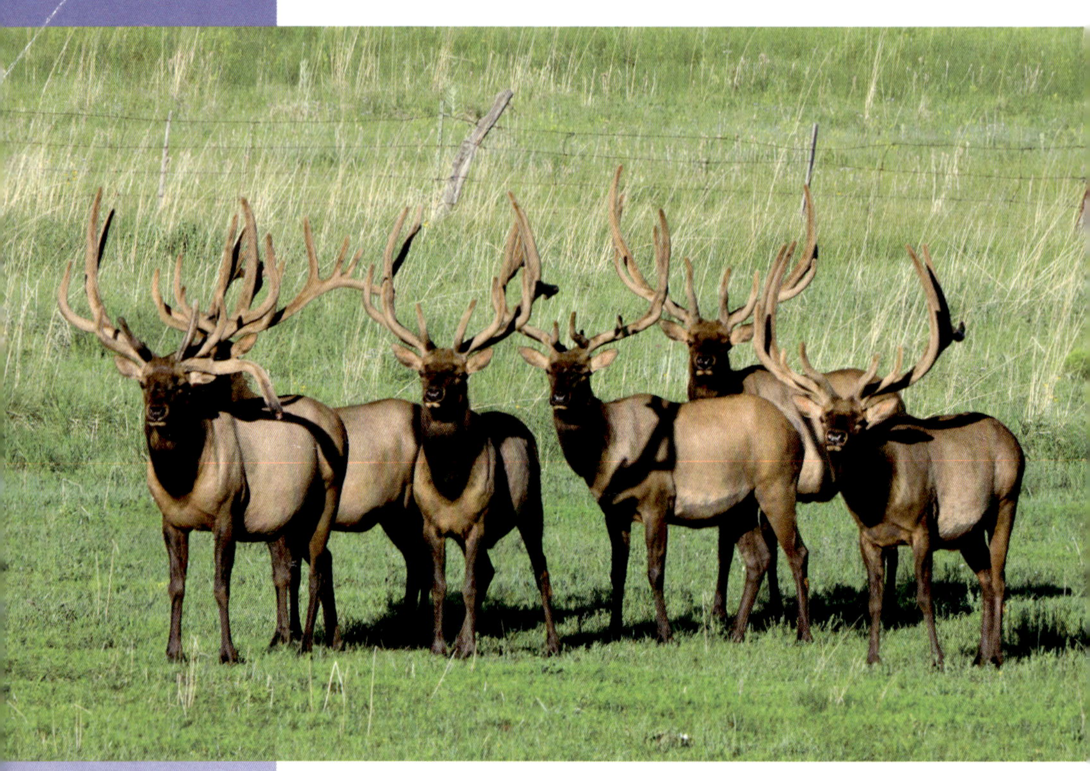

Bachelor herd of Rocky Mountain elk. Sony A77V, 400mm, ISO 400, f/11 @ 1/500 sec.

VIEW TIME
July to September

IDEAL TIME OF DAY
Early morning and late afternoon

VEHICLE
Any

HIKE
Easy

Intermingled among the intermittent wildflower-filled meadows and the occasional irrigated farmland and cattle grazing fields, pine and spruce trees dot the landscape along the section of the Coronado Trail spanning from the Nutrioso Reservoir to Luna Lake. At nearly 8,000 feet (2,430 m) in elevation, cool summer nights and tolerable daytime temperatures prevail. During the monsoon season, this mountainous terrain receives several inches of rain, making it one of the wettest places in the state. From diminutive horned lizards and spiders to giant Rocky Mountain elk and mule deer, this high elevation habitat provides not only a welcome relief from the summer heat, but also offers shutterbugs unique wildlife photography opportunities.

An early morning or late afternoon drive from Nelson Reservoir south to Alpine and then west to Luna Lake will prove you are in elk country, as some of the largest herds in the state roam here. Hot hang-out spots include the farm fields just north of Nutrioso Reservoir, the cattle grazing fields on the outskirts of Alpine, and the entire area surrounding Luna Lake. During the monsoon season, elk will linger either in large herds of mixed cows, calves, and young bulls or in bachelor herds of gigantic older bulls.

Depending on how close you get to the elk, pull out either a **medium or long telephoto lens (i.e., 100-400mm)**. You will photograph mostly from your vehicle, but if you decide to approach the elk around the Luna Lake area on foot, wear earth-toned clothing. But, be careful. Elk can be dangerous if you get too close. Keep some vegetation between you and the

DIRECTIONS:

From Springerville, travel south on AZ 190/180 (Coronado Trail) to Nelson Reservoir. To reach Alpine, drive south for an additional 17.1 miles (27.5 km). In Alpine, turn left to continue following AZ 180. Drive east for 3.9 miles (6.3 km) to Luna Lake.

The speed limit along AZ 190/180 ranges from 55 to 60 mph so be very careful when stopping to observe or photograph wildlife.

---

herd to block your approach so you do not spook the animals.

During the monsoons, many of the small roadside meadows fill with orange sneezeweed. Butterflies, moths, spiders, aphids, and many other types insects take advantage of the late-blooming flowers for food and shelter. Use a **macro lens** on a sturdy **tripod** for the sharpest image results. Because these critters are very small, consider focus stacking techniques (see page 196) to achieve an expanded depth of field. As you explore the blooms for insects and arachnids, move slowly while approaching or setting up the camera and tripod so you do not scare them away. Although the best light generally occurs during the early morning and late afternoon, consider visiting immediately after a rainstorm. At this time, the cold insects tend to sit still, and the overcast skies make for optimal diffused lighting to record increased color saturation.

While combing the countryside for elk and insects, the lucky photographer might run across a greater short-horned lizard. These well-camouflaged lizards move relatively slowly and can be easily captured. Be sure you possess an Arizona Game and Fish Department hunting license so you can legally handle reptiles. Gently pick up the lizard and position it where you can get an eye level shot. In aperture priority mode, set an f/13 aperture to maintain acceptable depth of field. Use fill flash when the ambient light is low.

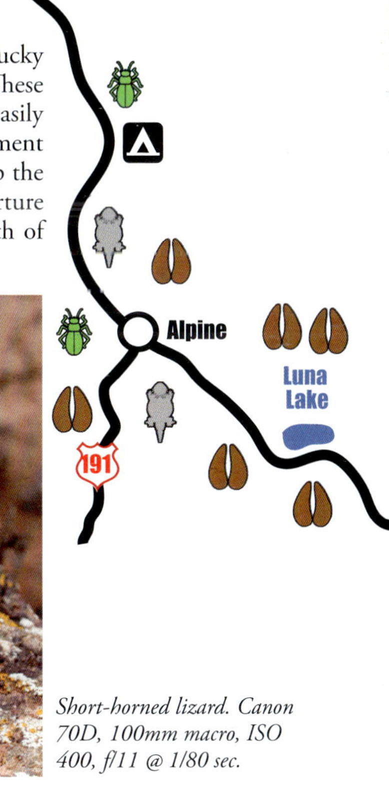

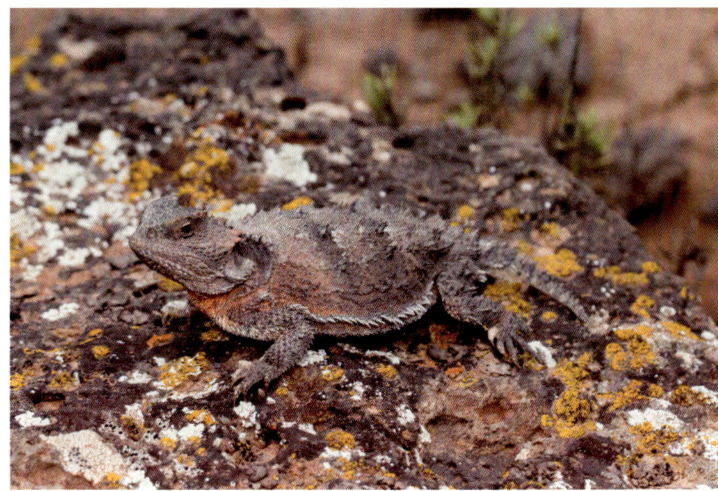

Short-horned lizard. Canon 70D, 100mm macro, ISO 400, f/11 @ 1/80 sec.

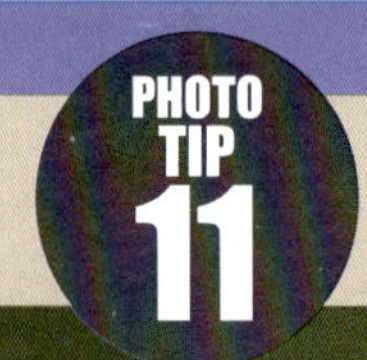

Macro Basics

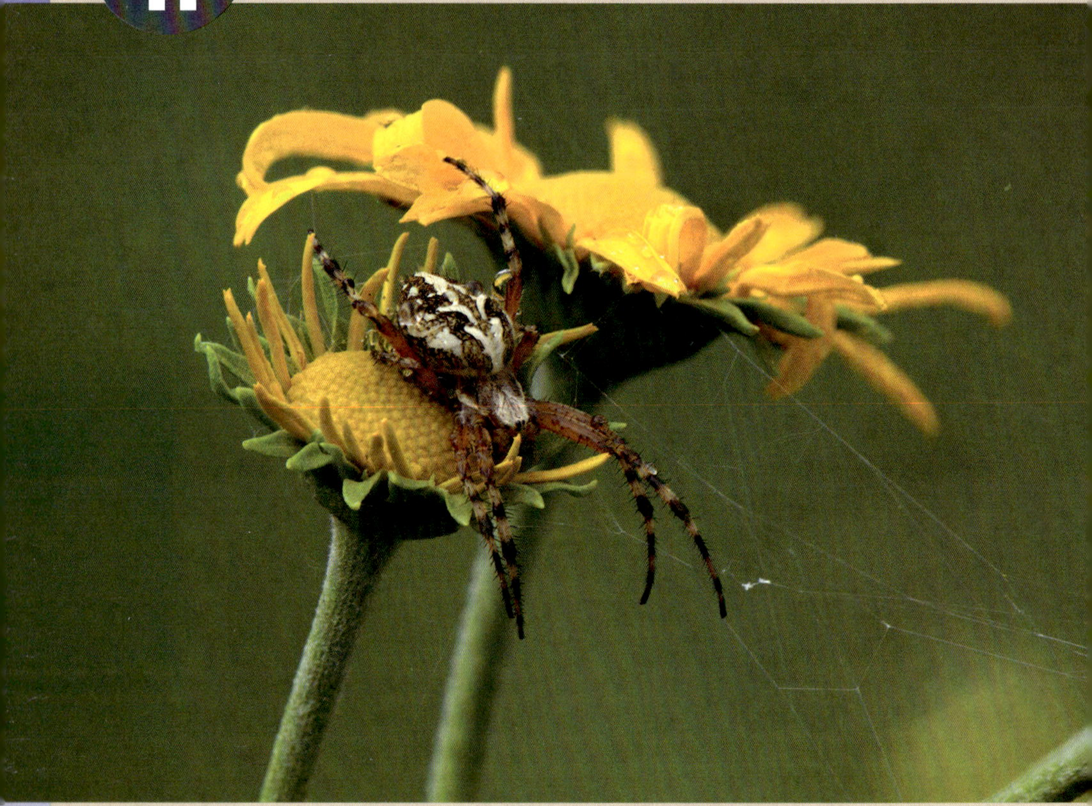

Orb weaver spider on orange sneezeweed. The image is a composite of seven images combined in Zerene Stacker. Canon 5DMIII, 180mm macro, ISO 1600, f/9 @ 1/640 sec.

A close-up photograph showcasing the beauty of a bee's eye or butterfly's wings can offer an exciting and different view than capturing a broader scene of an animal in its habitat. Called macro (or close-up) photography, the equipment and techniques differ from those used in most wildlife photography.

In order to get close enough to smaller subjects, either you need to use specialized macro lenses or make modifications to your "normal" lenses. Macro lenses (some camera manufacturers call them "micro" lenses) allow the lens to focus close enough to the specimen to render the subject at 1x (1:1, or full life-sized magnification ratio) or at 0.5x (1:2, or half life-sized magnification ratio). Canon makes a specialized lens that will allow up to 5x magnification, but takes a lot of time and practice to master.

For example, if you use a camera with a full size sensor (36mm wide) to take a photograph of a caterpillar that is 36mm long and the whole caterpillar fills the viewfinder, the image is considered 1x. However, if the caterpillar is only 18mm long and its body fills half of the frame then the image is considered to be 0.5x. For cropped frame cameras, even though the numbers change a little, the concept remains the same.

For those who do not possess a true macro lens, you can still achieve fine detail by using extension tubes, diopters (also called close-up lenses), and teleconverters. Place an extension tube between the camera and lens to allow the lens to obtain focus at a closer distance. The amount of magnification depends on the lens and the amount of "extension" from the camera. Close-up lenses are screwed into the filter threads of a lens and, similar to extension tubes, allow the camera to obtain focus at a closer distance. Finally, teleconverters can be placed between the lens and camera to enlarge the image without changing the focus distance.

Regardless of how the lens obtains close focus, the photographer must consider lens movement and depth of field. The more magnified the subject, the more obvious the camera and lens shake. The closer the lens gets to the subject, the greater the observed movement becomes. To remedy the effects of camera and lens movement in macro photography use a sturdy tripod and ball head. When possible, use Live View mode (if available) to focus. Live View mode employs mirror lockup which eliminates the movement caused by mirror vibration. If you do not have Live View mode, use the two-second timer and/or a cable release to eliminate the inherent movement caused when your finger pushes the shutter release.

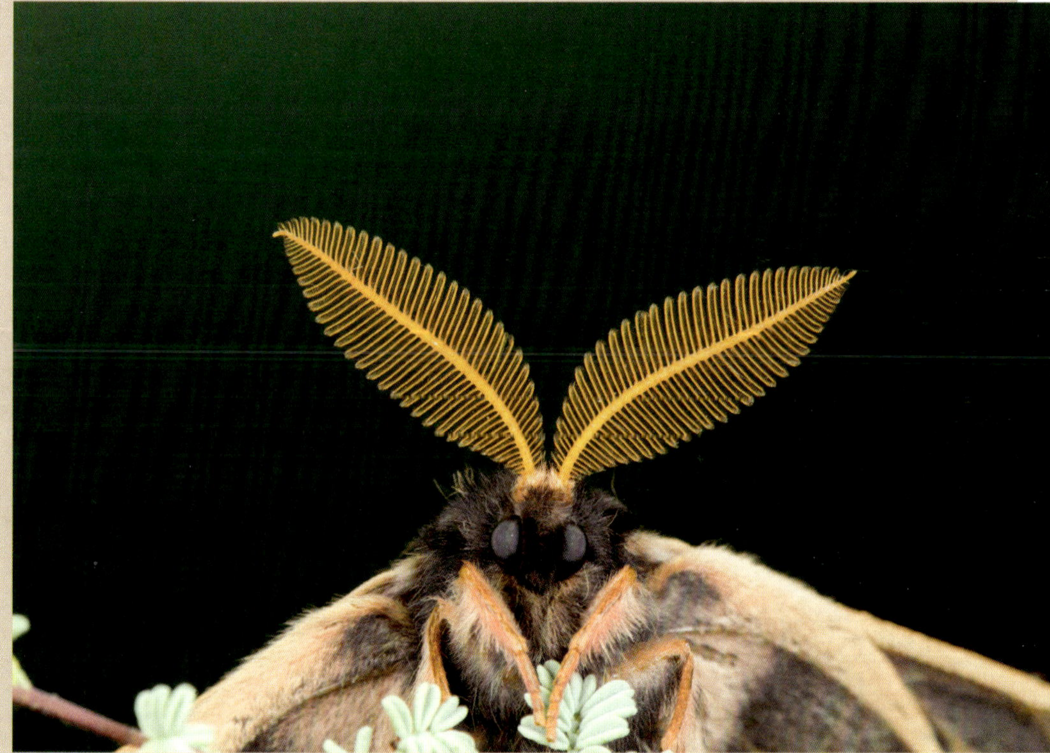

An Agapema homogena male shows off the massive antennae he uses to sense the location of female moths. Canon 5DMIII, 100mm macro, ISO 400, f/16 @ 1/200 sec., off-camera flash.

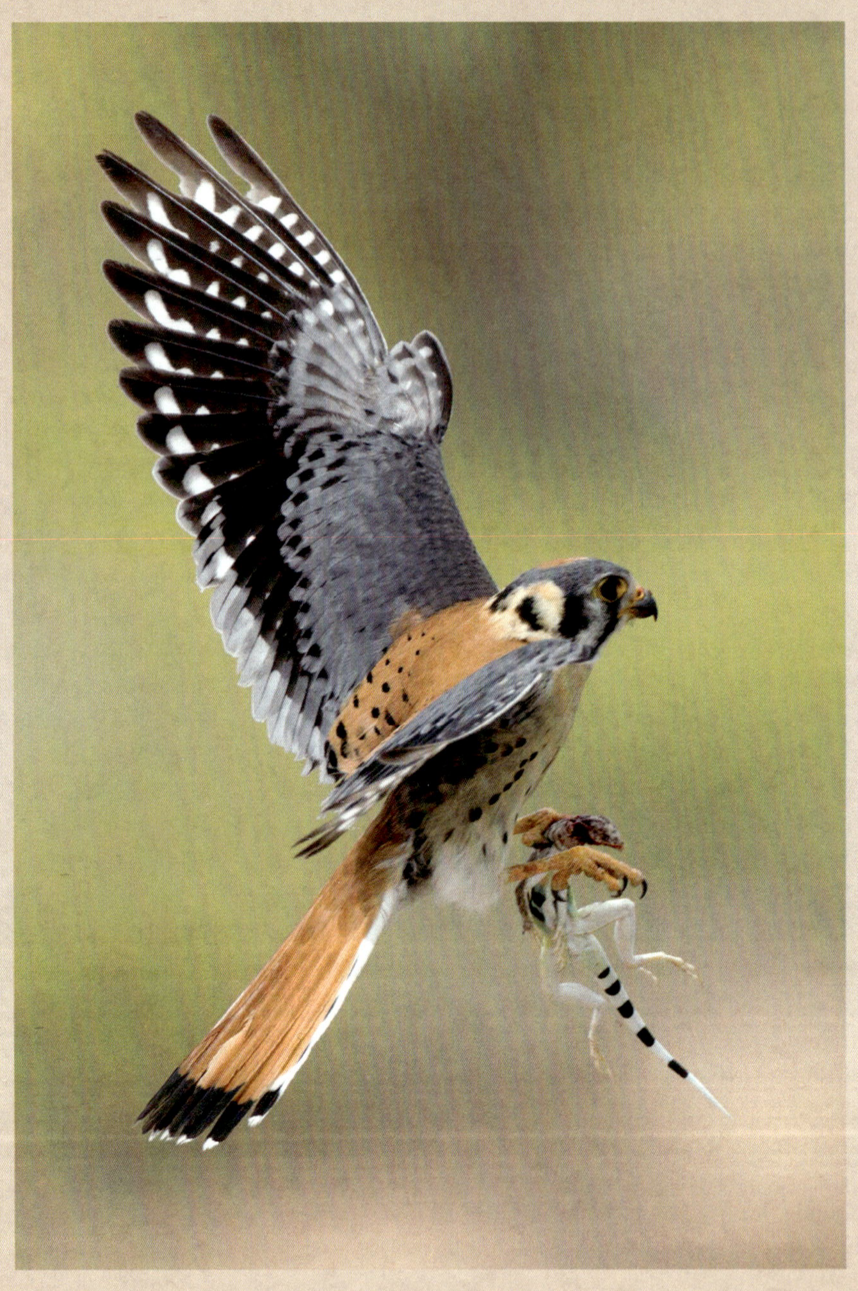

An American kestrel carries zebra-tailed lizard to babies in saguaro hole nest. Canon 1DMIV, 500mm, ISO 1250, f/6.3 @ 1/5000 sec., on-camera flash.

Wildlife of

SOUTHERN ARIZONA

S

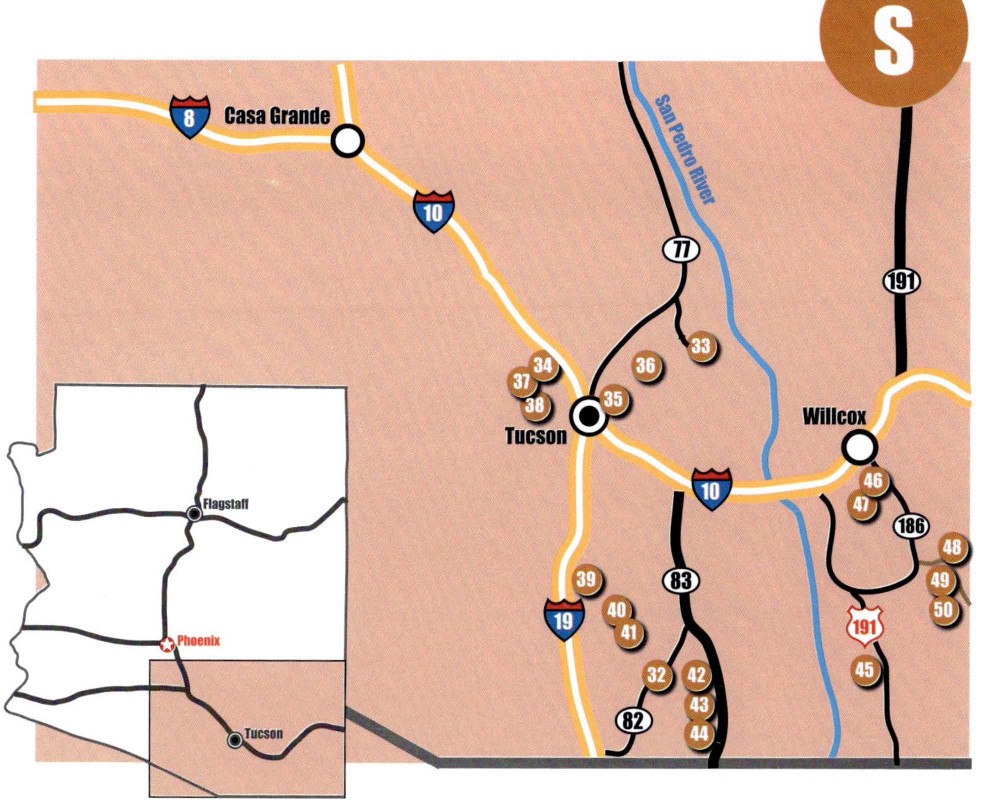

Casa Grande
8
10
San Pedro River
77
191
33
36
34
37
38
35
Tucson
Willcox
10
46
47
186
48
49
50
39
83
19
40
41
191
45
32
42
43
44
82
Flagstaff
Phoenix
Tucson

Sonoita

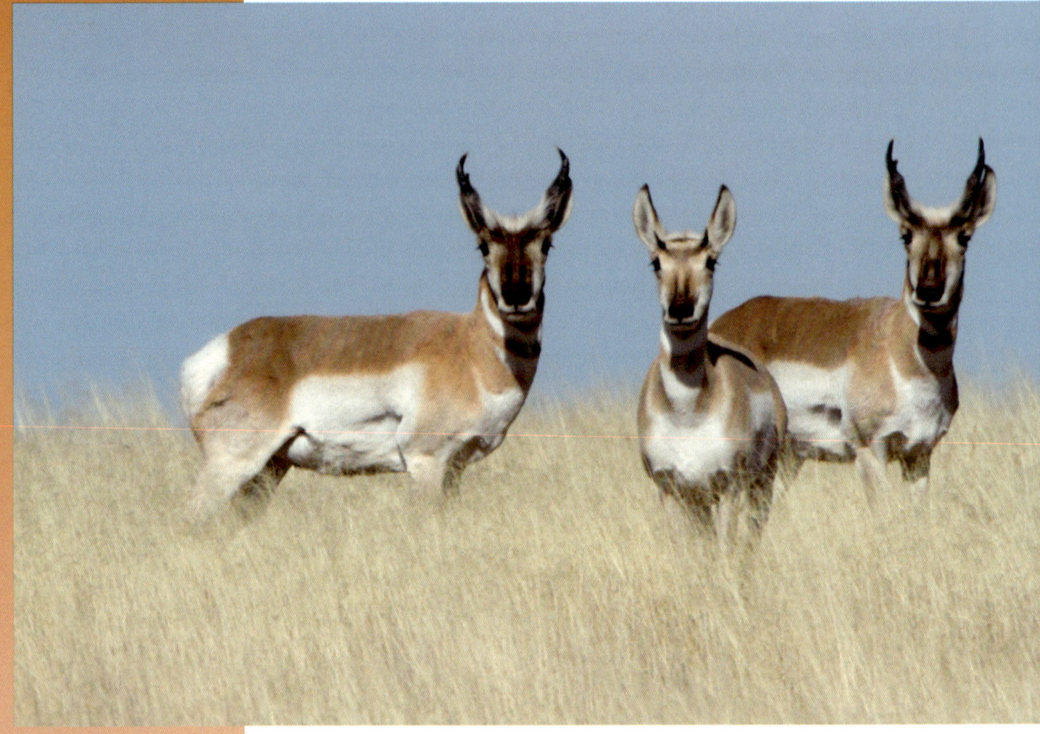

A small group of pronghorn stands in the brown, pre-monsoonal grass near Sierra Vista. Canon 7D, 500mm, 1.4x teleconverter, ISO 400, f/10 @ 1/2000 sec.

VIEW TIME
Year-round

IDEAL TIME OF DAY
Early morning and late afternoon

VEHICLE
Any

HIKE
Easy to moderate

The small, rural town of Sonoita sits at the crossroads of AZ 82 and AZ 83 in southeast Arizona. Surrounded by lush, rolling grasslands and visitor-friendly vineyards, dining, and wildlife-watching, Sonoita has an outdoor look that is reminiscent of Wyoming and Montana. For most of the year, the tall grass falls dormant and displays a golden brown color until the monsoon storms bring much needed rains which entices the yucca to bloom and the grass to turn brilliant green. Among the tall grass lives an abundant and very photogenic population of pronghorn.

To find pronghorn, all you need is a good pair of binoculars, a vehicle, and patience. Drive along the roads and look for the white rump patch of pronghorns. Alternatively, stop every mile (1.6 km) or so and use your binoculars to glass (meaning "to examine") both sides of the road for them. Frequently, pronghorn travel in herds.

When you locate a herd (or an individual), slow down and pull off the road. Shut your vehicle's engine off to reduce vibration, and turn on your **telephoto lens**' image stabilization (or vibration reduction) to keep your images sharp. Set a relatively wide aperture (e.g., f/5.6 to f/8) to blur the background. Tapping into high-speed continuous shooting mode will allow you to fire multiple frames quickly as the action happens.

If the herd appears off in the distance, get out of your car and move slowly towards them. Stop occasionally to take a few images and acquaint them with your presence. If possible, use the local terrain to camouflage your presence and mask your movements.

32

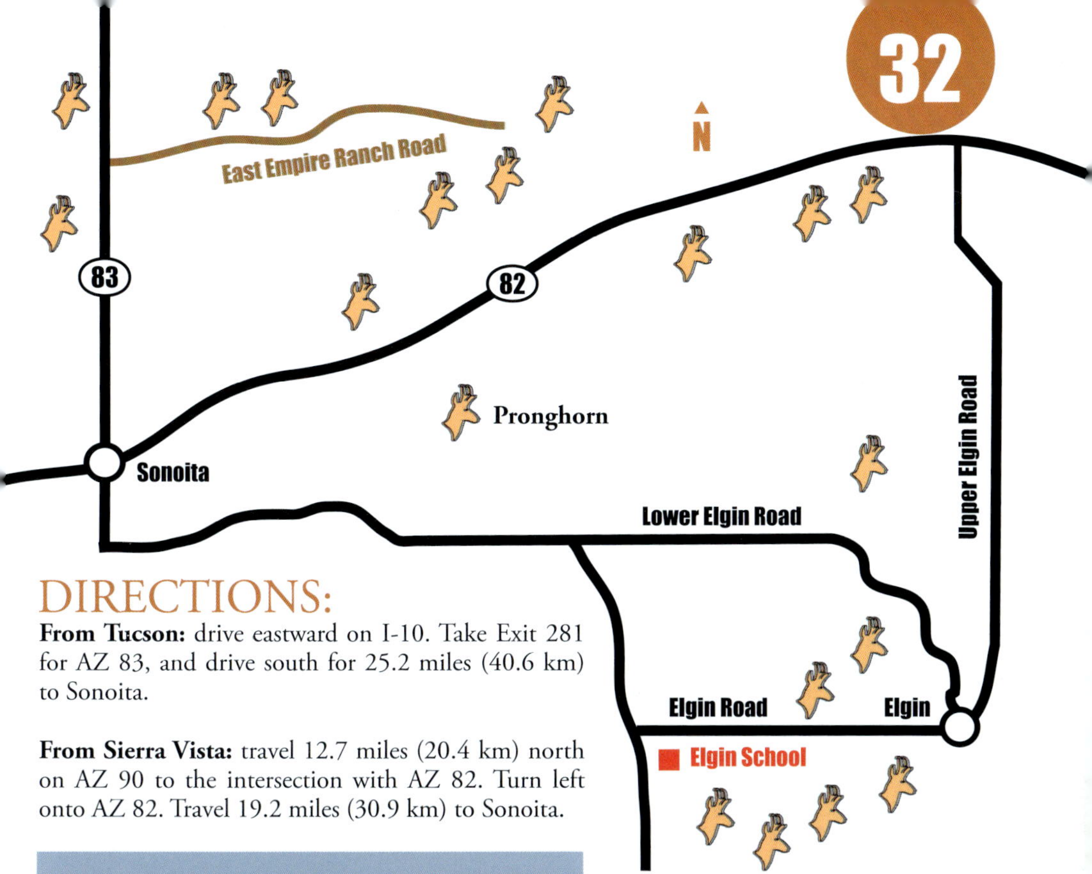

East Empire Ranch Road
N
83
82
Pronghorn
Sonoita
Lower Elgin Road
Upper Elgin Road
Elgin Road
Elgin
Elgin School

DIRECTIONS:

From Tucson: drive eastward on I-10. Take Exit 281 for AZ 83, and drive south for 25.2 miles (40.6 km) to Sonoita.

From Sierra Vista: travel 12.7 miles (20.4 km) north on AZ 90 to the intersection with AZ 82. Turn left onto AZ 82. Travel 19.2 miles (30.9 km) to Sonoita.

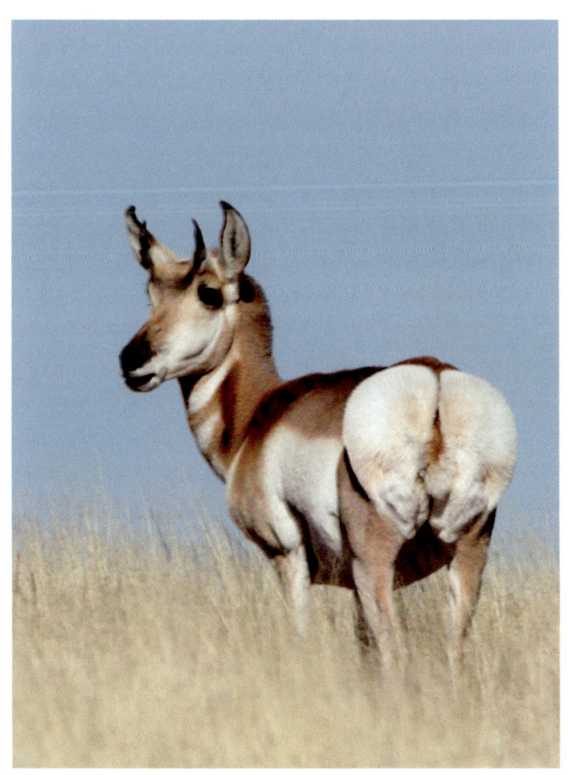

From Sonoita: travel north along AZ 83 and scan both sides of the road for the next two miles (3.2 km). Pronghorn commonly congregate near the windmill and a small wet area to the east of the road. Travel about 5 miles (8.1 km) further and go east on Empire Ranch Road. Watch both sides of the road for the next 3.5 miles (5.6 km). Return to the intersection of AZ 82 and AZ 83, watching carefully as you travel east for 9 miles (14.5 km) to Upper Elgin Road. Travel south to the town of Elgin and then drive west on Elgin Road. Watch both sides of the road as you near AZ 83, and be particularly vigilant near the Elgin School where a herd frequently assembles. From the intersection of AZ 83 and Elgin Road, travel south for three miles (4.8 km) as you search the roadside for pronghorn.

Pronghorn. Canon 7D, 500mm, 1.4x teleconverter, ISO 400, f/10 @ 1/1600 sec.

Making the Photo
7

Driving Along in My Automobile

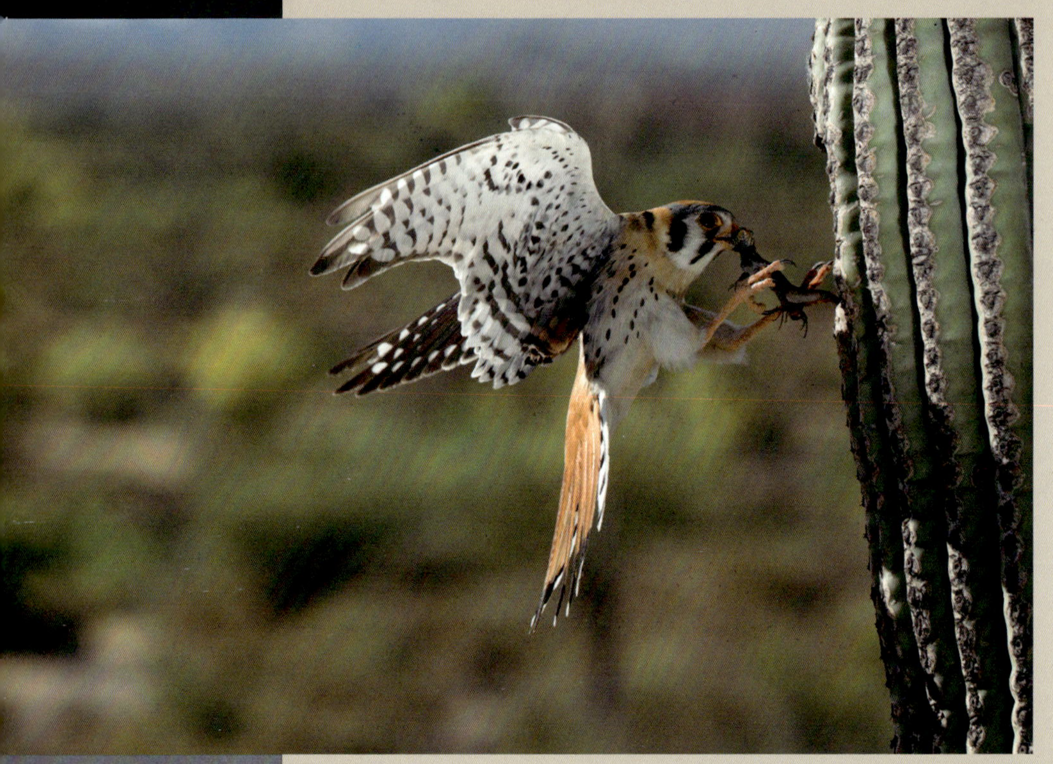

An American kestrel carries a lizard to its babies. Canon 5DMIII, 70-200mm at 122mm, 2x teleconverter,, ISO 1600, f/9 @ 1/3200 sec., on-camera flash.

Unlike Chuck Berry, I always have a "particular place to go" when I am driving along in my automobile. I am either returning from a late night shoot or driving to a new location to set up a blind or scout for photo opportunities. Scouting, in the traditional photography sense, indicates that the scout travels to a new place and searches for something to photograph. Those interested in animals might walk around looking at marshes, scanning the area with binoculars, or just sitting in one place to observe critters' behavior. A wildlife photographer always seeks an advantageous place to hide from their subjects or a place where attractants (such as food and water) exist. So how do I scout when driving along the highway at 45 miles (72.4 km) per hour or faster?

I look for noise in the environment—not the kind of noise that you can hear, but the type that you can see. As I drive along, the sky typically appears a shade of blue from horizon to horizon (at least in Arizona). If a flock of birds fly by, they create "visual noise" by breaking up the monochromatic silence of the sky. At night, frogs and snakes crossing the road stand out against the seemingly never-ending brown or black ribbons of highway. When I drive along the highways near Sonoita, I usually scan the light brown fields of grass for the white rumps or dark brown shoulders of pronghorn. If something stands out in the constant of a particular environment, or if I see movement, then I know a chance to photograph might occur.

Late one afternoon in May, I was driving home from an unsuccessful scouting for red-tailed hawk nests. About 50 feet (15.2 m) in front of my vehicle, I spotted a bird that looked out of place. It seemed a little larger than the rest of the flying birds, had backwards sweeping wings that flapped at an odd beat, and looked as though it was carrying something. Somewhere in the silence of my brain, a loud voice shouted "AMERICAN KESTREL!" Simultaneously, I realized that since it was May (the breeding season for kestrels), the bird was likely taking food to a nest to feed its young.

I pulled over to the side of the road to watch the kestrel's flight patterns. It kept flying and flying and flying until it almost disappeared out of sight, taking all my hopes of photographing the bird with it. At the last second, before I could no longer distinguish the bird from the dark blue sky, it landed on top of a giant saguaro cactus. At almost the same moment, his female mate flew to him, and I watched him transfer the food. In a split second, the female dropped straight down and disappeared from sight.

I began walking towards the cactus in hopes of finding a nest (or at least some indication that the nest existed). When I reached the cactus, I heard a faint noise coming from a hole about 12 feet (3.7 m) from the ground. Poking out of the hole were the tail feathers of the female kestrel; she was head down in the nest, feeding her young.

Over the next two years, I returned to this American kestrel nest over and over again. I set up my gear in the early morning light and took my photographs remotely from a blind about 300 feet (91.4 m) from the nest. After finding this gold mine, I was not about to screw up and offend the birds. I also wanted to live up to my pledge of only photographing undisturbed behavior of wild animals. With much enthusiasm, I made thousands of exciting, beautiful images of this family of kestrels.

During the first year, the adults fledged six young. How all of them fit into that tiny hole in the saguaro still makes me wonder! The next year, I had to leave the state before they fledged, but I later saw at least one fledging out of the nest and three young in the hole. Watching these wild raptors go about their lives was not only photographically exciting but also visually and mentally stimulating.

Today, I continue to scout from my vehicle hoping that another opportunity like the kestrel nest awaits just around the corner. Sometimes I find new things to photograph, but most of the time, I am merely practicing my looking for when it really matters.

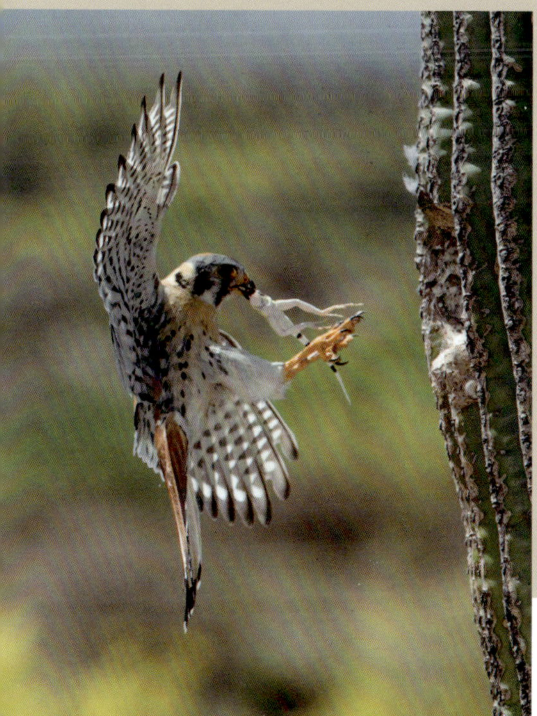

An American kestrel brings dinner home to its young. Canon 5DMIII, 70-200mm, 122mm, 2x teleconverter, ISO 1600, f/9 @ 1/3200 sec., on-camera flash.

San Pedro River Road

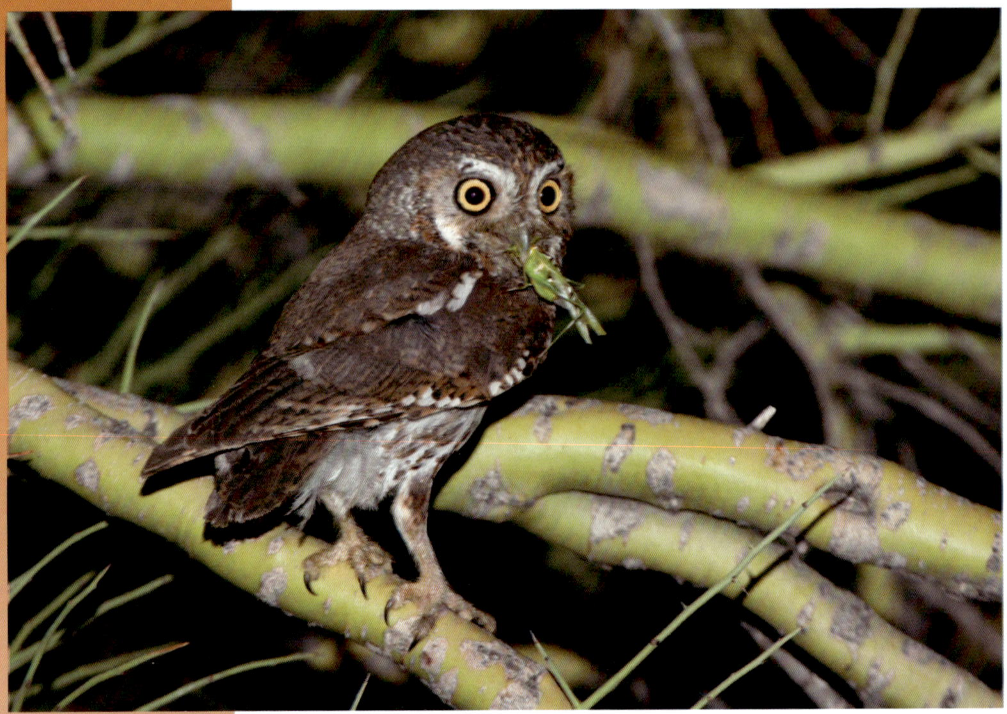

Elf owl with a katydid meal. Canon 40D, 100-400mm at 400mm, ISO 250, f/9 @ 1/250 sec., off-camera flash.

VIEW TIME
Late April to July

IDEAL TIME OF DAY
Night

VEHICLE
2WD high-clearance

HIKE
Easy to moderate

South of Mammoth, Arizona and adjacent to the San Pedro River, western screech and elf owls live among the dense saguaro cacti stands and palo verde and ironwood trees. Most owls are typically difficult to both find and photograph, but not these diminutive little birds. Arrive about 30 minutes before sundown to find large stands of saguaro cacti.

Shortly after dark, sit for a while and listen for the screech owl's hoot calls and the elf owl's puppy-dog yelp-like calls. If you have not heard any calls by 30 minutes past dark, play either owls' artificial call and listen for five more minutes. Keep playing their calls for a few more minutes. If you do not hear a response within 15 minutes or so, move on to another densely populated saguaro stand.

As soon as you hear one of the owls respond, walk towards the source. Keep playing the call while approaching the owl. When you are a few yards from the bird, place the call on the ground and use a headlamp (or flashlight) to find the owl.

Because of their light weight and ability to produce close-focus, frame-filling images, use a **telephoto zoom lens (in the range of 80-400mm)**. Tripods are too cumbersome for this type of desert photography. Take a friend with you, not only to hold the light on the owl as you focus and take pictures, but also for safety as rattlesnakes and sharp-spine cacti inhabit the area.

After snapping a few images, move closer (being as quiet as possible). Try framing the owl in both vertical and horizontal images. If the owl flies off, follow it and continue to play the calls.

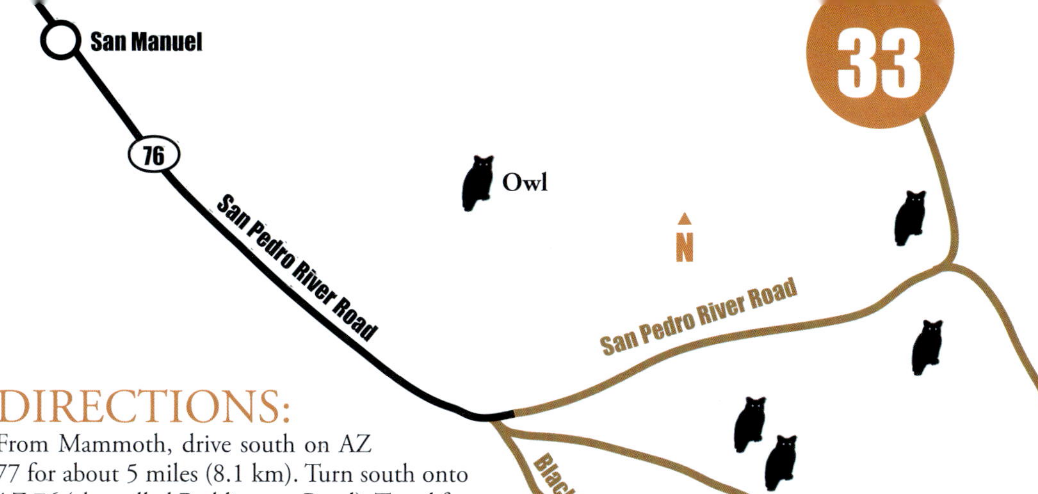

DIRECTIONS:

From Mammoth, drive south on AZ 77 for about 5 miles (8.1 km). Turn south onto AZ 76 (also called Reddington Road). Travel for 9 miles (14.5 km). At this point, the pavement ends at a junction of several roads. The San Pedro River/ Reddington Road goes to the left and two smaller roads continue.

The first option is to turn left on San Pedro River/ Reddington Road (the well-traveled of the three) and drive 3 miles (4.8 km) to the riverbed.

You can also drive the center unnamed dirt road for 1.8 miles (2.9 km).

Finally, the Black Hills Mine Road veers to the right. Drive along this dirt road for about 1.3 miles (2.1 km).

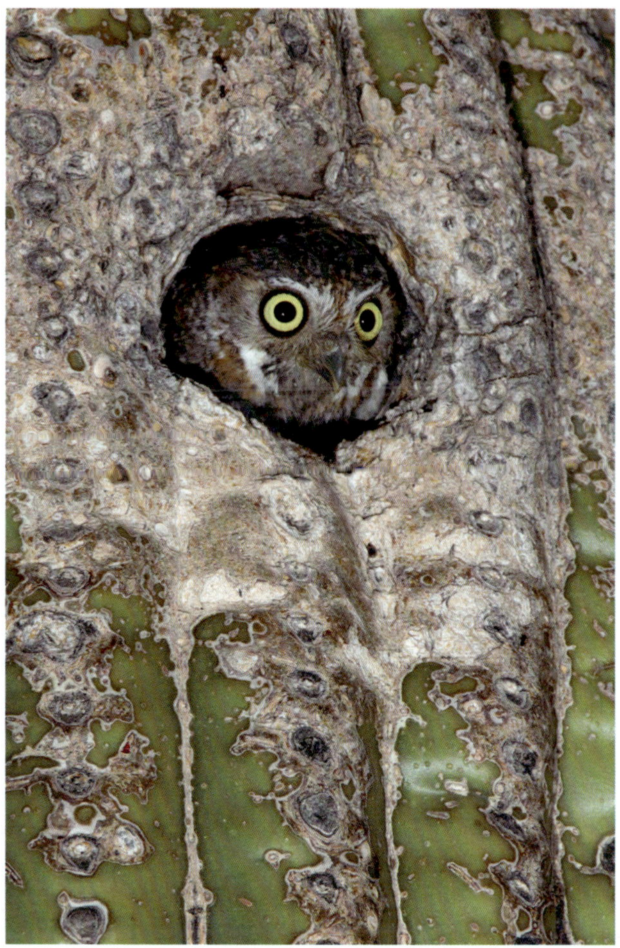

An elf owl peers from its nest hole in a saguaro cactus. Canon 7D, 500mm, 1.4x teleconverter, ISO 640, f/13 @ 1/200 sec., off-camera flash.

Making the Photo

8

Kangaroo Rats

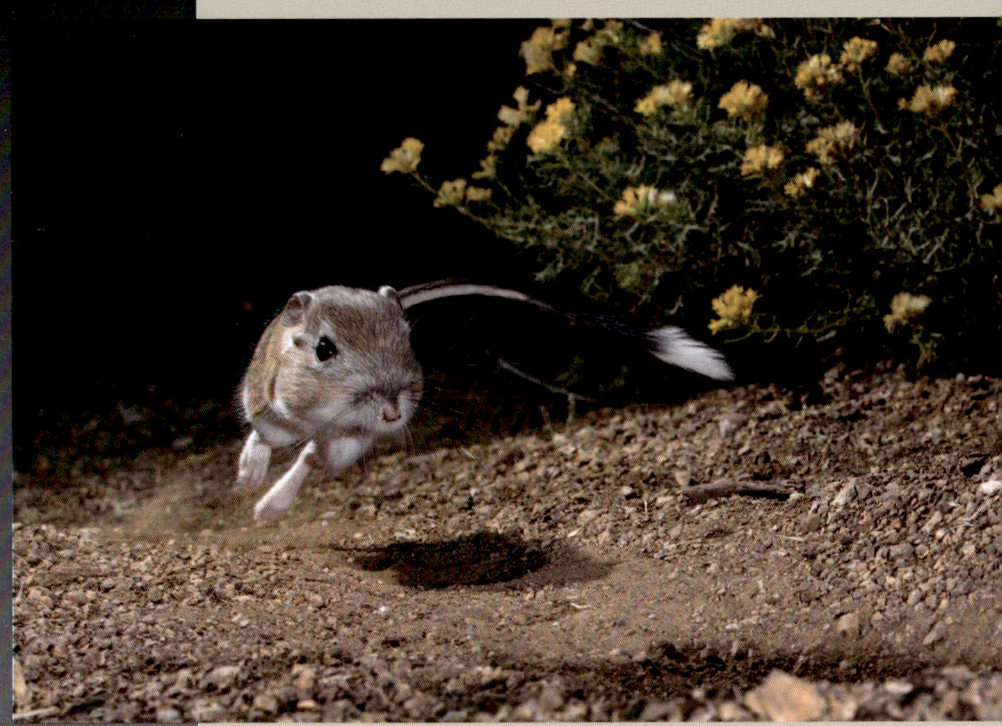

A banner-tailed kangaroo rat runs for its den. 50D, 70-200mm at 70mm, ISO 400, f/14 @ 2 sec., high-speed hummingbird set-up, Phototrap infrared trigger.

For several years, I wanted to photograph a running kangaroo rat. The vision of a kangaroo rat running at full speed with all four legs off the ground intrigued me. The problem? I had no idea how to accomplish the task. I knew nothing about kangaroo rat life history or distribution.

Fortunately, I owned a copy of Donald H. Hoffmeister's written treatise titled, *Mammals of Arizona*. I began investigating the chapter on these small rodents. I immediately learned that kangaroo rats never have to drink water! In order to maintain their body fluids, they uniquely metabolize water from the carbohydrates in the food that they eat (beneficial for an animal that lives in places where it may not rain for months or sometimes years).

In addition, Hoffmeister references the paper of a scientist who conducted population studies on kangaroo rats. After finding and reading the paper, I learned that kangaroo rats can smell seeds. Also, the experiments revealed that kangaroo rats enjoy brown rice, probably because it is high in carbohydrates and is easily metabolized by the rat into water.

But what did this have to do with photography? Everything!

From Hoffmeister, I knew where to find kangaroo rats, and from the article, I knew what they ate. I took my bag of brown rice to an area around Oracle where a friend of mine had seen many banner-tailed kangaroo rats. During the afternoon, I found a concentration of rat mounds and set up my security camera within six feet (1.8 m) of an active burrow (security cameras use infrared light to enable night vision). I then placed a pile of brown rice near the burrow, turned on my security camera and monitor, and waited. About an hour past sunset, a furry kangaroo rat exited its burrow and went directly to the brown rice. The rodent filled up its cheek pouches with seeds and scurried off to the burrow. A few minutes later, the rat returned for more free food until the pile disappeared. Now I understood how to get the rat where I wanted, when I wanted.

Next, I needed to determine the necessary equipment to take their photograph. From my extensive hummingbird photography experience, I knew that I could stop very fast objects (flapping hummingbird wings) and had no doubt that the same flash set-up could stop a running rat.

On my next trip to the rat burrows, I set flashes outside the mound in a position perpendicular to—and midway between—the rice seeds and the burrow entrance. I placed the camera on a low tripod and focused between the seed pile and the hole where I expected the rat to emerge. As predicted, the critter was hooked on the seeds. While I recorded some nice kangaroo rat portraits, the images failed to fulfill my initial vision of capturing a running rodent.

After the rat visited the seed pile a few times, I decided to speed the process. When the rat visited the seed pile, I shifted my feet to make noise and, as expected, the rat ran to the burrow, and it ran FAST. So fast that I could not react quickly enough even to get an image of the retreating tail. Bummer! Back home to the drawing board.

I recalled a technique I employ using a Phototrap infrared trip to take photos of flying bats. I point the infrared beam across an area where I believe the bats will fly and hook the automatic trigger to my camera and flashes. When the bat flies through the infrared beam, the trip sets off the camera and flashes and takes the bat's image. Considering this, I believed that if I could place the beam across the path of the running rat, the trip would work similarly. I bought more seeds, armed myself with my Phototrap, and drove back to the kangaroo rat mounds (two hours each way).

This time, I positioned the flashes and camera so when the kangaroo rat ran back to the nest from the seed pile, I had a face-on view. I placed the infrared beam so that the flashes would fire exactly where I had focused the camera. I crossed my fingers and waited for the sun to set.

On cue, the banner-tailed kangaroo rat dashed for its free meal. As soon as it filled up its pouches with seeds, I turned the Phototrap on, placed the camera on Bulb mode and shifted my feet. The rat beat feet for the burrow, crossed the infrared beam (which set off the flashes), and, much to my pleasure, I successfully made an image. I proceeded to repeat this process several times over the next couple of hours until the rat no longer reacted to my noises.

The images I recorded that night exceeded my expectations. The banner-tailed kangaroo rat was running directly in my direction (a face-on view), all four feet were off the ground, and the tail was tilted to one side to help counterbalance the rat as it ran. I felt a sense of pride for having taken the time and energy to develop new techniques to not only make a difficult image, but also achieve my original vision.

Sweetwater Wetlands Park

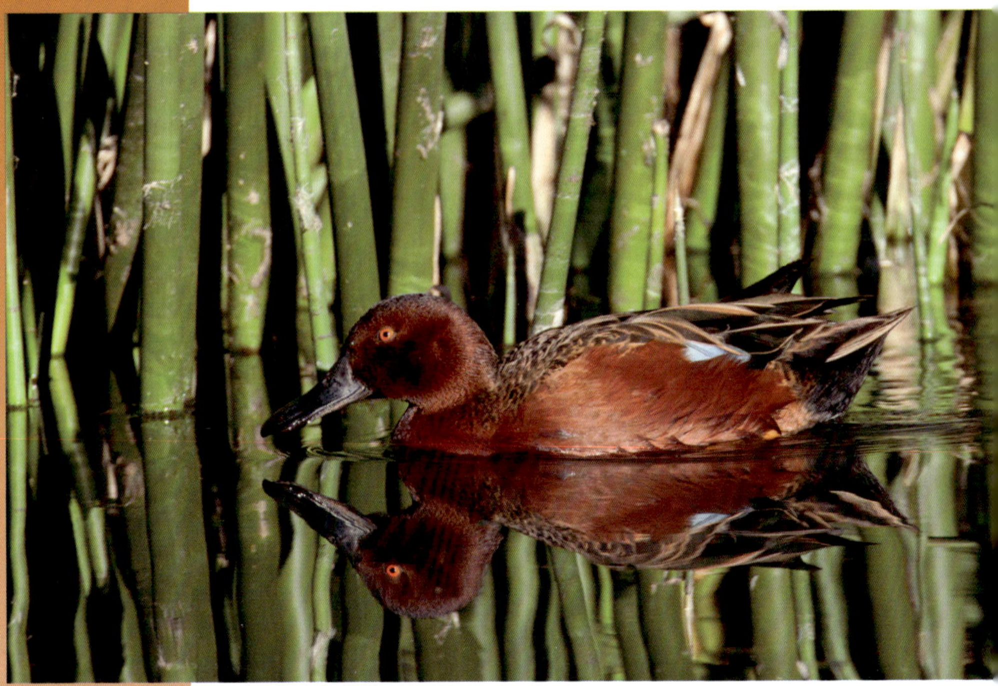

Cinnamon teal. Canon 50D, 500mm, 1.4x teleconverter, ISO 200, f/8 @ 1/200 sec.

VIEW TIME
October to June

IDEAL TIME OF DAY
Early morning and late afternoon

VEHICLE
Any

HIKE
Easy to moderate

If you love photographing waterbirds and waterfowl, then grab your **telephoto lens** and **tripod**, and head to the Sweetwater Wetlands Park in Tucson. During the winter months, thousands of ducks migrate to these wetlands to escape the frozen north and find food and shelter. A slow moving stream and a series of culverts and pipes connect the sixteen or so small ponds (some with, and some without, islands). The water level varies, but all of the cattail-lined ponds normally stay full. A series of trails provides access to each pond, but in some areas, the vegetation is so thick that photography is not possible. In addition, two lookout platforms overlooking the largest ponds offer an ideal perch for shutterbugs.

From the west end of the parking lot, walk west for 300 feet (9.4 m) or so along West Sweetwater Drive. Walk into the park at the small pump house and note the four small, elongated ponds. During the winter months, a pair or two of cinnamon teal normally call these ponds home. After locating the teal, sit down and be quiet. A patient and still photographer will not scare off the normally skittish ducks.

During March and April, red-winged blackbirds breed in these ponds. The males use the cattail perches to defend their territories and attract females. Watch for a male to fly to the top of a cattail. When it defends its area, the blackbird will sing and display its bright red feathers on the front of the wing. If you see this behavior, find a good place to take a seat, ensuring your background is distant enough to record an out-of-focus backdrop using a wide aperture (e.g., f/4 or f/5.6).

The larger ponds to the south and west end of the park host northern

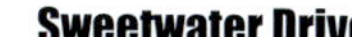

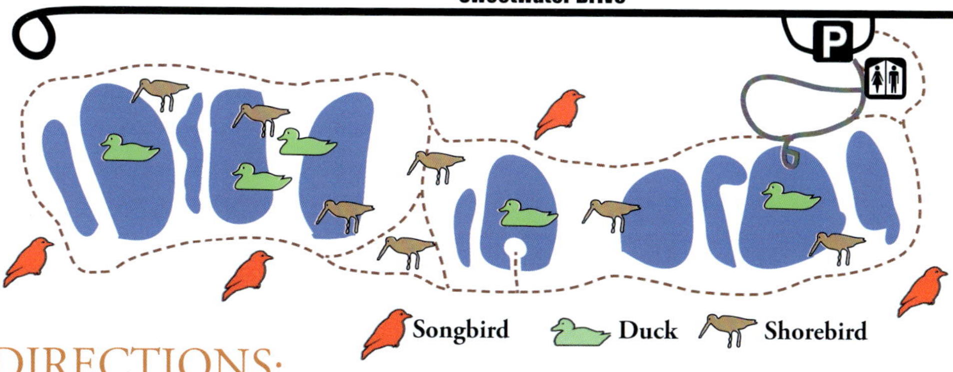

DIRECTIONS:

From downtown Tucson, travel north on I-10, and take Exit 254 for Prince Road. Turn left onto Prince Road. Drive 0.3 miles (0.5 km) until the road ends. Turn right onto Business Center Drive and continue for an additional 0.2 miles (0.3 km). Turn left on River Park Road which loops around the business park and becomes Commerce Drive. Take the first left (which has no street sign), and follow it to the dead-end at Sweetwater Drive. Turn left and continue 0.2 miles (0.3 km) to the Sweetwater Wetland Park parking lot on the left.

For up-to-date information on bird locations and abundance at Sweetwater Wetlands Park, visit the Tucson Audubon Society's website at **www.tucsonaudubon.org/what-we-do/visit/planning/locations/212.html**.

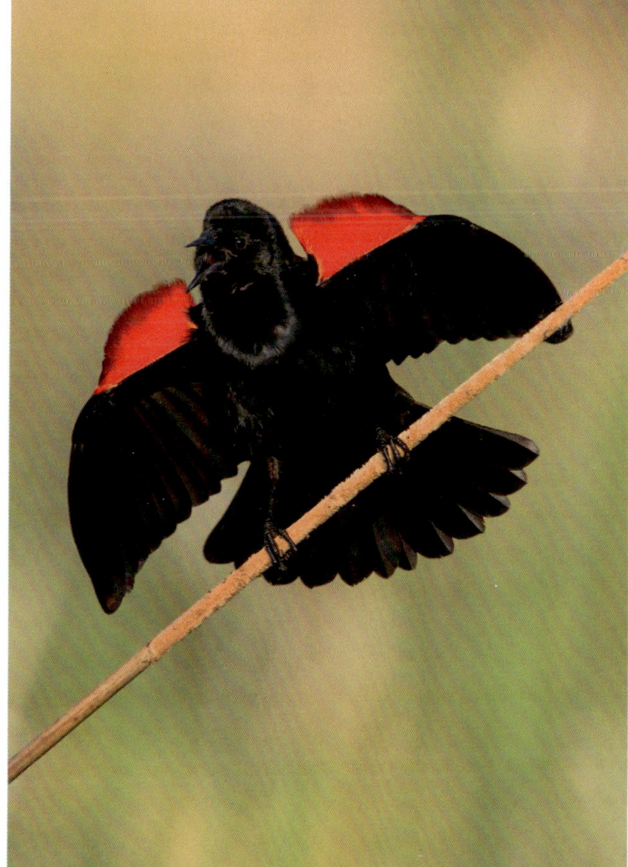

shoveler, grebe, mallard, pintail, teal, merganser, bufflehead, scaup, wigeon, and the difficult to spot sora. Walk the interconnecting trails looking for openings in the marsh where the ducks appear close enough to make photographs.

Try to time your visit for windless days so the calm waters reflect the green or tan colors of the surrounding vegetation. Get as low to the water as possible to ensure the best angle for the reflective backdrops. Use tripods whenever you can to keep your camera steady and your images sharp.

A red-winged blackbird aggressively defends its territory at Sweetwater Wetlands Park. Canon 70D, 500mm, 1.4x teleconverter, ISO 400, f/8 @ 1/500 sec.

Making the Photo
9

Snowfall

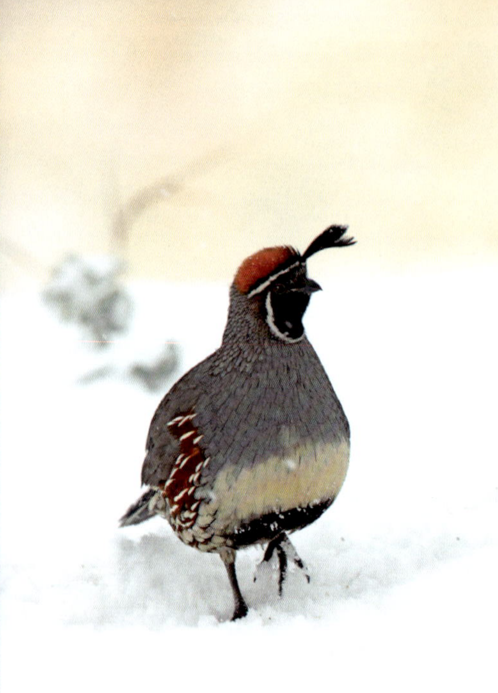

Gambel's quail in a rare Oracle snowstorm. Canon 5DMII, 500mm, ISO 800, f/8 @ 1/800 sec.

I left the Midwest for Arizona, eager to leave the winters behind. Climbing up Bascom Hill at the University of Wisconsin during heavy, wet snow wearing tennis shoes, or sampling for fish during a snow blizzard on Lake Sangchris in central Illinois, was not my idea of fun. But you know the old adage, you never know what you miss until it's gone!

After several years in my new desert home, I thought I would search for that rare desert snow storm and take images of desert animals in the snow. The trick was to figure out exactly when and where such a fleeting event would occur.

Months of waiting for snow at low elevations turned into years, and I began to wonder why I did not pick an easier goal. I chased snow all over the desert, but by the time I arrived, it had disappeared. I should have realized that (unlike more northern climates), even though the snow fell in the cold air temperatures, the ground temperature remained well above freezing. As a result, the rare (typically bi-annual) desert snows only lasted a few hours at best.

Then I got smart (at least I thought I did), and started to chase the snows before they arrived. But how accurate are the weather forecasters? When it comes to predicting snow in the desert, they are not very accurate at all. I was not used to the frustration of going on a photo shoot and coming back with nothing. I told my sad story to anyone who would listen, hoping to get some advice on what to do.

Then, my good friend from Oracle, Bud Bristow, called me at midnight to inform me it was snowing hard there. I put my photo blind and all of the necessary equipment in the car and drove to Bud's house. My luck held strong because Bud was an avid bird feeder and had some hunting dogs (more on them later).

I arrived about 4 a.m. and set up my blind where I thought the birds would appear in the morning. The snow continued to fall, and the wind picked up enough that I had to hold on to the top of my blind so it would not blow away.

Despite the storm's fury, first came the Gambel's quail, and then cactus wren, mourning dove, curve-billed thrasher, canyon towhee, and other hot-desert birds. They fluffed their feathers to keep the cold out, and they ate furiously to stoke their internal fires to keep warm. It is possible none of them had ever seen snow before and were perplexed.

While in manual camera mode, I continuously monitored my light meter so it did not dictate the exposure I recorded in the final image. Light meters are geared to render neutral, middle grey tones no matter the object's luminance. Because I did not want to create an image where the snow appeared gray and unattractive, I needed to add between 1.5 and 2 stops of light. This allowed me to maintain whiteness and texture without blowing out any highlights.

What happened next made Bud's hunting dogs my friends for life. Whenever Bud fed the dogs, a portion of their food was strewn outside of their chain link kennel. Occasionally, a coyote or two wandered around the fence looking for scraps to eat. On this already exciting morning, the mom and her pup stopped by to scavenge.

I watched them move around the kennel and walk within 50 feet (15.2 m) of me. The birds flew away from the predators, and the coyotes stopped to watch. I took a few frames and hoped for the best. Then, in a second, both coyotes trotted off. I looked at my images on my camera's playback mode. There she was: a coyote looking into the distance with snow falling around her. Snow covered the barrel and prickly pear cacti, adding to my vision of snow in the desert. I could not put my excitement into words.

The snow stopped and began to melt. By 11 a.m., only a few patches of snow in the shade of trees and shrubs remained—the magic moments gone after a single morning. Over the past ten years, I have been unable to repeat this event. So much for my hating snow ever again.

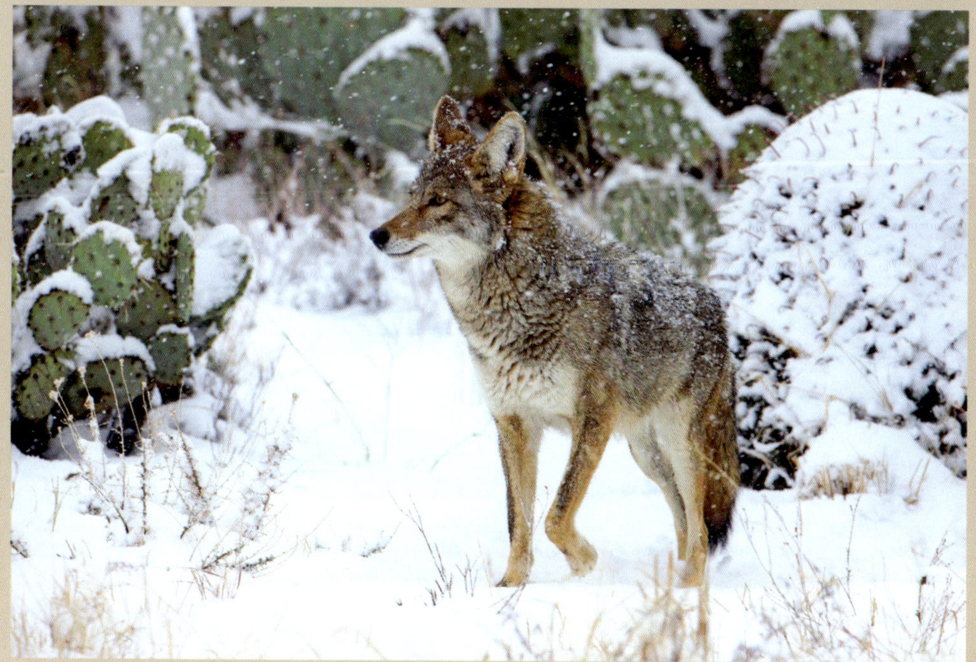

Coyote among snow-covered barrel and prickly pear cacti. Canon 5DMII, 500mm, ISO 800, f/9 @ 1/800 sec.

SOUTHERN ARIZONA

Tucson Botanical Gardens

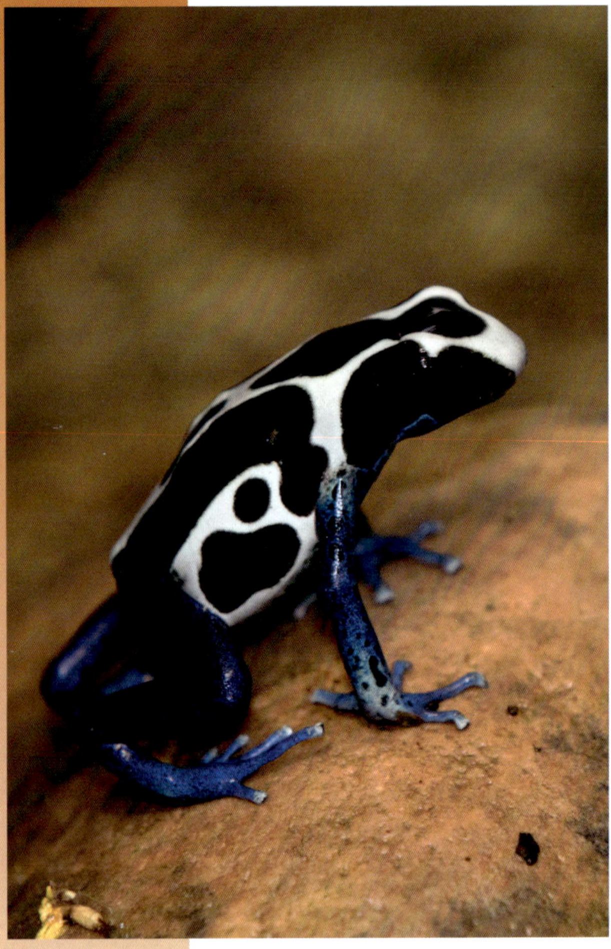

A poison dart frog waits for an ant meal at the gardens. Canon 7DMII, 70-200mm at 130mm, 2x teleconverter, ISO 800, f/7.1 @ 1/80 sec., on-camera flash @ -2/3 FEC.

In 1974, the Tucson Botanical Gardens joined with Bernice Porter to both protect the Porters' home (including her botanical gardens) and establish a non-profit organization to promote the appropriate use of plants and water in a desert environment. In cooperation with Cox Communications, the garden developed the beautiful and photogenic Cox Butterfly and Orchid Pavilion in 2006.

Because the change in humidity from the dry desert air to the indoor humid pavilion can fog your lens, place your **long telephoto lens (e.g., 300 mm or longer)** on your camera before you enter and wait a few minutes while your camera lens defogs. Changing lenses while your camera equipment is still adapting to the higher humidity could fog the camera's sensor and render it useless for the duration for the shoot.

Roam the small pavilion and look for butterflies sitting on plants or the abundant orchids. Set your camera on aperture priority, ISO 400, and an f/8 aperture. Because the backdrops appear relatively close to the butterflies, a narrow depth of field is necessary to keep the backdrop blurred. Watch your backgrounds to keep them simple and free from distractions. Then, keep the wings of the butterfly parallel to the camera body and perpendicular to the long axis of the lens to maximize your depth of field and wing sharpness. Hand-holding your camera will give you the utmost flexibility in your pursuit, but turn on image stabilization (or vibration reduction) and use continuous autofocus.

Although the pavilion produces diffused lighting ideal for capturing rich colors, use an **on-camera flash** on TTL set to minus one or two stops to add a little fill flash. Once you locate a subject, focus on the butterfly head and hold your camera steady. Breathe out, hold your breath, and gently push the shutter button.

VIEW TIME
October to May

IDEAL TIME OF DAY
Early morning to late afternoon

VEHICLE
Any

HIKE
Easy

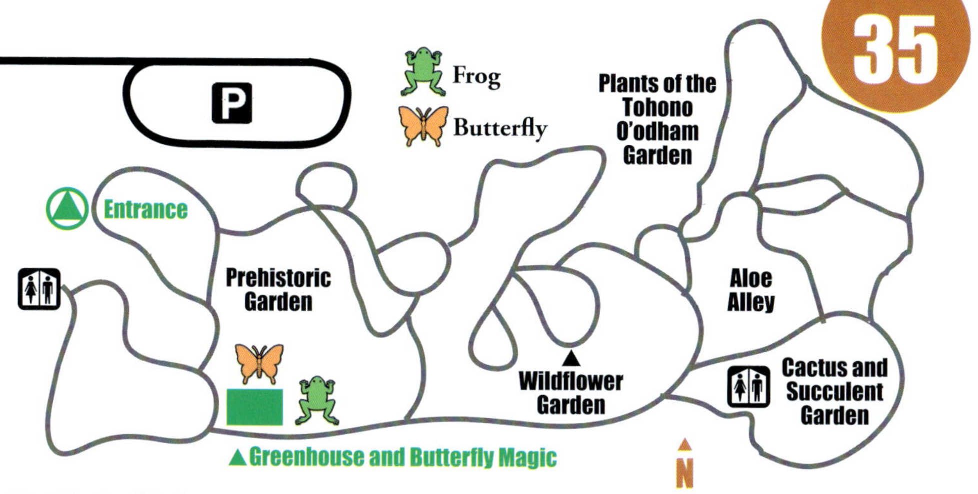

DIRECTIONS:

From downtown Tucson, follow I-10 to Grant Road (exit 257). Turn east and drive 4.8 miles (7.7 km). Turn south onto Alvernon Way. Drive 0.1 miles (0.2 km), and then turn east into the small gravel parking lot. If this lot is full, visitors are permitted to park in the shopping center parking area to the north.

Tucson Botanical Gardens opens its gates every day, year-round from 8:30 a.m. until 4:30 p.m., except for Independence Day, Thanksgiving Day, Christmas Eve and Day, and New Year's Day.

From October to May, the butterfly greenhouse is open from 9:00 a.m. to 3:30 p.m.

The park charges a small admission fee unless you have purchased an annual membership pass. For more information, visit **www.tucsonbotanical.org**.

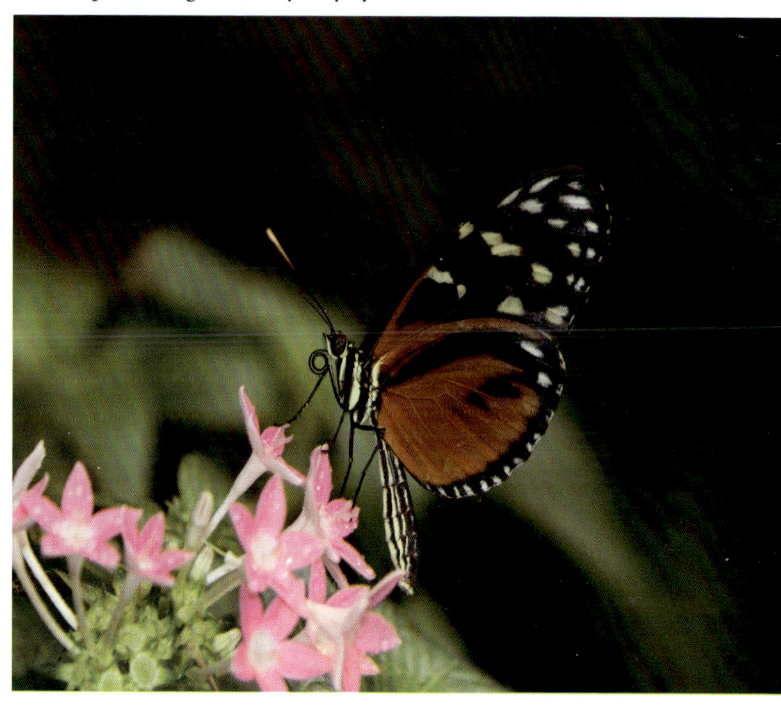

A spotted tiger glassywing sips nectar from a flower. Canon 7DMII, 70-200mm at 119mm, 2x teleconverter, ISO 800, f/8 @ 1/60 sec., on-camera flash.

Just as exciting as the butterflies and orchids are the many poison dart frogs that inhabit the pavilion (and are normally found in Costa Rica and along the Amazon River in South America). Scan the ground and the plants for bright green, yellow, or blue frogs. The frogs are used to people and might sit still long enough to allow you to make a portrait image of them. Get low to the ground to shoot at eye level. Use the same exposure settings as for the butterflies, but boost the ISO up a little to maintain a fast shutter speed in the added darkness on the floor of the pavilion.

SOUTHERN ARIZONA

Mount Lemmon

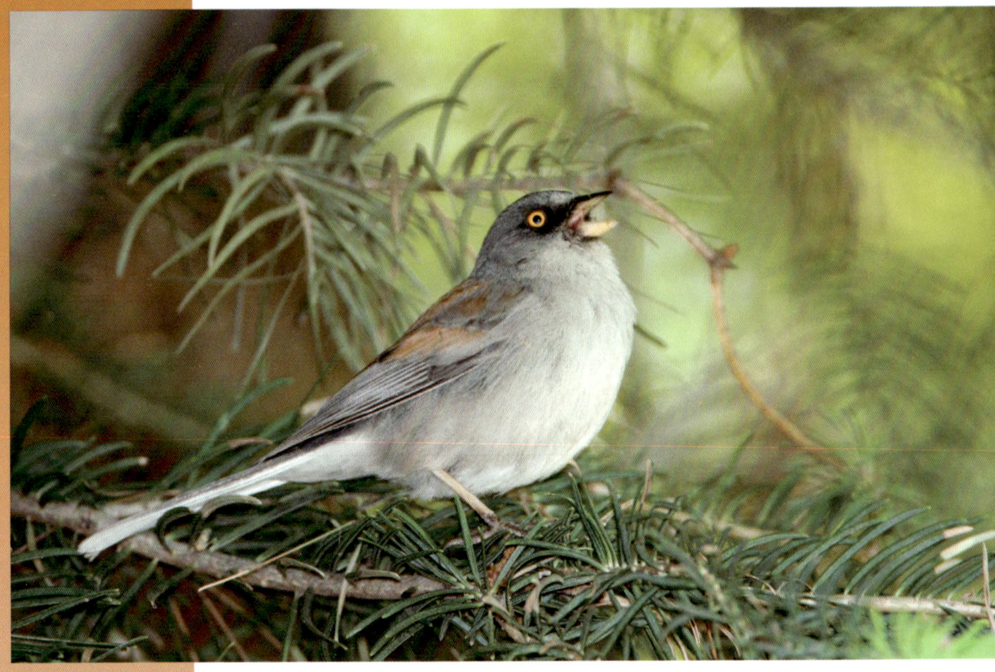

A yellow-eyed junco calls to attract a female. Canon Digital Rebel XTi, 500mm, 1.4x teleconverter, ISO 200, f/5.6 @ 1/60 sec., on-camera flash.

VIEW TIME
April to June

IDEAL TIME OF DAY
Early morning or late afternoon

VEHICLE
Any

HIKE
Easy to moderate

Mount Lemmon Highway (also known as the General Hitchcock Highway or the Catalina Highway) climbs for 28 miles (45.1 km) from the desert floor at 2,000 feet (609 m) in elevation to the small town of Summerhaven near 9,100 feet (2,774 m). Along the way, the two-lane road twists and turns through deep canyons, oak woodlands, mountain forests, and vast vistas of the surrounding lands. Travelers visit Mount Lemmon to camp in the cool summers and ski in the snowy winters (Mount Lemmon Sky Valley is the southernmost ski area in the United States). However, bird watchers and photographers gather here for the abundance of hermit thrush, American robin, olive warbler, red-faced warbler, mountain chickadee, yellow-eyed junco, and three species of jay.

Explore the trails originating at the Marshall Gulch Trailhead just south of Summerhaven. Walk up the canyon and stop frequently to look and listen. Red-faced warblers are particularly abundant in this area and easy to locate. Work this location early in the breeding season (i.e., April and May) with a **400mm or longer telephoto lens**. If you are not there on a diffused cloudy day, find a place where the light does not fall directly on your subject (such as shady glen or canyon). If shadows appear too harsh and direct, use a pop of **on-camera fill flash**.

When you locate a subject, place your artificial bird call near low hanging branches and start the call for your choice of species. Normally the birds might be nervous about your presence but if you arrive during breeding season, they should respond to the calls with calls of their own and approach the speaker to look for their rivals. Keep the shutter speed fast to freeze their motion. Place the fill flash on high-speed sync. Set your lens on

DIRECTIONS:

From the Tucson area, head east on Tanque Verde Road. Take a left onto the Catalina Highway which turns into the Mount Lemmon Highway. Drive about 29 miles (46.7 km). Veer left towards Summerhaven when the Mount Lemmon Highway splits. Continue another 1.4 miles (2.3 km), passing several parking areas on the left, before reaching the last dirt parking area. The unmarked Sunset Trail starts on the southernmost end of this parking area.

The formal Marshall Gulch Trail starts on the northwest side of this same lot. In addition, an unmarked trail (informally referred to as the Marshall Gulch Trail as well) begins behind the pit toilets and follows the creek.

A federally issued "America the Beautiful" annual pass or a Coronado Recreational Pass is required. For more information, visit the Coronado National Forest website at **www.fs.fed.us/r3/coronado**.

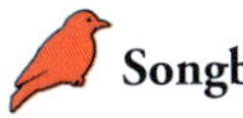

continuous autofocus mode.

As soon as you establish focus, take several frames in fast succession on high-speed continuous shooting mode. Watch your histogram and highlight alert to insure adequate exposures. Under these conditions, light areas in the backdrop can often appear overexposed.

Continue to walk up and down the canyon and around the picnic area. Thrushes, robins, and more warblers await your calls.

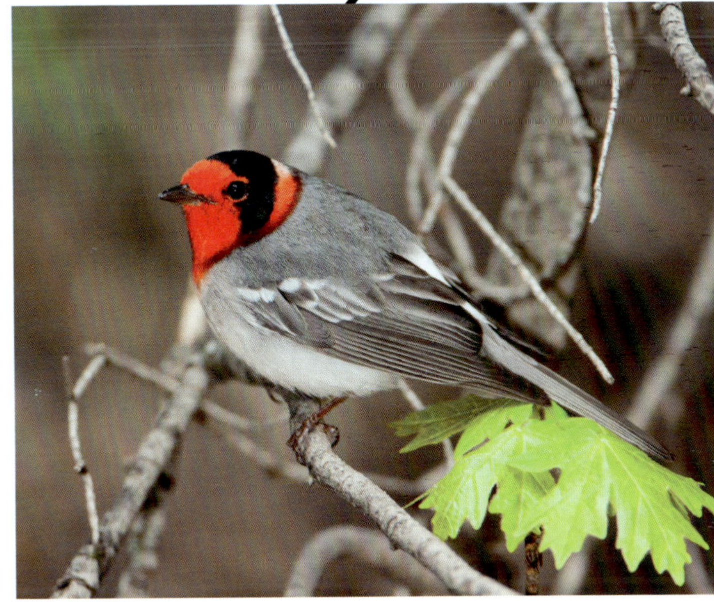

Red-faced warbler. Canon Digital Rebel XTi, 500mm, 1.4x teleconverter, ISO 400, f/5.6 @ 1/60 sec., on-camera flash.

Saguaro National Park–West

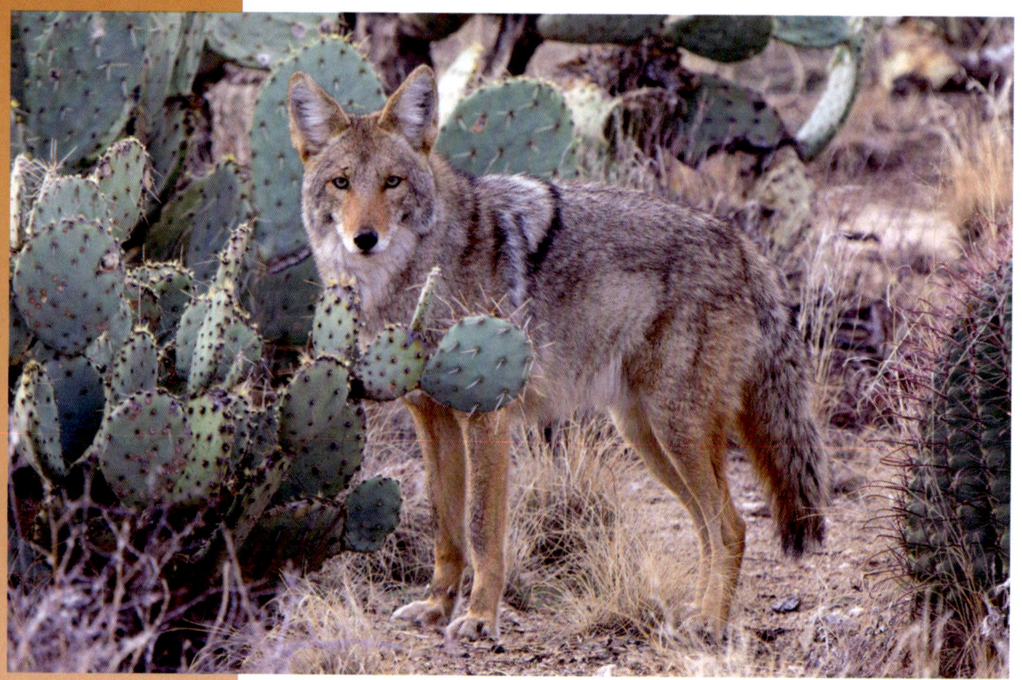

ABOVE: The wily coyote surveys for danger (or maybe food). Canon 50D, 100-400mm at 400mm, ISO 400, f/8 @ 1/200 sec.
RIGHT: Bullock's oriole on saguaro cactus flower. Canon Digital Rebel XTi, 500mm, 1.4x teleconverter, ISO 200, f/7.1 @ 1/2000 sec.

VIEW TIME
May to July

IDEAL TIME OF DAY
Sunrise to late morning

VEHICLE
Any

HIKE
Easy to moderate

Saguaro National Park protects thousands of giant saguaro cacti—one of the most iconic living symbols of the Sonoran Desert. The park consists of two districts (Rincon Mountain to the east of Tucson and Tucson Mountain to the west) totaling over 80,000 acres (32,374 hectares). The grand scenery found in the western section appears so pristine upon entering the park, you may feel as if you are going back to a time before humans changed the Arizona landscape.

The park's vast terrain and wildlife distribution make photographing in the park different than many other places highlighted in this book. To maintain the "wild" in "wildlife," the park does not set out feeders to attract animals. In addition, with such a broad space to roam, most creatures here are not habituated to human presence and therefore cannot be easily approached.

However, if you understand the cycles of the saguaro, even a novice photographer can take wonderful images of the unique desert fauna. In May, the saguaro begins to produce massive numbers of blossoms. These blooms (full of both nectar and pollen) offer a long-awaited food source for a variety of the desert creatures. Almost every desert bird feeds on the saguaro flower. Gila woodpecker, gilded flicker, cactus wren, curve-billed thrasher, Bullock's oriole, Gambel's quail, white-winged dove, mourning dove, verdin, and the like use the tops of saguaro cacti as their dinner tables. More often than not, several birds will land on the cactus arm and begin feeding on its flowers.

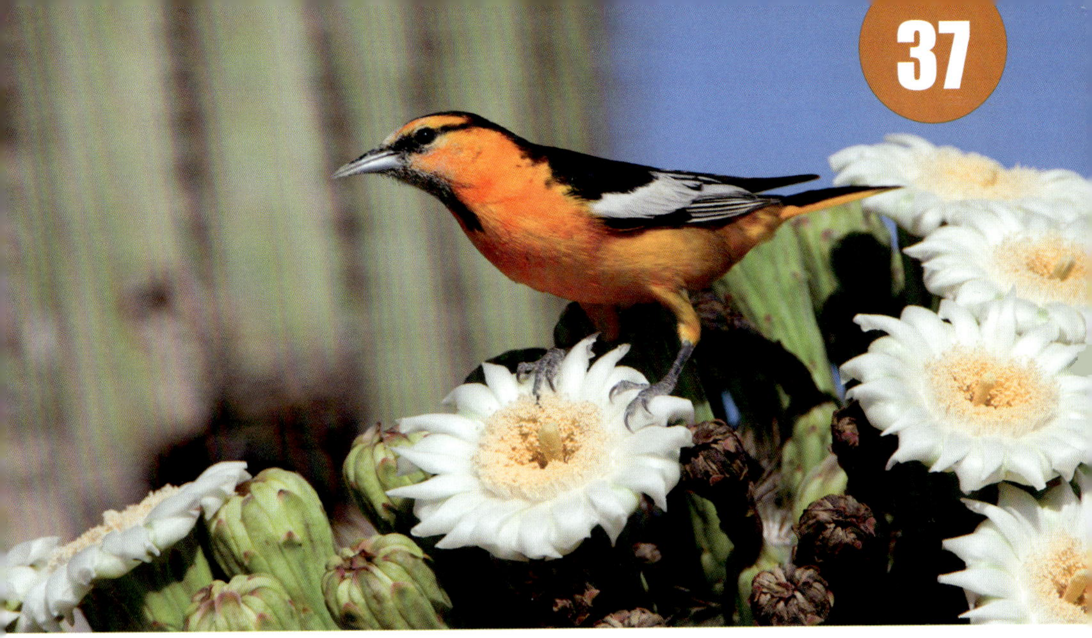

Arrive just before sunrise in the cooler temperatures. Scan the desert for a low-hanging saguaro arm displaying several blooms. Place a **telephoto lens** on a sturdy **tripod** and sit or stand quietly. Although early morning offers the coolest time to photograph, many birds continue to feed on the flowers until they close their petals just after noon.

As the blooms die, they form fruit. With its thousands of seeds and rich, red pulp, local birds and other wildlife munch on this tasty treat. Coyote, javelina, several species of small mammals, desert tortoise, and other ground-based critters enjoy the fruit once it falls to the ground. The strategy for photographing this type of activity remains the same: arrive at the park early in the morning with your telephoto lens in hand and wait for the feast to begin. The blooms and fruit will also attract several species of bees, hornets, ants, and other desert insects.

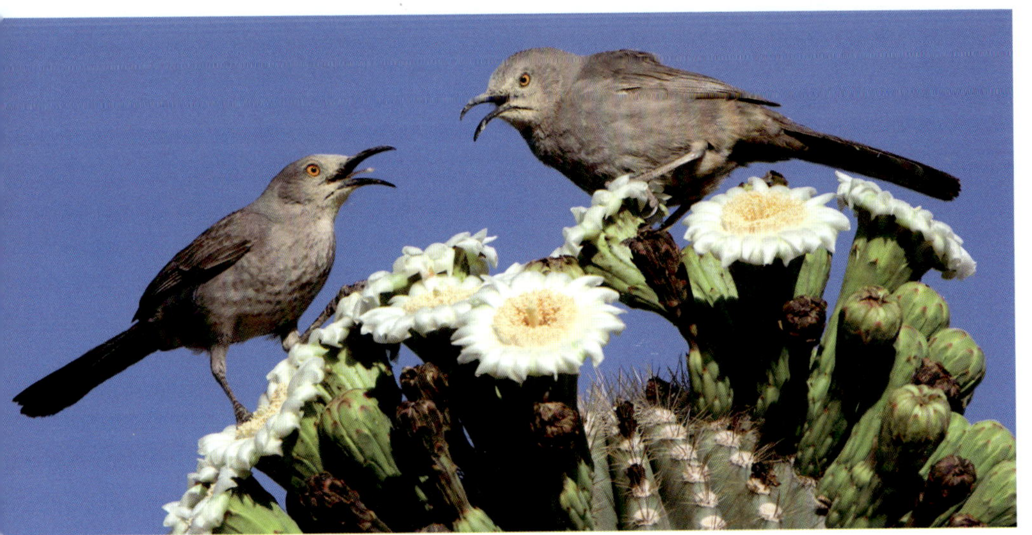

A pair of curve-billed thrashers have a discussion. Canon Digital Rebel XTi, 500mm, ISO 200, f/9 @ 1/500 sec.

Saguaro National Park–West

DIRECTIONS:

From central Tucson, take Speedway Boulevard west toward the nearby Tucson Mountains. Travel approximately 4.5 miles (7.2 km) from the I-10 highway, and then veer left onto Anklam/Gates Pass Road when the road splits. Drive another 4.7 miles (7.6 km) on Gates Pass Road up and over Gates Pass. Where Gates Pass Road ends on the far side of the mountains, turn right onto Kinney Road. Drive about 4 miles (6.4 km), passing the Arizona-Sonora Desert Museum on the left, until the road splits. Veer right to continue following Kinney Road.

The Saguaro National Park Visitor Center will appear on the right hand side of the road. Stop here to first pay your entrance fee or show your "America the Beautiful" annual pass issued by the National Park Service.

Wildlife appears most abundantly in the wash behind the visitor center as well as along the Golden Gate and Hohokam roads. To access Golden Gate and Hohokam roads, take Kinney Road west of the visitor center for 1.5 miles (2.4 km). At this point, Hohokam Road is to the right and Golden Gate Road is straight ahead (where Kinney Road ends).

PHOTO TIP 12

Remote Equipment

When photographing many species of wildlife, getting to the right place at the right time takes skill. Skittish animals may not allow the photographer to get close enough to take an image, even when the human sits in a blind. Nocturnal mammals and birds hide themselves well in the dark, making it difficult to see their exact position in order to take a photo. Animals may visit a location so infrequently that the photographer either gets bored or tired waiting for them. In all these instances (and more), remote equipment can act as a surrogate for the photographer's presence.

The most basic piece of remote equipment is the wired or wireless remote release. These units plug into the remote control socket on the camera and operate the shutter button from as close as a few feet to in excess of 0.5 mile (0.9 m to 0.8 km) depending on the make and model. For a few dollars more, you can purchase a wired or wireless intervalometer. When using an intervalometer the photographer can determine how many shots they want to take, when to start taking photos, the duration between images, when to stop taking images, and the like.

At night, you control the light. However, oftentimes with nocturnal wildlife, you cannot see your subject well enough to focus. An infrared remote release can help. An infrared release sets off the camera when the infrared light is either interrupted or reflected back to a sensor. To use this tool, place the beam from the trip where you expect your subject is likely to travel. In addition, position two or three flashes configured as slaves to the camera. Then, manually focus on the spot where the animal will likely interrupt the trip, and wait. Wildlife (and humans) cannot see infrared light, and, therefore, are not afraid of it. When the animal crosses the beam, it will trigger the camera and flashes to take a photo.

Oftentimes, a simple security camera connected to a small (8 inches (20.3 cm) or so) television screen can help you monitor nests or trails at night. Place the camera close to where you expect the animal to be, stretch the connection to your blind, and watch the screen.

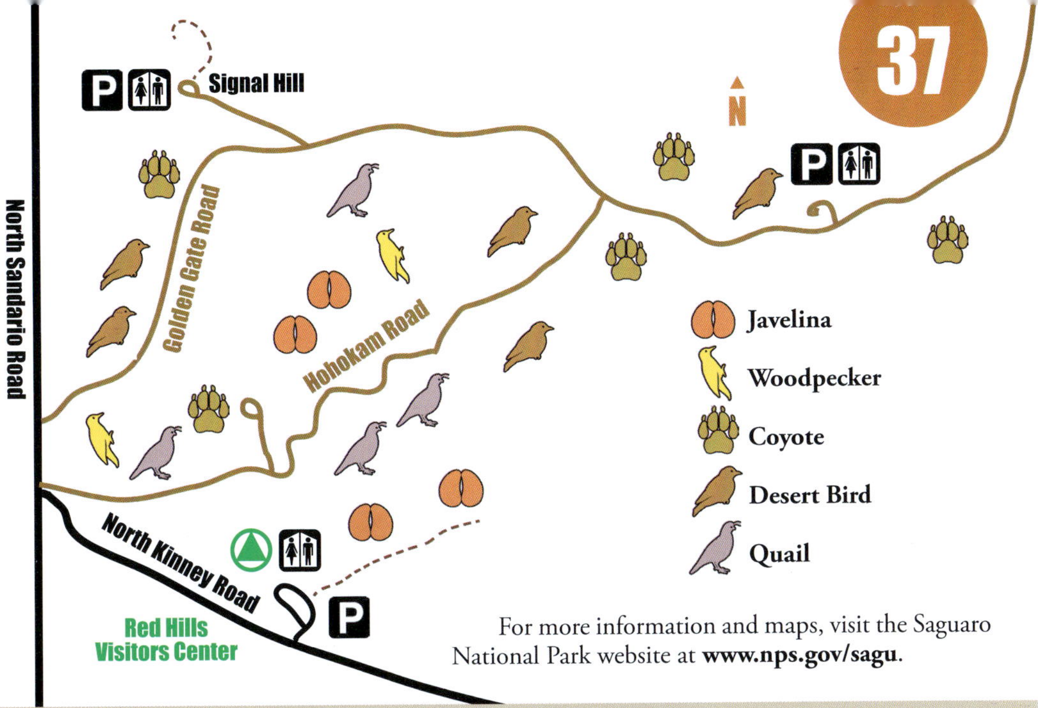

37

Signal Hill

Golden Gate Road

Hohokam Road

North Sandario Road

North Kinney Road

Red Hills Visitors Center

N

Javelina

Woodpecker

Coyote

Desert Bird

Quail

For more information and maps, visit the Saguaro National Park website at **www.nps.gov/sagu**.

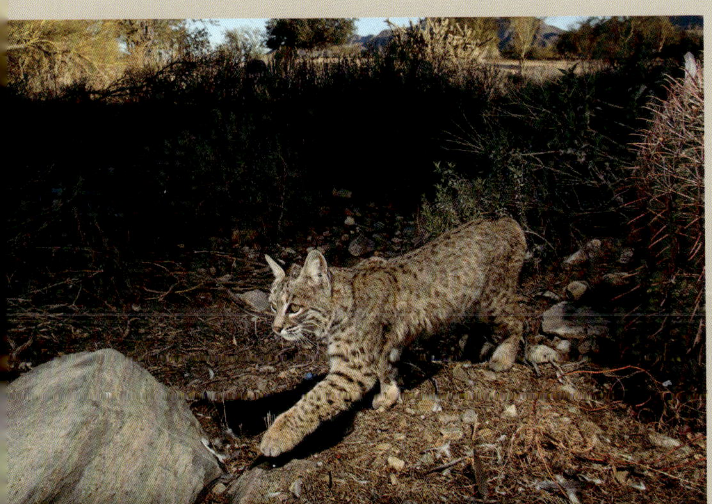

A bobcat slinks along the edge of a golf course. Canon 5DMII, 17-35mm at 24mm, ISO 400, f/13 @ 1/200 sec., remote off-camera flash, Phototrap infrared trigger.

You might want to place your camera close to the subject but do not want to continuously take photographs using an intervalometer nor sit next to the camera. In these situations, sit in a blind positioned so you can watch the subjects with binoculars. Then, trip the camera using either a wired or wireless remote connected to the remote slot on the camera. When the subject comes into range, you can trigger the camera, sometimes from as far away as 1,500 feet (457.2 m).

Handy devices exist that enable the photographer to take the image, manipulate the exposure, change the focus point, modify the white balance, and most other settings on the camera from a remote location (usually within 300 feet (91.4 m) or so). The photographer can see the scene before and after taking the image. You can also download your photograph for a preview on a monitor. Some devices require tethering to a computer while others use Wi-Fi to communicate with the camera. Manufacturers of wireless camera controls include Vivitar, CamRanger, and Canon EOS Utilities.

Arizona-Sonora Desert Museum

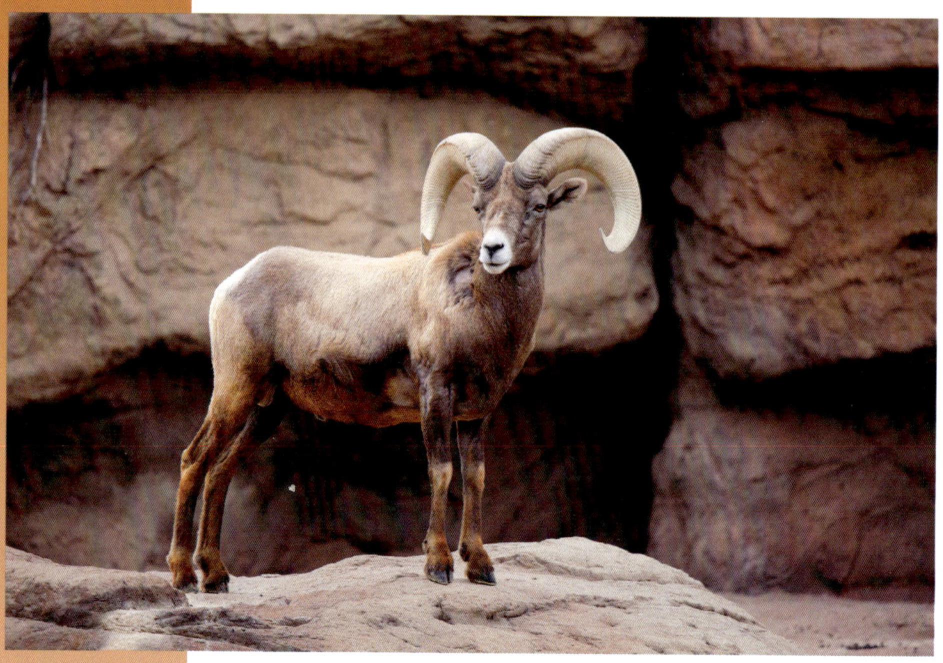

ABOVE: A desert bighorn sheep ram shows off his horns. Canon 5DMII, 70-200mm at 200mm, ISO 800, f/4 @ 1/640 sec.
RIGHT: Endangered Mexican gray wolf. Canon 20D, 500mm, 1.4x teleconverter, ISO 200, f/7.1 @ 1/50 sec.

VIEW TIME
Year-round

IDEAL TIME OF DAY
Early morning and late afternoon

VEHICLE
Any

HIKE
Easy to moderate

Just two miles (3.2 km) from the pristine Tucson District in Saguaro National Park, the Arizona-Sonora Desert Museum showcases over 230 species of native wildlife and 1,200 species of plants spread across 21 acres (8.5 hectares). With approximately two miles (3.2 km) of crisscrossing paths through a maintained desert landscape, this museum belies the classic definition of a white concrete building simply holding and displaying a collection of stuff. Although caged wildlife presents most of the photographic opportunities here, the world-class enclosures will give your image a more wild feeling.

Cat Canyon features excellent displays for gray fox, porcupine, and ocelot, but the manmade canyon walls surrounding the bobcats provide the best photo opportunities. Going about their daily business in full view, most of the time, these small cats sleep and roam only a few yards from the top of the exhibit. While a **short telephoto lens (e.g., 300mm)** will help record frame-filling portrait shots, you may need a **smaller-focal-length lens** to take a full body image. The museum permits **tripods** and **monopods**, so bring one along to keep your camera steady while you photograph many of the critters.

The popular Mountain Woodland exhibit houses endangered Mexican gray wolves, white-tailed deer, Merriam's turkeys, mountain lions, and black bear (all in separate natural-looking habitats). The Mexican gray wolves move throughout their enclosure and stop once in a while to gaze at the humans outside of their fence. Their soft fur and glossy brown eyes give more of a puppy dog appearance than a vicious predator. The mountain lion comes down for water on occasion but, for the most part, sits atop a large

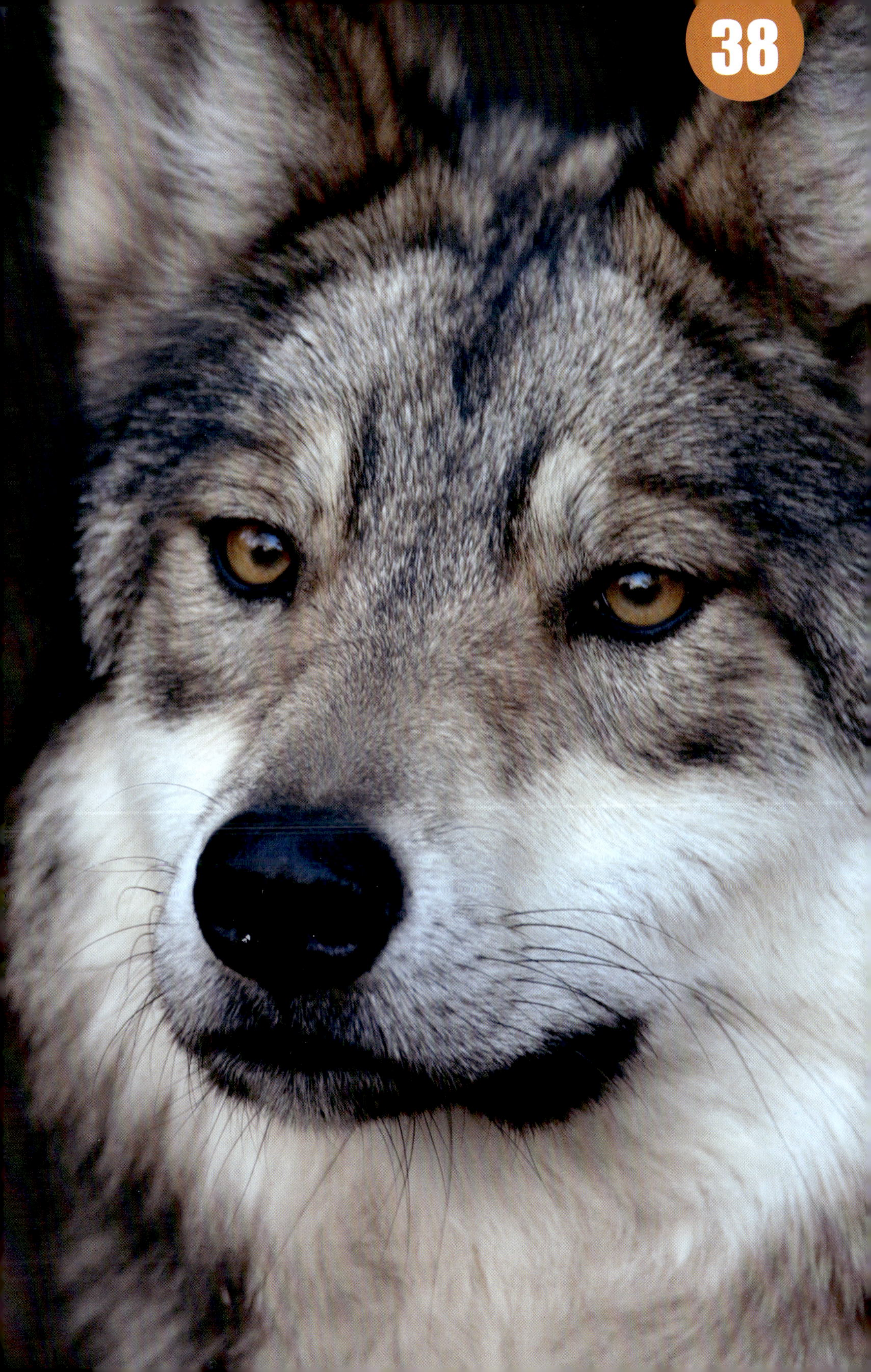
38

Arizona-Sonora Desert Museum

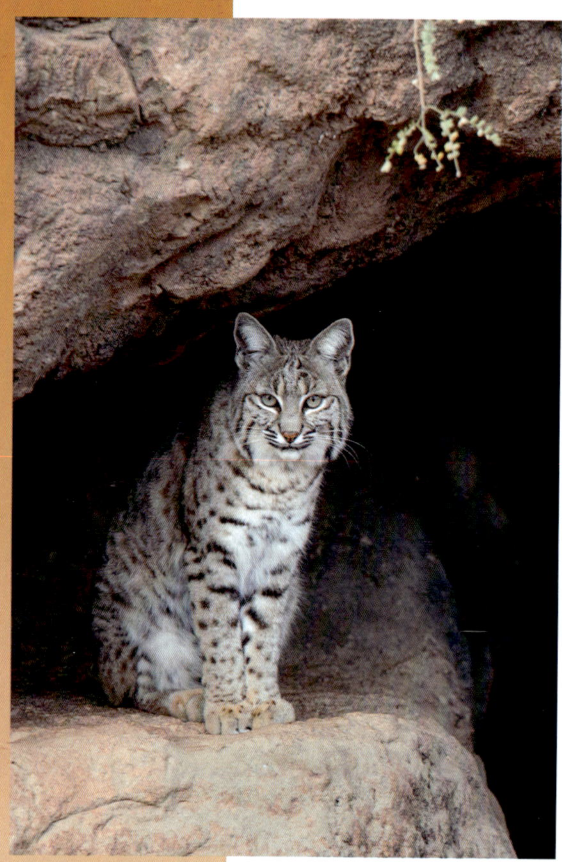

A bobcat lazily watches the visitors at the Arizona-Sonora Desert Museum. Canon 5DMII, 70-200mm at 200mm, ISO 1600, f/5 @ 1/500 sec.

boulder offering a classic western scene. The black bear generally lies in a corner trying to stay out of the heat, while the deer and turkeys are seldom seen.

In the Riparian Corridor, bighorn sheep sit on top of manmade rock. The majestic-looking ram seems very proud of his huge horns and fondly stands in front of his ewes. A short telephoto zoom lens will help record this spectacle. Because of the bright sun, set your tripod aside, and hand-hold your camera instead. In addition to the sheep, river otter play in simulated ponds, while coatimundis and beaver hide behind the vegetation. Several native fish swim lazily in aquarium displays. Use an off camera flash to light the fish when taking their photographs.

The Desert Grassland exhibit houses a small colony of black-tailed prairie dogs. They breed, feed their young, call to each other, and run around—apparently having loads of fun. Their silly antics make them a crowd favorite. Plexiglass surrounds the open air enclosure. Because many dirty-fingered children (and maybe adults too) have left fingerprints, take some cleaning wipes when visiting this display to clean the plexiglass before attempting to shoot. Place your telephoto zoom lens on your tripod, and watch the prairie dogs as they gallivant around their space.

If you have extra time, swing by the hummingbird exhibit. Although spring offers the best time to photograph nesting and other activities, the birds actively buzz around the aviary year-round. The small area does not lend itself for tripod use (fold it up when you enter the exhibit), so flip on your image stabilization (or vibration reduction) and steadily hand-hold your camera. Even though these birds will fly in close proximity to you, follow the action through your telephoto zoom lens.

Before you leave the museum, do not miss the raptor free-flight demonstration (held twice a day in winter). Park rangers guide red-tailed hawks, great horned owls, barn owls, Harris's hawks, and prairie falcons in flight around the grounds within the range of small telephoto lenses. These raptors do not wear ankle bracelets or other adornments, so they will appear wild despite their captive environment. To record sharp images of these flying birds, enable continuous autofocus and high-speed continuous shooting mode. Also, keep your shutter speed in excess of 1/1500 of second.

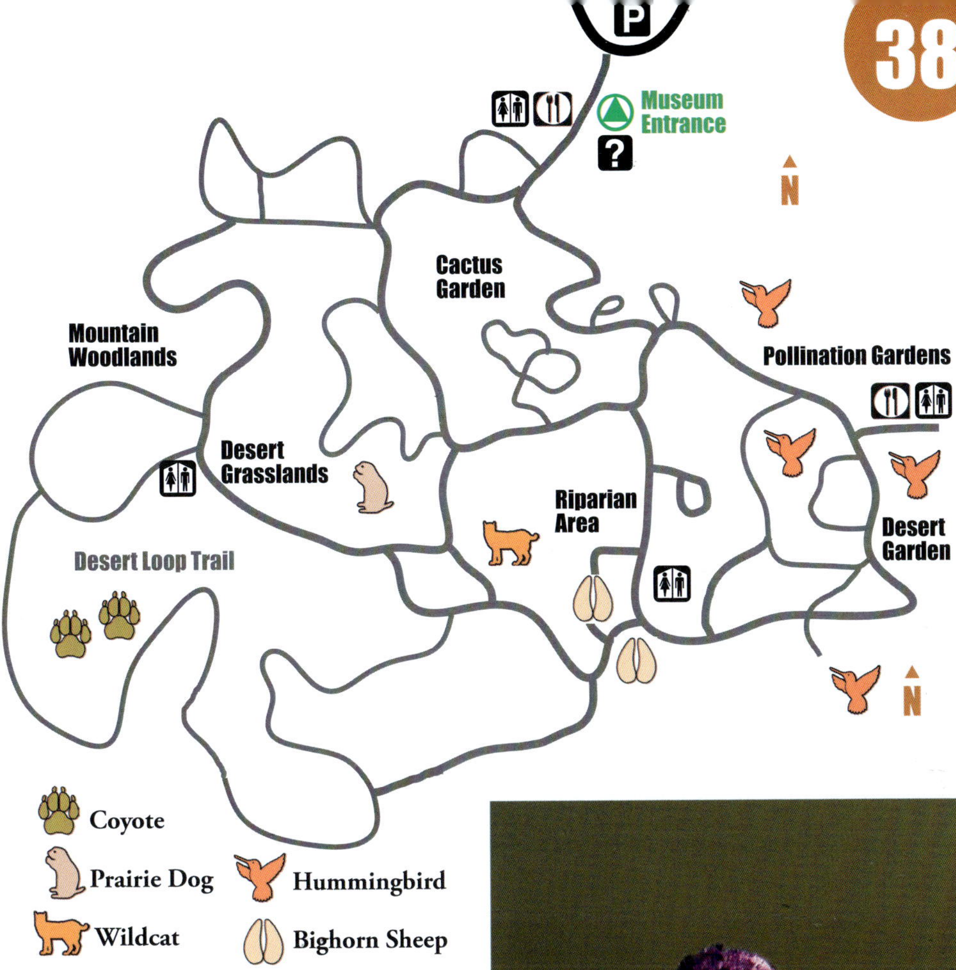

Museum Entrance

N

Cactus Garden

Mountain Woodlands

Pollination Gardens

Desert Grasslands

Riparian Area

Desert Garden

Desert Loop Trail

N

Coyote

Prairie Dog

Hummingbird

Wildcat

Bighorn Sheep

DIRECTIONS:

From Tucson, take Speedway Boulevard west toward the nearby Tucson Mountains. Travel approximately 4.5 miles (7.2 km) from I-10, veering left onto Anklam/Gates Pass Road when the road splits. Drive another 4.7 miles (7.6 km) up and over Gates Pass on Gates Pass Road. When the Gates Pass Road ends on the west side of the mountains, take a right onto Kinney Road. Drive about 4 miles (6.4 km), turning left into the museum's parking area.

Visitors must pay an entrance fee. Museum members are admitted for free. For more information and park hours, visit **www.desertmuseum.org**.

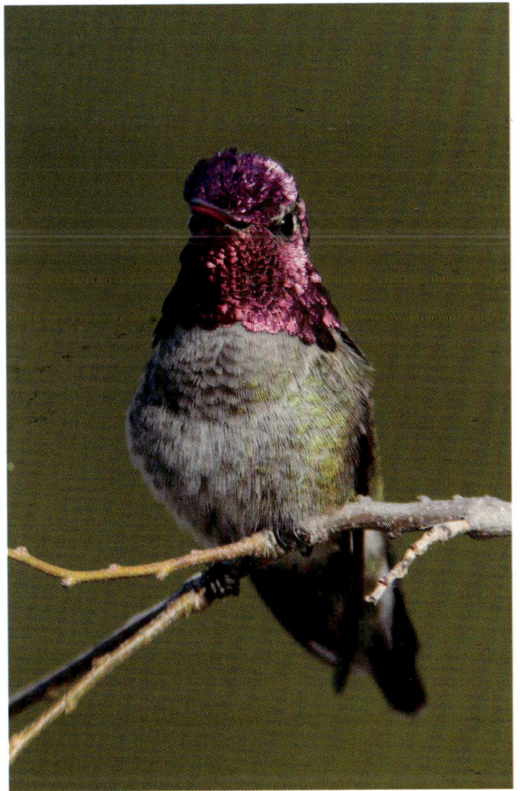

Anna's hummingbird flashes its brightly colored head to ward off intruders. Canon 20D, 500mm, 1.4x teleconverter, ISO 200, f/7.1 @ 1/400 sec.

Collaboration

Even though I have spent most of my adult life researching and managing wildlife in Arizona and around the world, my field expertise with wildlife is relatively limited and narrow in scope. I know a lot about the animals that I have photographed, but there are others where I have less knowledge.

To fill the gap and find new photo adventures, I talk to the experts. If I am attempting to hone my skills in insect photography, I will not only research on the web, but also talk with specialists in the field of insect biology. If I am unfamiliar with a certain location, I look at maps to become familiar, but I will also find someone that has previously visited that spot—perhaps a hunter, camper, research biologist, or the like.

A few years ago, while discussing bird nest locations with a friend, I discovered that Dr. Courtney Conway at the University of Arizona was conducting research on nesting birds throughout southeast Arizona. Dr. Conway had hired several students and ornithologists to survey different habitats in order to locate bird nests and determine the specific habitat requirements, success rates, and other biological points of interest. I contacted Dr. Conway and asked if he would allow me to have access to the nest sites. In return, I offered to let him use any of my images for his research papers, website, or for whatever he needed them. We agreed, and I began to accompany his researchers into the field to locate nests.

His workers were masters at finding bird nests. In late May, we were searching for nests near Colossal Cave Mountain Park when one of the biologists located a Bell's vireo nest. As usual for the species, the nest sat in a thicket of thorny vegetation perched on a flimsy limb of a mesquite tree. Four babies occupied the nest, and both the male and female parents actively fed them. I had all of my camera equipment with me except for a blind. Since I did not know when I would return to the area, I decided to stay and photograph the site. The hatchling Bell's vireos leave the nest in 10-12 days, so time was of the essence.

Unfortunately, although shade covered the nest, I had to stand in the glare of the hot May sun to achieve the best angle. The adult birds seemed oblivious to my presence and continued to feed the young the entire time I baked in the hot sun.

The nest rested about 20 feet (6.1 m) from my location. In order to take frame-filling images, I used my 500mm lens and a 1.4x teleconverter mounted on my sturdy tripod and gimbal head. Given the amount of shade and mottled light on the nest, I decided to use some light fill flash.

The large and heavy nestlings caused the nest to droop at a bad angle for the birds, but at an angle that opened up the entire clutch to my camera. The parent birds flew quickly in and out of the nest, offering me only a frame or two each time they visited. Just once over those three hours that I cooked in the baking sun did both attend the nest at the same time. The male brought in a food item and handed it to the female, while the hungry hatchlings waited with gaping mouths for the morsel. I only recorded three frames before the male left, two of which appeared soft and out of focus

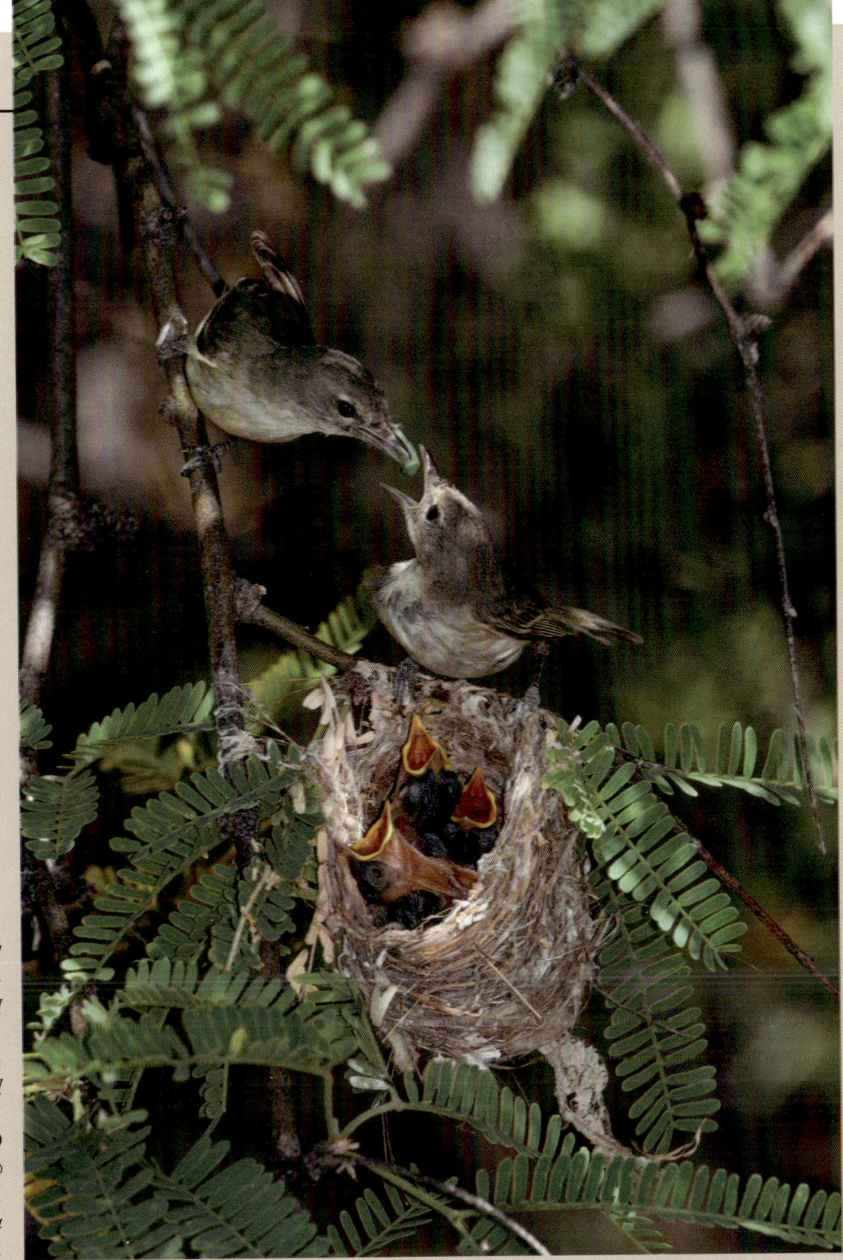

Mother and father Bell's vireo feed their young. Canon Digital Rebel XTi, 500mm, ISO 200, f/10 @ 1/200 sec., on-camera flash.

due to movement of the flimsy branch the nest rested upon. However, one frame appeared well framed, properly composed, perfectly exposed, and depicted a natural occurrence few people have seen.

Although I took hundreds of images of several bird species, this image represented the best I was able to make during the two months that I spent with Dr. Conway's workers. Even more important to me, simply spending time in the field with these hard-working biologists and learning just a fraction of what they knew about bird nests and biology will stay with me forever.

SOUTHERN ARIZONA

The Pond at Elephant Head

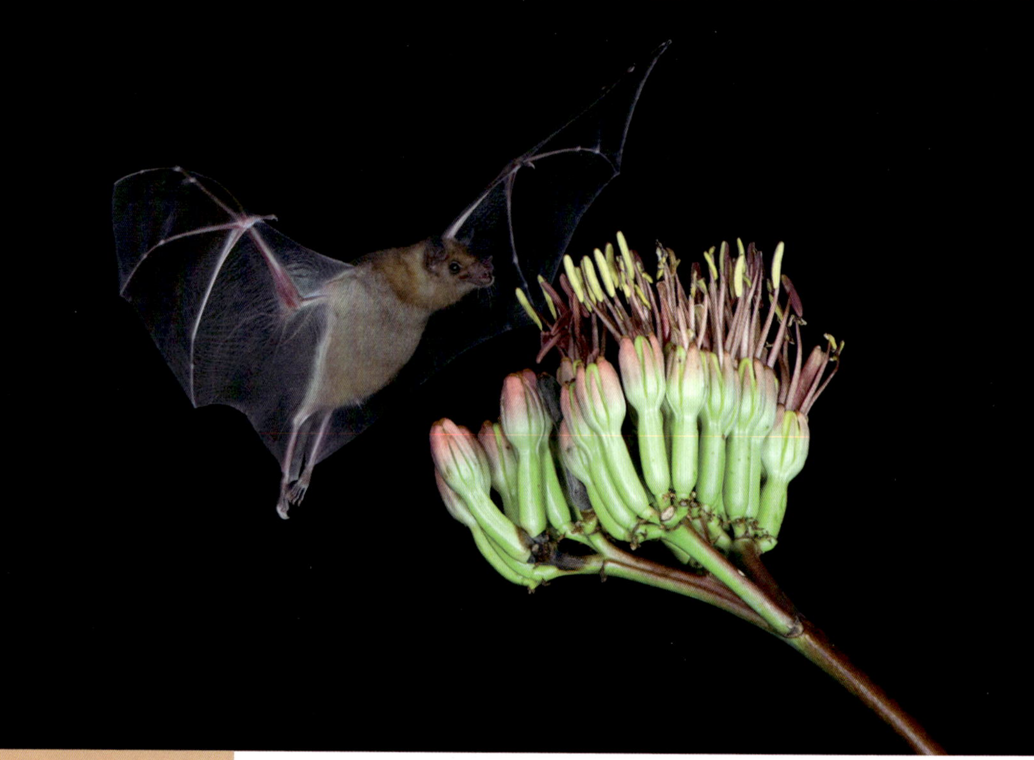

A lesser long-nosed bat feeds at agave flower. Canon 1DMII, 100-400mm at 310mm, ISO 200, f/13 @ 22 seconds, Bill Forbes' bat set-up.

VIEW TIME
May to early October

IDEAL TIME OF DAY
Night

VEHICLE
Any

HIKE
Easy

Bill Forbes purchased this property near Amado with the intent of developing a wildlife sanctuary just for photographers. The pond, blinds, perches, hummingbird feeders, remote trips, and everything else is there only for photographers. Set in an upper Sonoran desert habitat, his property features all of the desert birds, a friendly roadrunner, a couple of small mammals, and a lot of bats.

Unless you are an expert on finding bats or can work with someone in possession of special permits, bat photography is an ordeal at best—unless you visit with Bill. Formally called the Pond at Elephant Head, this is the only facility in Arizona set up for bat photography and available for general public usage.

You bring your **telephoto zoom lens**, **camera**, sturdy **tripod**, and a locking **cable release**, and Bill supplies the rest for you. He offers all of the high-speed flashes, infrared trips, comfortable seating, and even a television screen on which to watch the action.

From May through late August, several species of bats come to drink water out of the pond. After you position your camera, set it on manual mode. Set your ISO speed to 400, your aperture to f/10, and your shutter speed to 20 seconds.

After dark, lock down the shutter button on your cable release, which triggers the camera to take 20-second exposures over and over again. When

DIRECTIONS:

From Tucson, travel south on I-19 approximately 35 miles (56.3 km). Take Exit 56 for Canoa Road. Turn left and go under the freeway, and then turn right at the stop sign on the frontage road. Travel 2.5 miles (4 km), and then turn left on Elephant Head Road. Continue 2 additional miles (3.2 km) until the road ends, where you will make a sharp left turn onto Canoa Road. Travel 0.5 miles (0.8 km), going through the S-curve. Travel 0.5 miles (0.8 km) and then turn right on West Dove Way. Go one mile (1.6 km). The paved road will turn to gravel. Turn left at 1500 West Dove Way, and veer to the left until the driveway ends.

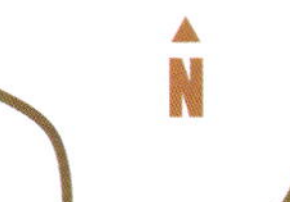

The Pond at Elephant Head requires a formal appointment to visit. For more information, visit **www.phototrap.com/pond.htm**.

the bat takes a drink from the pond, it will pass through an infrared trip that will set off the flashes, and because the camera's shutter remains open, the illuminated bat will appear in your exposure.

In late August, both lesser long-nosed and Mexican long-tongued bats begin to visit the pond, but they are not there to drink water. As the only nectar-feeding bats in Arizona, they feed mainly on the nectar of agave plants. Bill grows agave flowers in his yard and sets up infrared trips around them. When the bat comes to feed from the flower, they break the beam and set off several high-speed flashes enabling the moving creatures to appear in your frame.

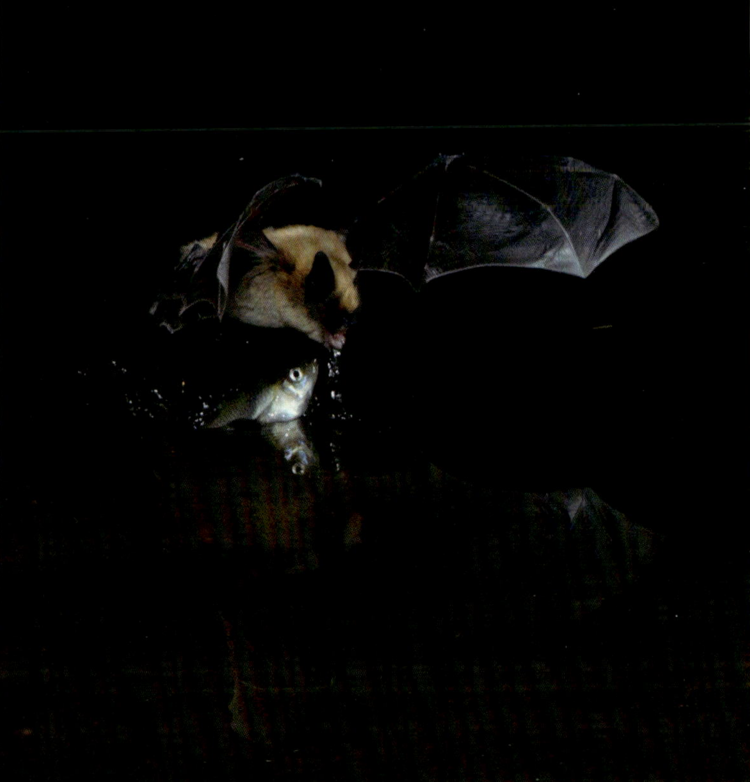

A myotis bat drinks and scares a fish. Canon 1DMIV, 24-105mm at 105mm, ISO 400, f/18 @ 20 sec., Bill Forbes' bat set-up.

SOUTHERN ARIZONA

Santa Rita Lodge

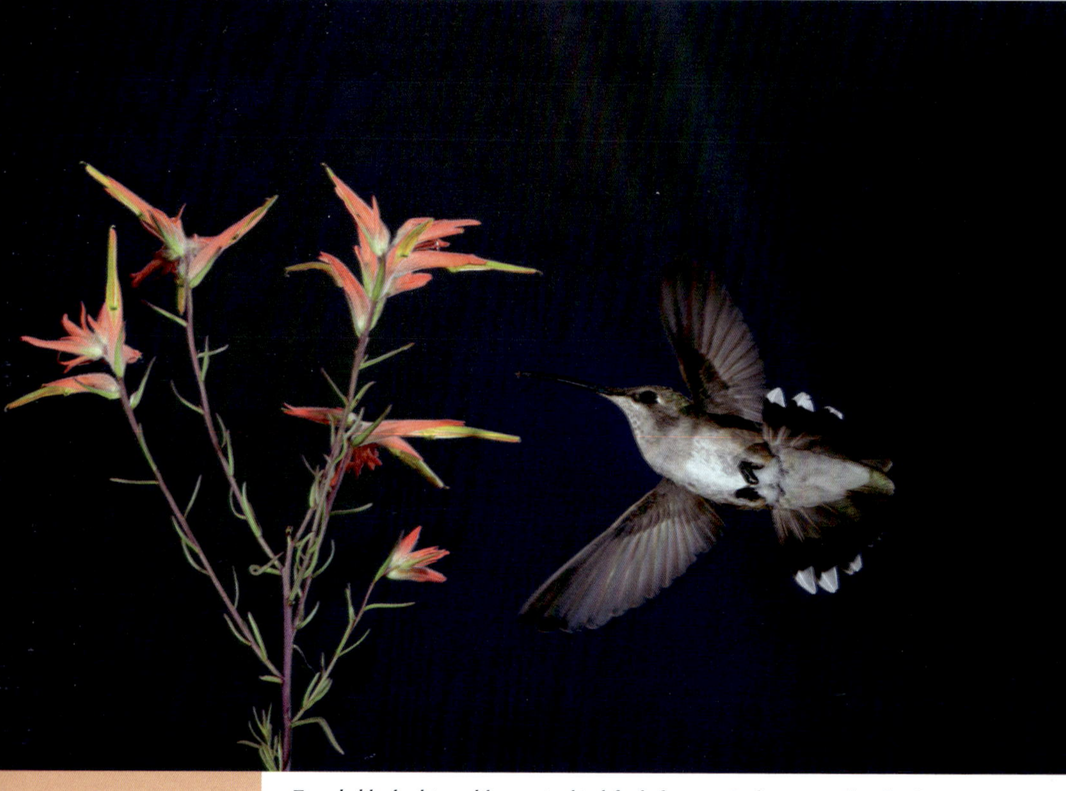

Female black-chinned hummingbird feeds from an Indian paint brush. Canon 1DMIV, 70-200mm at 102mm, 2x teleconverter, ISO 400, f/20 @ 1/320 sec., high-speed hummingbird set-up.

VIEW TIME
March to September

IDEAL TIME OF DAY
Sunrise to late morning; early afternoon to sunset

VEHICLE
Any

HIKE
Easy

Shutterbugs can stay and play at the Santa Rita Lodge! C.R. Dusenberry built the Santa Rita Lodge in 1922 so Tusconians could get out of the heat and enjoy horseback trips to the top of the Santa Rita Mountains. The property caters specifically to birders, bird photographers, and nature lovers. Five rustic cabins and eight casitas sit peacefully adjacent to Madera Creek and next to some of the best wildlife habitat in the Santa Rita Mountains.

Jam-packed with bird seed feeders, hummingbird feeders, suet feeders, thistle feeders, a small water feature, and plenty of shade, the lodge attracts a variety of birds like Gould's turkeys, woodpeckers, jays, lesser goldfinches, screech owls, white-breasted nuthatch, bridled titmouse, and many more. In addition, a pair of nesting elf owls perch themselves frequently atop the telephone pole next to the parking lot. During the spring migration and early summer breeding season, the number of bird species skyrockets to include warblers, painted redstarts, several species of hummingbirds, wrens, flycatchers, grosbeaks, and other sought-after birds of a feather.

However, as the number of animals increases, so does the number of human visitors. The area in front of the feeders fills up with hopeful birders and photographers, and rooms become difficult to book, so plan your outing well in advance.

Offering unmatched hummingbird photography opportunities, feeders surround each cabin and casita. The lodge owners allow guests to use whatever high-speed equipment necessary to obtain the best images. Bring your lightning-fast **flashes**, cameras, and backdrops (see "Hummingbird Photography" on page 154).

Prepare to shoot at the crack of dawn, when these hungry jewels of the bird world arrive for their first drink. In a good year, expect to see up to seven species of hummingbirds visit the feeders. Work at the feeders until around 10 a.m., when the hummingbirds take a siesta or travel to the many flowers in and around the canyon. After your own siesta, begin photographing again around 3 p.m. Stay at the hummingbird station until almost dark, as they come in for their last drink before finding a bush in which to sleep the night away.

The feeding area attracts the greatest number and diversity of birds, but because of its positioning, bring a **long telephoto lens (e.g., 500 or 600mm lens**, if available, and a **teleconverter**). Occasionally birds will land on the surrounding trees and pose for the observant photographer which yields more natural-looking images than birds sitting on feeders.

From March until July, Gould's turkey toms (the boys) intently attract the hens (the girls) to their flock—and the toms will take a lot of hens! Photograph the tom turkeys as they strut and gobble in front of the seemingly uncaring hens. Get within about 10 feet (3.1 m) of these show-offs with your telephoto zoom lens. When the turkeys roam the oak tree forests surrounding the lodge, watch for the hot spots created by the mottled light. Fill flash will help even out the lighting, or simply wait for the flock to move into better light.

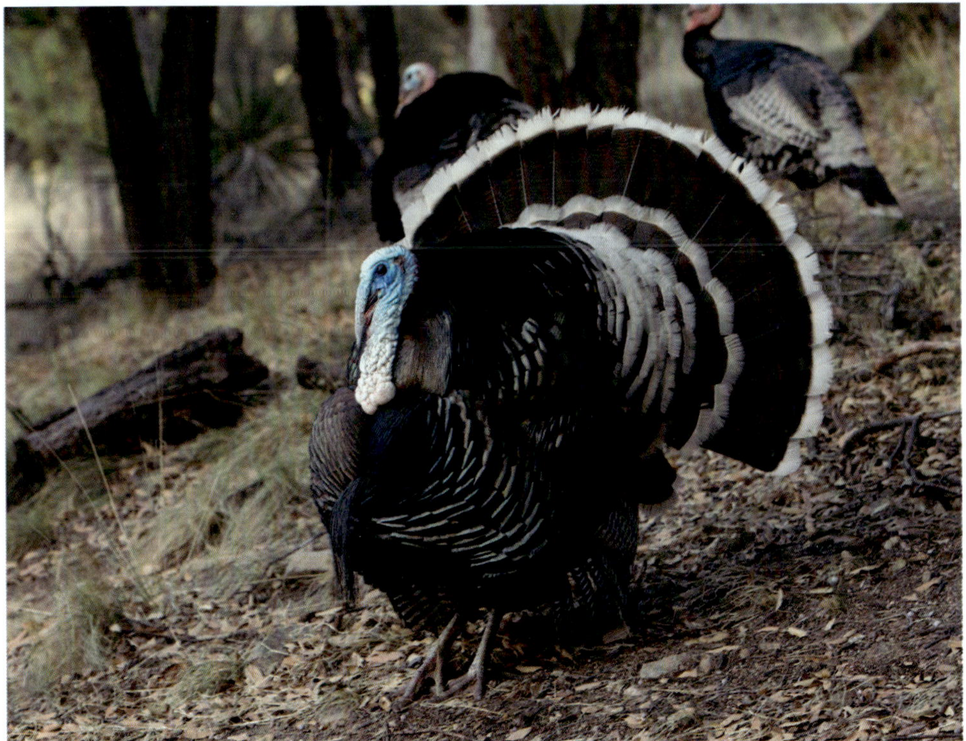

A Gould's turkey tom struts his stuff to attract hens to his harem. Canon 1DMIV, 70-200mm at 102mm, ISO 800, f/5.6 @ 1/160 sec.

Santa Rita Lodge

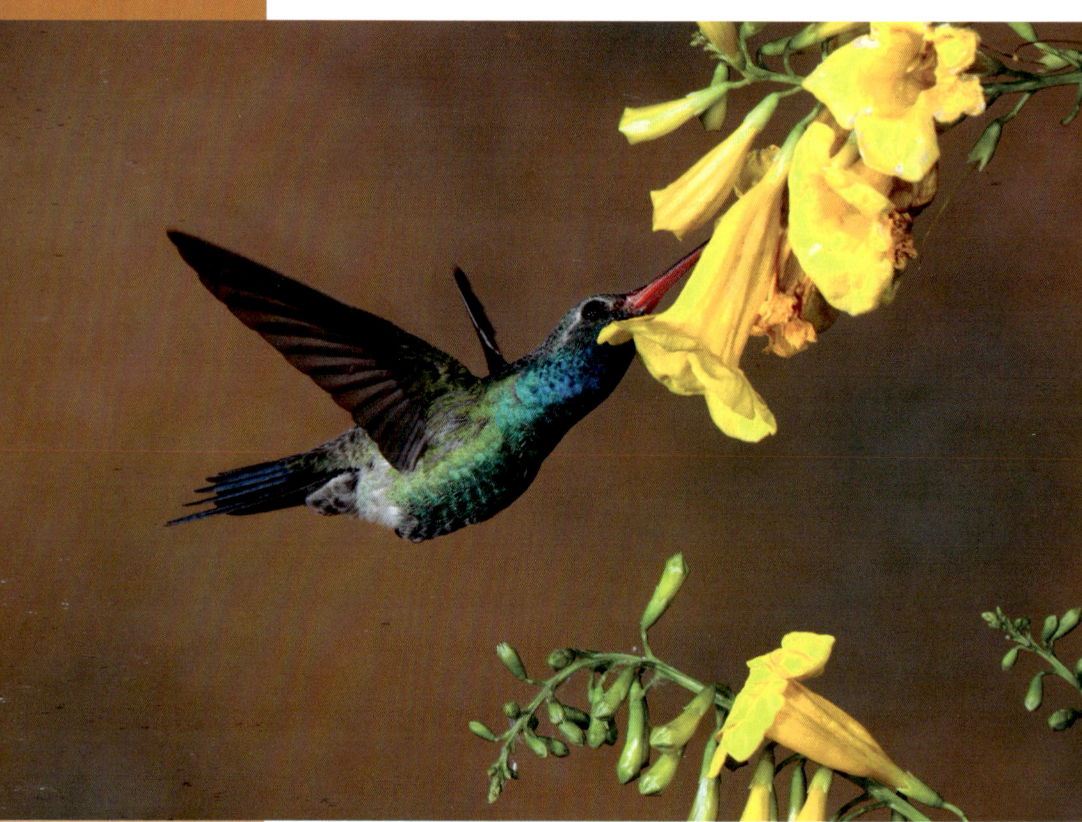

A broad-billed hummingbird feeds from a tecoma flower. Canon 50D, 100-400mm at 285mm, ISO 200, f/16 @ 1/200 sec., using high-speed hummingbird set-up.

Just after sunset, the resident pair of elf owls starts moving from their nest in the telephone pole to the surrounding oak trees. Easy to see but difficult to photograph, use an elf owl call to mimic their territorial songs. Work the calls as described in the "Bird Calls" tip (see page 74) to get close to them. Because the owls are not very vocal during or after the nesting season, the best month to photograph elf owls is in April when they first arrive in Arizona.

When you find a photogenic subject, use your telephoto zoom lens (e.g., 100 to 400mm) to fill the frame with the animal. In manual camera mode, start with an aperture of f/8, and set the shutter speed to match your flash's maximum sync speed. Have a friend hold a flashlight so that you can focus on the owl using autofocus. Alternatively, use a headlamp or suspend a light below the lens so as to illuminate the bird for focusing. A wary owl holds little patience. When the opportunity arises, shoot quickly on continuous high-speed shooting mode as much as the owl will allow.

Western screech owls also inhabit the lodge property. To photograph this species, use the same techniques as outlined above for elf owls (including the time of year). Western screech owls respond more aggressively to the calls and, therefore, are easier to locate and photograph.

41

DIRECTIONS:

From Tucson, follow I-19 south to Exit 63 for Continental Road. Turn left (east) and travel 1.1 miles (1.8 km). Turn right onto Whitehouse Canyon Road. After 0.9 miles (1.5 km), Whitehouse Canyon Road merges with Madera Canyon Road so continue on Madera Canyon Road. Travel another 12.2 miles (19.6 km) to the Santa Rita Lodge.

For reservations and information, visit **www.santaritalodge.com**.

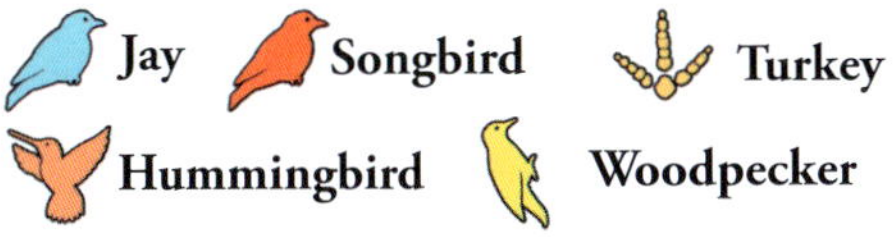

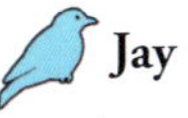 **Jay** 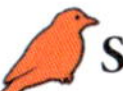 **Songbird** 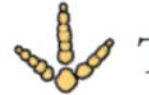 **Turkey**

Hummingbird 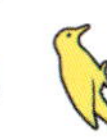 **Woodpecker**

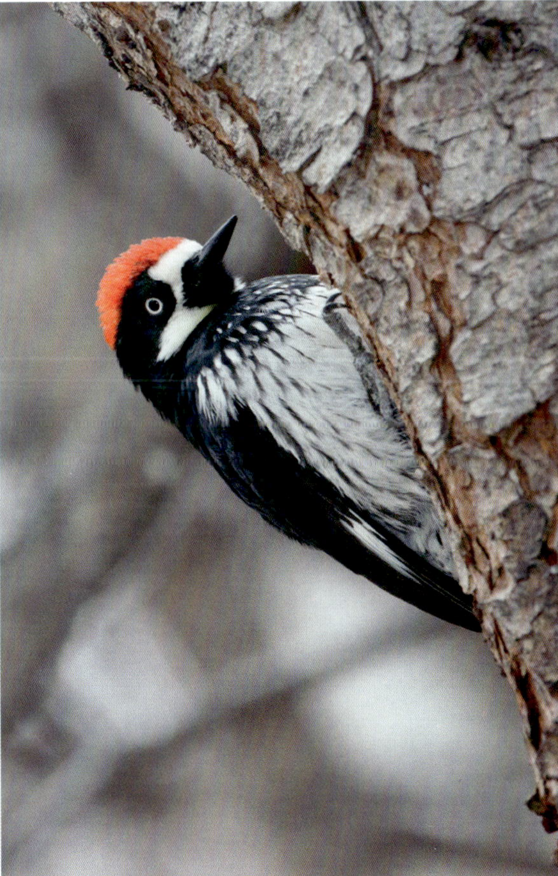

The comical clown face of an acorn woodpecker. Canon 1DMII, 500mm f/4, ISO 400, f/9 @ 1/40 sec., on-camera flash.

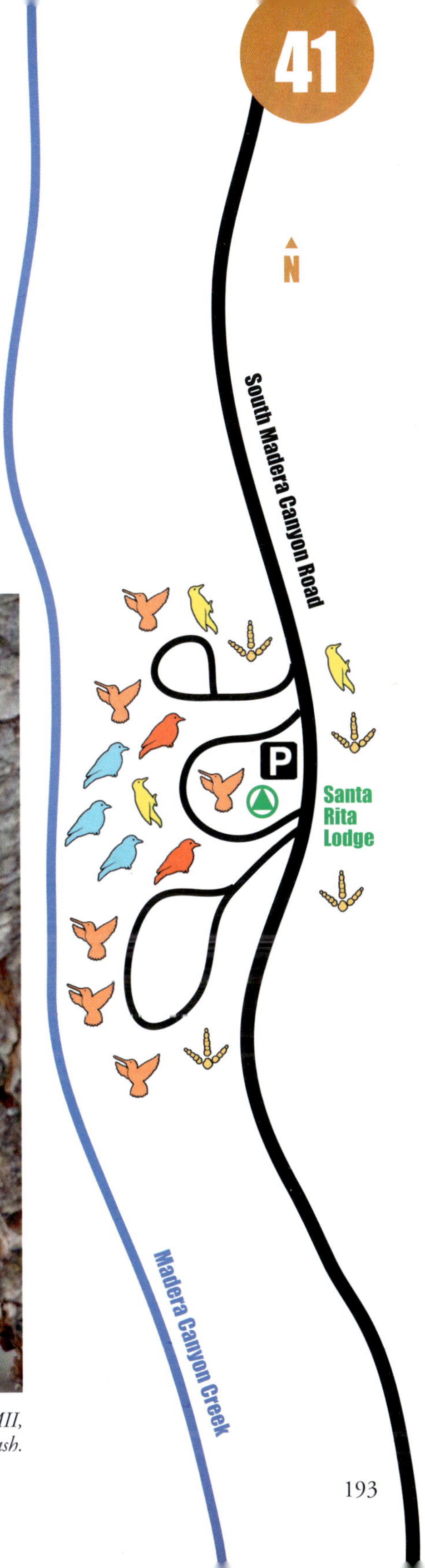

N

South Madera Canyon Road

P

Santa Rita Lodge

Madera Canyon Creek

Madera Canyon

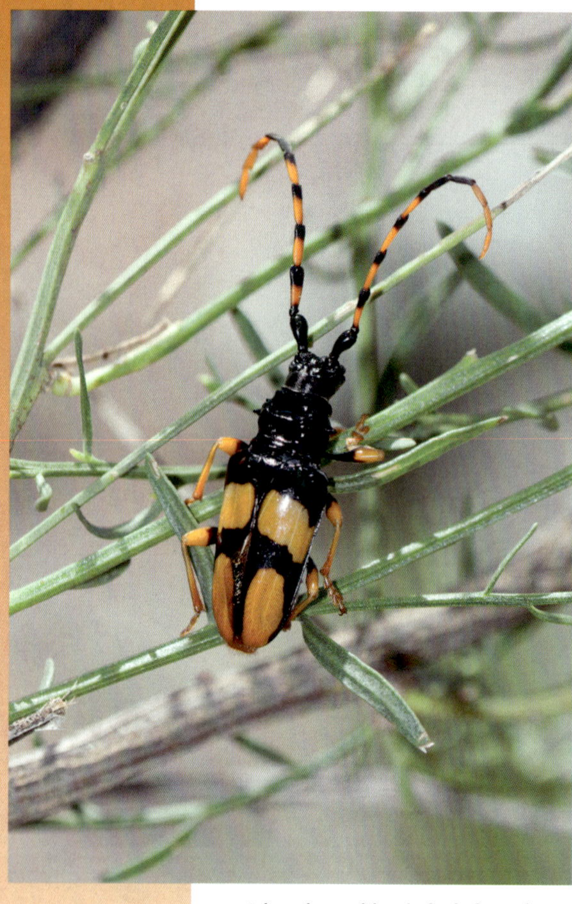

A long-horned beetle feeds from the sap of a desert broom. Canon Digital Rebel XTi, 100-400mm at 135mm, 500D macro, ISO 200, f/10 @ 1/200 sec., off-camera flash @ -2/3 FEC.

VIEW TIME
April to October

IDEAL TIME OF DAY
Sunrise to sunset; night

VEHICLE
Any

HIKE
Easy to moderate

Known as one of the Sky Islands of southeastern Arizona, this scenic location flaunts a large number of endemics and a high diversity of species. Photographers, bird watchers, insect enthusiasts, and others interested in the natural world come to Madera Canyon to see and enjoy the unique habitats and animals.

From April to June, bird photography takes center stage. At this time, migrant warblers, hummingbirds, several species of flycatchers, elf owls, elegant trogons, black-headed grosbeak, painted redstarts, white-winged dove, and more travel through the Santa Rita Mountains on their way to northern summer breeding grounds (or return there from their winter in the south to set up home). In addition, many of the resident birds, such as Gould's turkey, Mexican jays, ladder-backed woodpeckers, Arizona woodpeckers, acorn woodpeckers, Western screech owls, white-breasted nuthatch, bridled titmouse, and many more feathered creatures, become interested in reproduction, nesting, and raising their young here.

To make the most of photographing birds in Madera Canyon, take your **longest telephoto lens**, **fill flash**, and sturdy **tripod**. Use your binoculars to scout for birds and try your bird calls to either bring them in range or pinpoint their location. If the creek appears dry, scout the area to determine if some birds have a different drinking or bathing spot. This type of photography pits the skills of man against the evolutionary behavior of birds. A successful photography outing takes all of the photographer's skill at bird behavior, sneaking, and camouflage.

When the monsoons begin, the canyon comes alive with moths, butterflies, and all manners of insects. During the day, wander around the foothills checking out desert broom plants for bugs. At night, use a black light or mercury vapor lamp to attract flying insects. If you turn on the lights and place a white bed sheet on the ground next to the light, one can draw even more flying bugs. On a good night, hundreds of insects including praying mantis, moths, walking sticks, wasps, and hordes of prehistoric-looking creatures will settle on your bed sheet.

Move the insects to some nearby vegetation before photographing them, or better yet, collect them to photograph at a later time in a controlled setting.

To photograph small insects, hand hold the camera with a **macro lens**, and either use fill or full flash (depending on shutter speed). If shooting at night, place the camera on manual mode, and set the aperture at f/16 and the flash sync speed for your camera. Use continuous autofocus. Turn your flash to TTL. Diffuse your flash to reduce the harsh shadows between the insect and the setting on which you have placed it.

During the day, set the camera on aperture priority mode, and then change your aperture to f/16 and the flash sync speed. Set the lens on continuous autofocus. Then, put your fill flash on TTL high-speed sync and underexpose by one to two stops.

DIRECTIONS:

From Tucson, take I-19 south for approximately 23 miles (37 km). Take Exit 63 for Continental Road. Turn left and drive 1.1 miles (1.8 km). Turn right onto White House Canyon Road. After 0.9 miles (1.5 km), White House Canyon Road merges with Madera Canyon Road. Travel another 6 miles (9.7 km) to Proctor Road the base of the Santa Rita Mountains.

Visitors need a federally-issued "America the Beautiful" annual pass or a Coronado Recreational Pass. For more information, visit the Coronado National Forest website at **www.fs.fed.us/r3/coronado**.

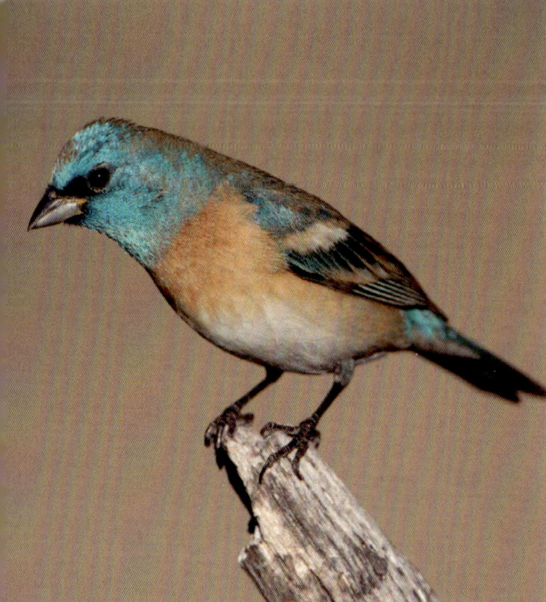

Lazuli bunting. Canon 50D, 500mm, 1.4x teleconverter, ISO 200, f/6.3 @ 1/1250 sec.

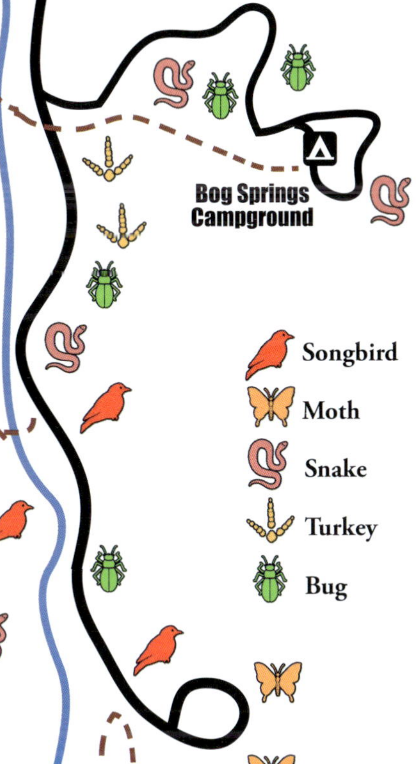

Focus Stacking

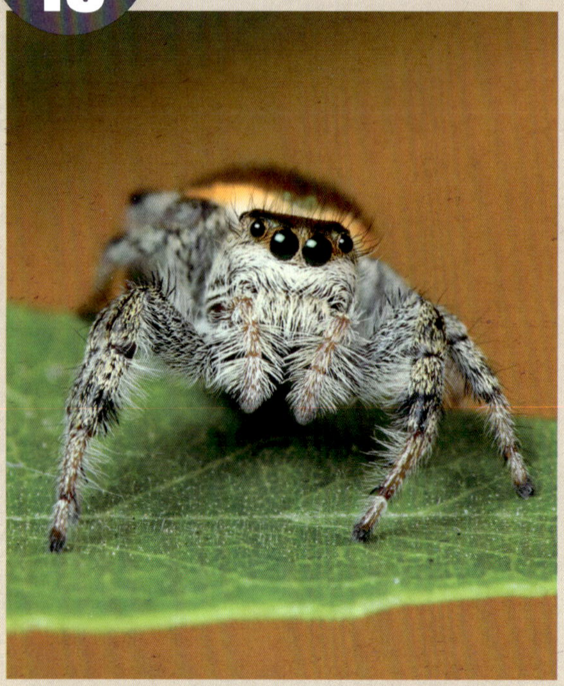

Final image of emerald jumping spider created using ten stacked images. Base exposure: Canon 70D, 180mm macro, 1.4x teleconverter, ISO 400, f/11 @ 1/200 sec., off-camera flash.

When photographing macro subjects, obtaining just the right depth of field can challenge any photographer. Because of the lens type and the close focusing proximity, depth of field ranges appear tiny. For example, if you use a 100 mm macro lens on an APS-C camera and you focus at 12 inches (30.5 cm) with an aperture of f/16, the depth of field is 0.16 inches (0.4 cm) from 11.92 to 12.08 inches (30.3 cm to 30.7 cm). Handy tools for quickly determining depth of field is the Depth of Field (DOF) Master website (**www. dofmaster.com**) or depth of field apps available for purchase and download to your smartphone.

To keep as much of the subject in sharp focus as possible without relying upon small apertures, focus on the important elements of the subject (e.g., for wildlife, the animal's eye), and position your lens parallel to that object. For example, if you want the entire body of a butterfly in focus, arrange your camera such that the butterfly's open wings are perpendicular to the camera lens. By having the body of the subject parallel to the body of your camera, you can sharply focus on both of the wingtips simultaneously.

However, if you want only the eye of a lizard in sharp focus, for example, and you do not care about whether the rest of the lizard appears in focus, then do not worry about the camera's position relative to the lizard's body. Simply focus on the eye, check your depth of field with the depth of field preview button (if available) and shoot.

When using apertures greater than f/16, watch out for diffraction. Diffraction is the bending of light as it passes through a small opening. When using very small apertures (f/32 for example) the bending of the light can case the image to loose focus.

When a photographer cannot obtain the desired depth of field using classic photographic techniques, multiple-image focus stacking can help extend the apparent depth of field in a final image. With this method, a photographer takes several photos of a subject at various focus points and then later combines all of the photos using specialized computer software like Zerene Stacker and Helicon Focus.

For instance, perhaps you wish to take a photograph of a jumping spider. In doing so, you want at least the area from the front legs to the eyes in sharp focus. The spider (a very large one)

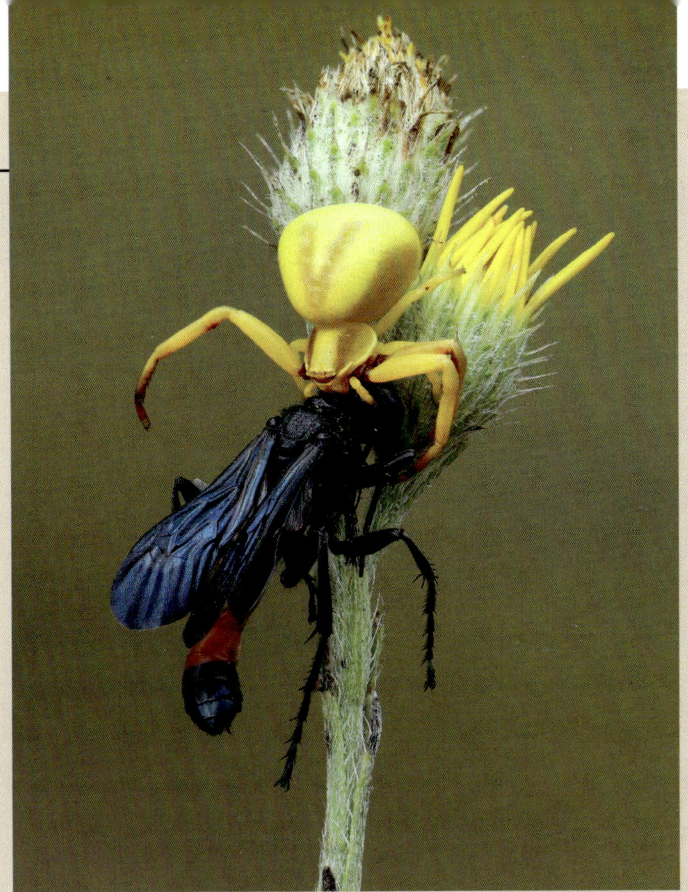

This crab spider held on to its wasp prey and stayed still long enough for me to take 36 images for the focus stacking process. Base exposure: Canon 7DMII, 180mm macro, ISO 200, f/7.1 @ 1/200 sec.

is about 0.5 inch (1.3 cm) long. The distance from the front of the legs to the eyes is 25% of that distance, or about 0.15 inch (0.4 cm). To obtain the appropriate field of view, you use a 100 mm lens on a cropped frame camera and an aperture of f/16. According to DOF Master, your close-focus falls about 12 inches (30.5 cm) away, so the effective depth of field equals about 0.16 inch (0.4 cm). If you focus on the front leg, the eyes will not appear in focus. If you focus on the eyes, the front legs will not appear in focus.

With the focus stacking technique, take one photo with the front legs in focus, one photo focused about halfway between the front legs and the eye, and one photo with the eyes in focus, and then combine the three images in image stacking software. The final image will yield sharp focus from the front legs to the eyes. To get more of the spider in focus, take more photos for the software to manipulate.

When using this technique, use a solid tripod setup. If you are not using a flash, and if your shutter speed is below about 1/250 of a second, use both mirror lockup and the two-second timer to help prevent movement from mirror slap. Also, set a two-second timer or remote to help reduce movement resulting from your finger pressing the shutter button. If you are using a flash, then mirror lockup and a two-second timer are not as important. When you use a flash, your effective shutter speed equals the duration of the flash (between 1/500 and 1/1500 of a second).

In the case of extremely small subjects (like insect faces or eyes), using the focusing ring on the lens does not work (the changes in focus distances are too small). For this type of focus stacking, tap into focusing rails, where you can achieve precise movements in between shots. In addition, plan to take in excess of 100 images to produce the best final image with the desired depth of field.

Fort Huachuca

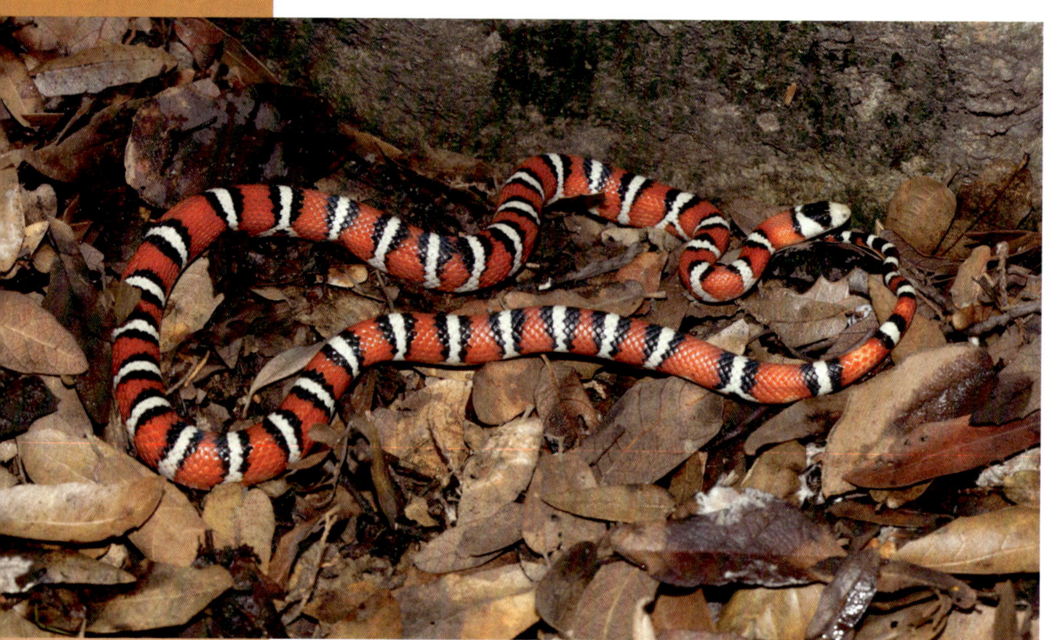

Sonoran mountain kingsnake nestles in the leaves in a wash near Fort Huachuca. Canon 5DMII, 100mm macro, ISO 500, f/13 @ 1/200 sec., on-camera fill flash.

VIEW TIME
April to October

IDEAL TIME OF DAY
Early morning and late afternoon

VEHICLE
2WD high-clearance

HIKE
Easy to moderate

The army established Fort Huachuca in the late 1800s to protect settlers and travelers from the local Apache tribes. Today, the fort serves as a major military installation housing the United States Army Training and Doctrine Command and the Army Intelligence Center. Located at the base of the Huachuca Mountains, the habitat ranges from pristine grasslands to oak woodland and pine forest. The high biodiversity of the Huachuca Mountains provides for some of the greatest wildlife photographic opportunities in Arizona.

The list of butterflies fluttering around Garden Canyon is extremely long, but expect to find Arizona sister, two-tailed swallowtail, black swallowtail, southern dogface, several skippers, American and painted ladies, orange sulphur (and other sulphurs), hairstreaks, white admiral, and many more. Begin looking for the butterflies at the base of Garden Canyon in the prairie grassland, and continue exploring past the third campground.

Hand-hold your camera with a **telephoto zoom lens** with a **flash** attached (in case the butterfly flies into the shade or positions itself with the sun to its back). Set the camera on aperture priority or program mode (one of the few times I use the program mode is when I use fill flash in bright sun). Put the flash on high-speed sync. Keep the shutter speed above 1/1000 of a second and your aperture around f/10 or wider. Get low to the ground to place your subject against the blurred and out-of-focus background to draw greater attention to your butterfly.

In Huachuca Canyon, spot Coues white-tailed deer, sulphur-bellied flycatchers, elegant trogons, Gould's turkey, Mexican jay, woodpeckers, flycatchers, and several warblers. Drive the lower portion of Huachuca Canyon (the paved portions of the road) and keep an eye out for deer,

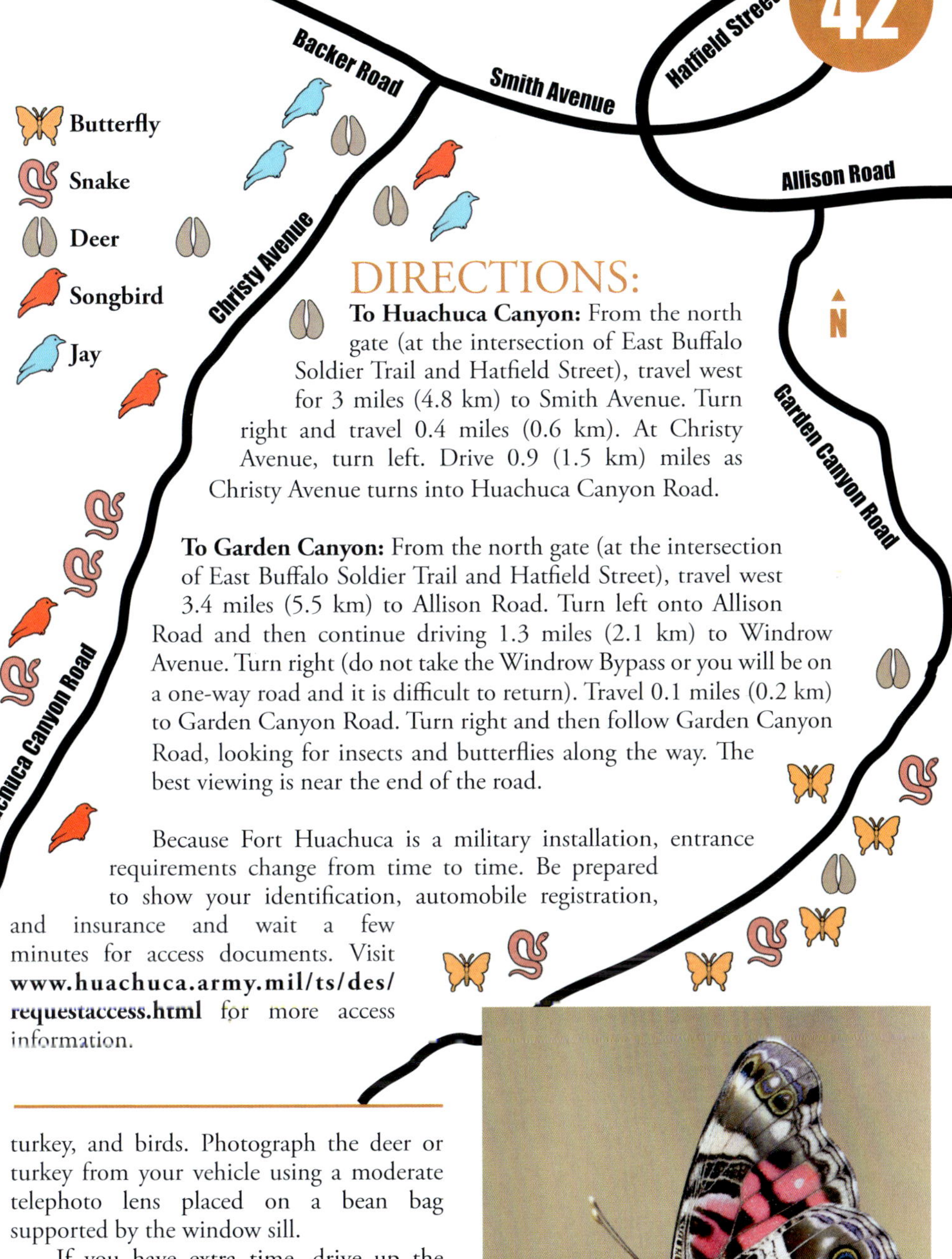

DIRECTIONS:

To Huachuca Canyon: From the north gate (at the intersection of East Buffalo Soldier Trail and Hatfield Street), travel west for 3 miles (4.8 km) to Smith Avenue. Turn right and travel 0.4 miles (0.6 km). At Christy Avenue, turn left. Drive 0.9 (1.5 km) miles as Christy Avenue turns into Huachuca Canyon Road.

To Garden Canyon: From the north gate (at the intersection of East Buffalo Soldier Trail and Hatfield Street), travel west 3.4 miles (5.5 km) to Allison Road. Turn left onto Allison Road and then continue driving 1.3 miles (2.1 km) to Windrow Avenue. Turn right (do not take the Windrow Bypass or you will be on a one-way road and it is difficult to return). Travel 0.1 miles (0.2 km) to Garden Canyon Road. Turn right and then follow Garden Canyon Road, looking for insects and butterflies along the way. The best viewing is near the end of the road.

Because Fort Huachuca is a military installation, entrance requirements change from time to time. Be prepared to show your identification, automobile registration, and insurance and wait a few minutes for access documents. Visit **www.huachuca.army.mil/ts/des/requestaccess.html** for more access information.

---

turkey, and birds. Photograph the deer or turkey from your vehicle using a moderate telephoto lens placed on a bean bag supported by the window sill.

If you have extra time, drive up the canyon on the unpaved sections. Stop at the different picnic areas and look for lizards and snakes. Also, listen for the soft calls of the elegant trogon.

American lady butterfly. Canon 7D, 70-200mm at 135mm, 2x teleconverter, ISO 400, f/8 @ 1/250 sec., on-camera flash @ -2/3 FEC.

BEATTY'S Miller Canyon Guest Ranch

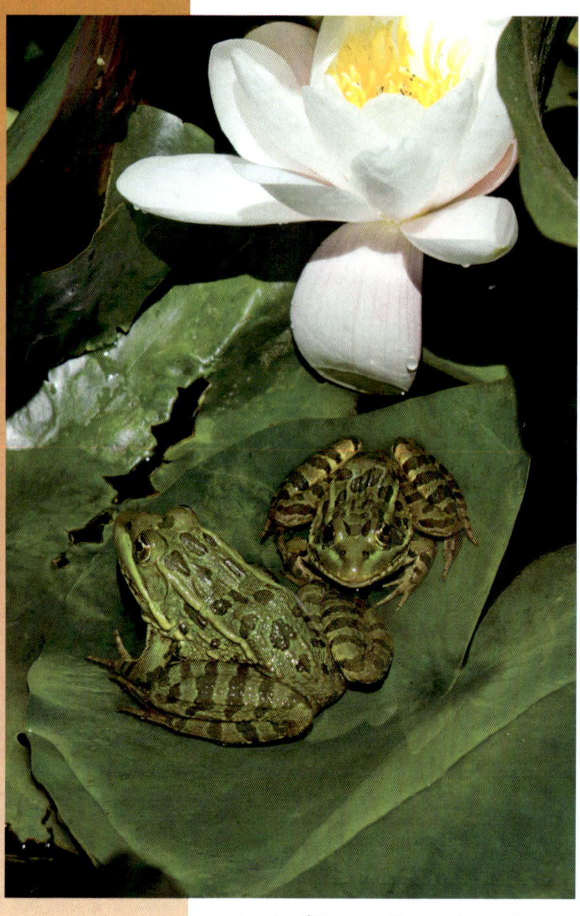

A pair of Ramsey Canyon leopard frogs sit on a lily pad at Beatty's Miller Canyon Guest Ranch. Canon 1DMIV, 70-200mm at 115mm, 2x teleconverter, ISO 400, f/13 @ 1/400 sec.

VIEW TIME
April to August

IDEAL TIME OF DAY
Early morning

VEHICLE
Any

HIKE
Easy

At the end of the road to Miller Canyon, a quaint little apple orchard, some cabins for rent, and a hummingbird extravaganza await at Beatty's Miller Canyon Guest Ranch. Tom Beatty knows all there is to know about the area and can direct you to local bird nests and other interesting wildlife photographic opportunities.

When you visit Beatty's, leave your high-speed hummingbird photography gear behind, as Tom does not allow photographers to use it. Cabin guests have full run of the property which increases your chance for unique close-up images of these fast-moving birds. Pack your **telephoto lenses** and a **tripod**. Find a location where the hummingbirds like to perch and manually focus your camera on the spot where they sit. When one perches, take several shots using high-speed continuous shooting mode. Some finicky hummers fly off at the sound of a whirling shutter, so if your camera has a silent shutter mode, use it.

Visitors that do not rent one of the cabins are restricted to the "grandstand" seating at the parking lot. From these seats, photographers can observe hummers feeding, fighting, and perching.

Thanks to Tom's involvement in a national effort to bring back the Ramsey Canyon leopard frog (now called the "Chiricahua leopard frog"), his ranch offers the best location in the state to take frame-filling portraits of this amphibian. Ramsey Canyon leopard frogs were almost extirpated until hard working citizens (led by Tom) converted their properties to frog sanctuaries.

The small pond at the gazebo near the snack house contains several leopard frogs. During the summer months, the pond also showcases several white water lilies, making for a colorful addition to the frogs' persona. If the white lilies appear in your composition, watch your histogram so you do not overexpose them. Most light meters tend to overexpose whites when the overall scene is of average luminance. Any moderate telephoto lens in the range of 300mm works well for this scene.

43

DIRECTIONS:

From Sierra Vista, drive about 9 miles south (14.5 km) on AZ 92. Turn right on Miller Canyon Road. Travel 2.5 miles (4 km) to the parking lot at the end of the road. Walk past the stop sign onto Beatty's Ranch.

For reservations and more information, visit **www.beattysguestranch.com**.

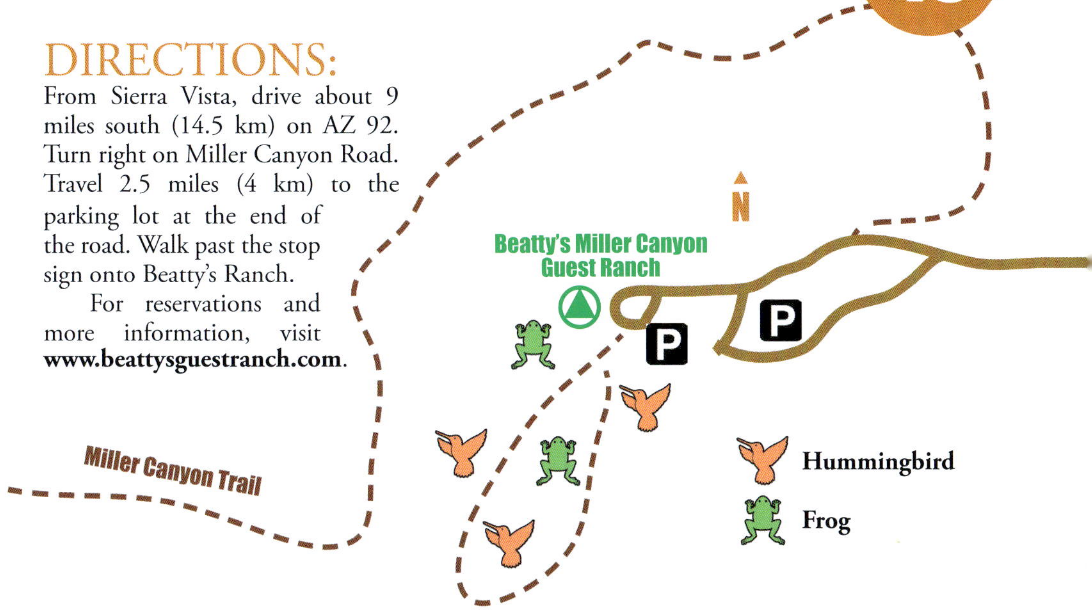

N

Beatty's Miller Canyon
Guest Ranch

P

P

Miller Canyon Trail

Hummingbird

Frog

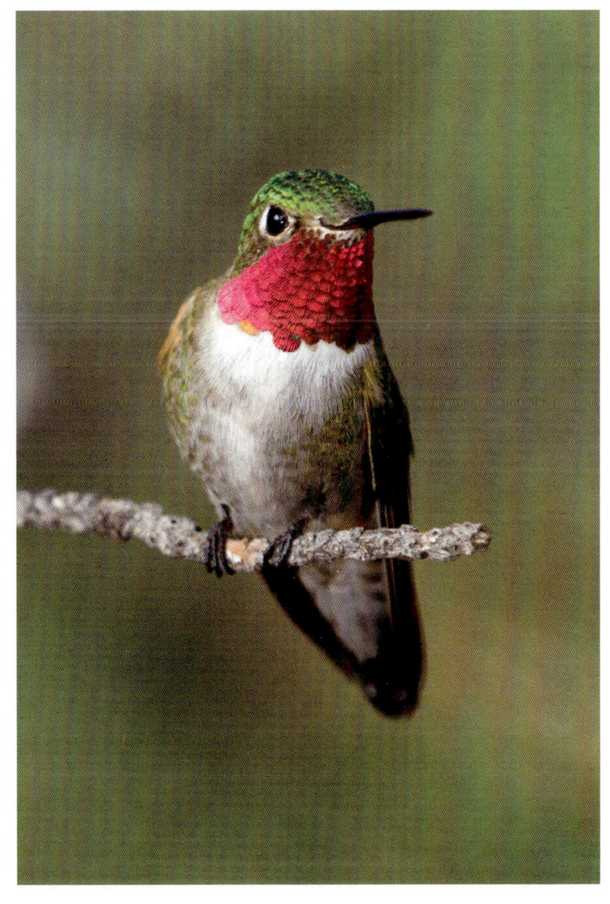

Broad-tailed hummingbird. Canon 7D, 70-200mm at 200mm, 1.4x teleconverter, ISO 400, f/6.3 @ 1/160 sec., off-camera flash @ -1/3 FEC.

Battiste Bed, Breakfast, and Birds

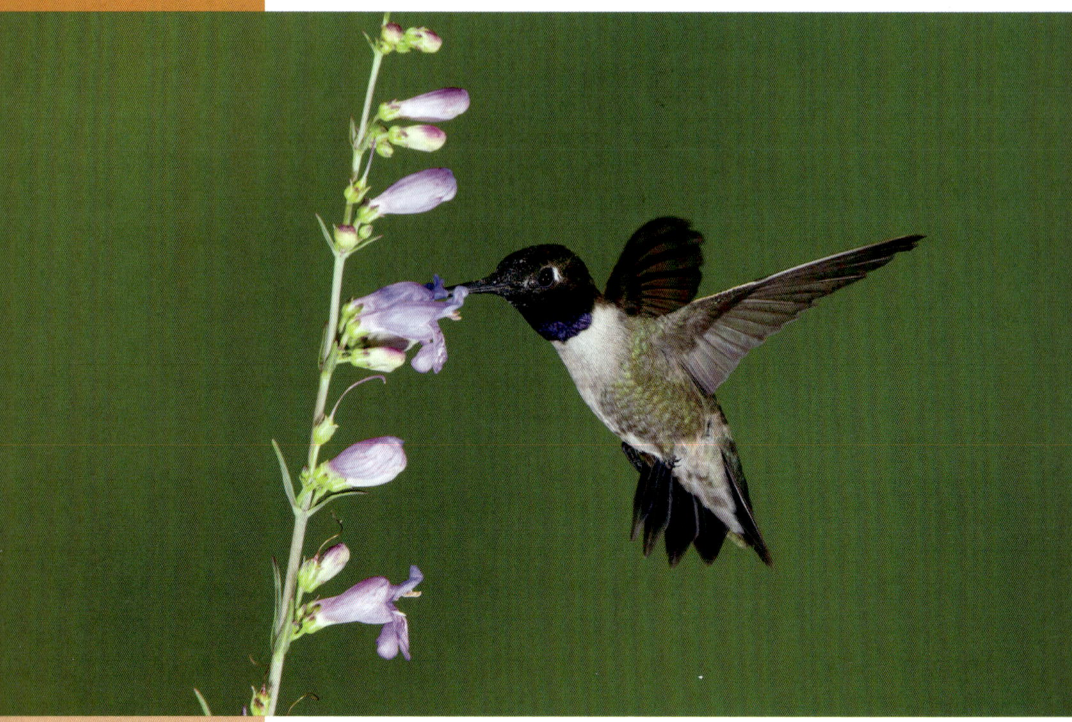

Black-chinned hummingbird male. Canon 1DMIV, 70-200mm at 146mm, 2x teleconverter, ISO 400, f/18 @ 1/320 sec., high speed hummingbird set-up.

VIEW TIME
March to August

IDEAL TIME OF DAY
Early morning, sunset, and night

VEHICLE
Any

HIKE
Easy

In the foothills of the Huachuca Mountains, Tony Battiste runs a bed and breakfast business that caters to both birders and bird photographers. A bird blind, small pond, plenty of bird food, and a lot of perches adorned with flowers and local plants offer the lucky photographer an opportunity to shoot thousands of images in a single morning. Oriole, finch, woodpecker, cowbird, pyrrhuloxia, and many other species come to feed here each day.

The best equipment for photographing at the lodge (and at the remote Mearns's quail location) is a **100-400mm telephoto zoom lens** mounted on a cropped frame camera (if shooting a **full-frame sensor camera**, use a **1.4x or 2x teleconverter**). When photographing from the blind at the pond, set your camera on aperture priority and dial in an aperture around f/8. In addition, set a fast enough ISO speed to allow a shutter speed of 1/1000 of a second or faster. High-speed continuous shooting mode will help fire multiple frames once you spot a subject.

At the remote Mearns's quail site, bring **flash** to overcome the dappled shade from the overhanging oak trees. Use either a bracket to raise the flash off the camera or place the flash on a **light stand** a couple of feet from the camera and use a TTL remote flash assist. The quail start to enter the area at sunrise when the light appears dim. Set your aperture to f/5.6 and the ISO speed fast enough to allow for a shutter speed of 1/125 of a second or faster. Place your flash on high-speed sync in case you increase your shutter speeds above your flash's maximum sync speed as you shoot. When the sun rises higher in the sky, modify the shutter speed (and ISO speed, if

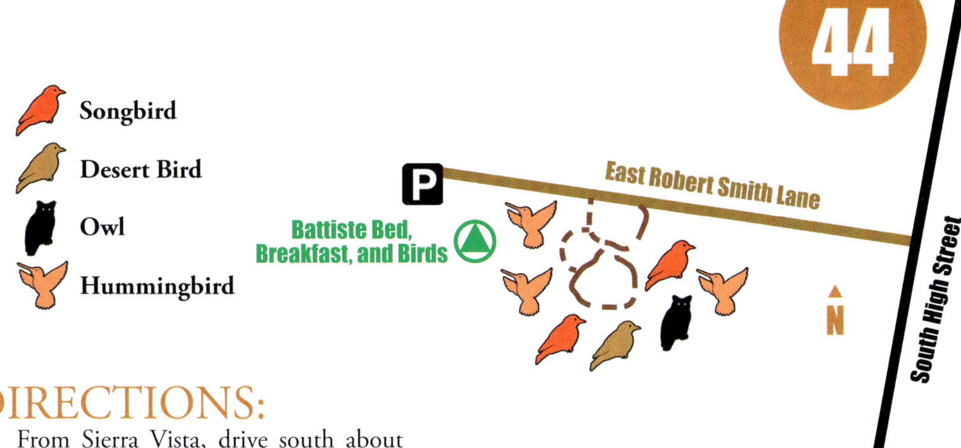

DIRECTIONS:

From Sierra Vista, drive south about 9.3 miles (15 km) on AZ 92. Turn west on East Vista Grande Road. Drive 0.2 miles (0.3 km) to High Street. Turn left, and travel another 0.4 miles (0.6 km) to Robert Smith Lane. Turn right and park at the second house on the left (4700 E. Robert Smith Lane).

For reservations and information, visit **www.battistebedandbirds.com**.

---

needed) to keep the shutter speed at 1/250 of a second or faster. A tripod is mandatory when shooting at these relatively slow shutter speeds. Use continuous autofocus to help you keep the moving quail sharp in your final image.

From late April until late June, a pair of elf owls nest in an old telephone pole within the birding area. When photographing the elf owls at night, the flash will serve as the only light source. After locating the owl, use a small flashlight or headlamp to focus using autofocus. Switch your ISO speed to 100 and your aperture to f/10 while keeping your shutter speed matched to your flash's maximum sync speed. Add or subtract power from the flash so the owls look naturally illuminated.

Although a vertical image seems more natural for an owl sitting in a tree, do not forget to try a horizontal orientation as well. When photographing owls, always leave a little more space at the bottom of the frame to allow the bird room to fall into—not out of—the scene.

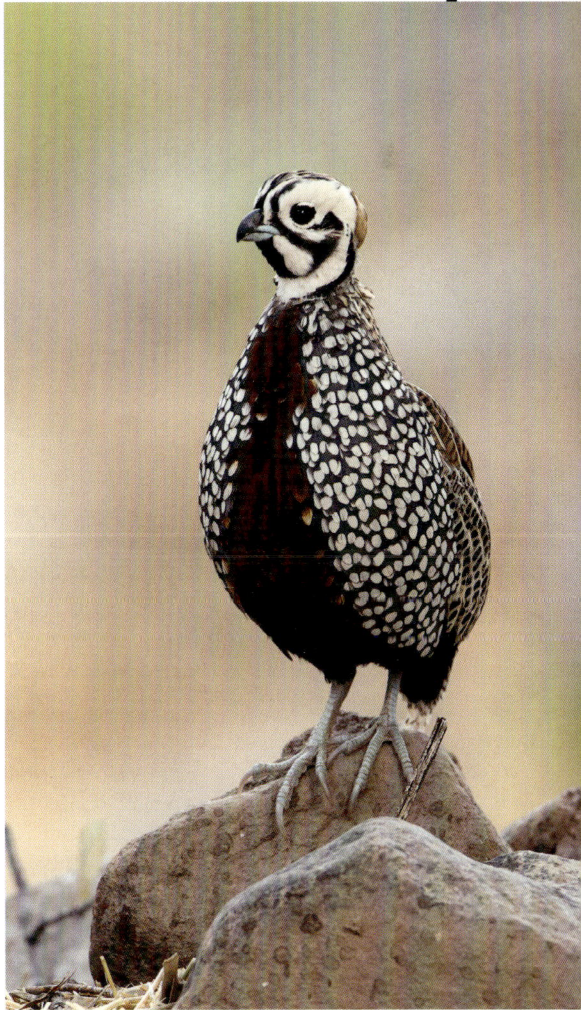

Montezuma quail male. Canon 5DMIII, 500mm, ISO 400, f/9 @ 1/80 sec., remote off-camera flash @ +1/3 FEC.

Whitewater Draw Wildlife Area

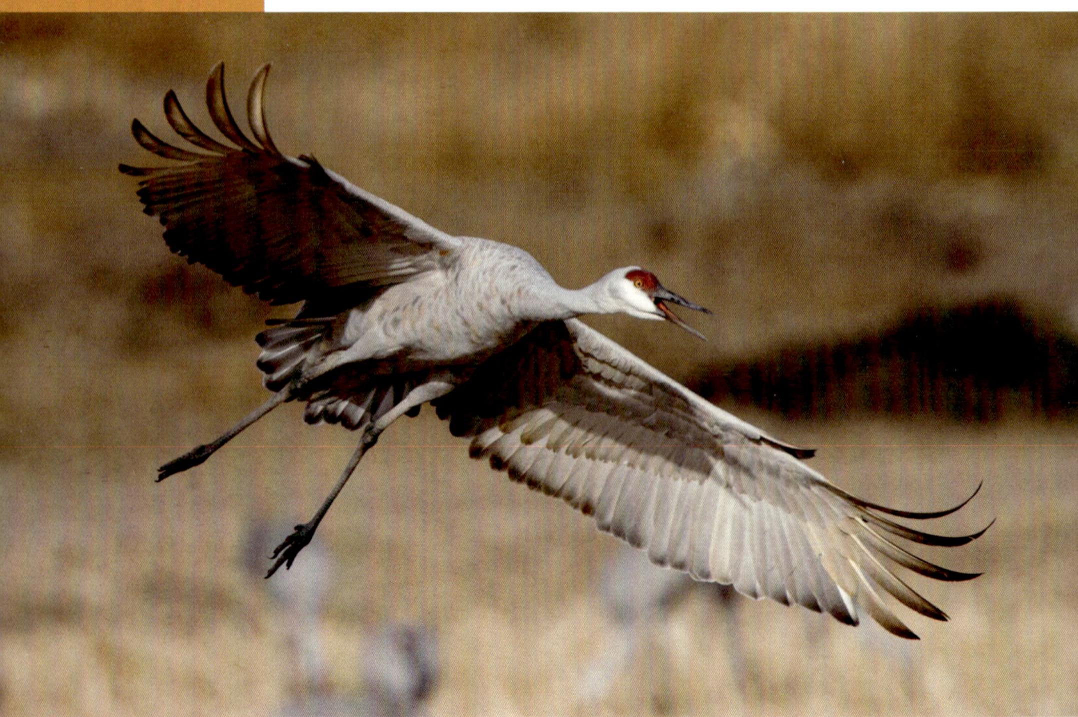

A sandhill crane sings a prehistoric song as it prepares to land. Canon 1DMII, 500mm, 1.4x teleconverter, ISO 400, f/6.3 @ 1/2500 sec.

VIEW TIME
October to March

IDEAL TIME OF DAY
Sunrise to sunset

VEHICLE
Any

HIKE
Easy to moderate

Set in the midst of Sulphur Springs Valley's agricultural lands, no other place in Arizona presents the same opportunity to view and photograph sandhill cranes. As many as 35,000 cranes roost in the several ponds at the wildlife area between October and February. In addition to the hordes of cranes, other birds such as ducks, water birds, sparrows, egrets, herons, flycatchers, vireos, doves, quail, phoebes, and wrens also make appearances here. The Arizona Game and Fish Department (AZGFD) purchased the Whitewater Draw in 1977. Since then, the AZGFD has managed the landscape to enhance wetland species and provide watchable wildlife—and an abundance of photography opportunities.

Telephoto lenses in the range of 400mm and longer on a **cropped frame camera** atop a **tripod** provides the bird photographer the optimal set-up. Despite the sandhill cranes large size, they usually appear a long way off, and any lens smaller than a 400mm will not do the job. Images taken of the smaller water birds, ducks, doves, and the like also benefit from the use of long lenses.

Arrive at the wildlife area at least 45 minutes before sunrise. Starting just before sunrise, and lasting until a half-hour after, thousands of sandhill cranes leave the ponds in unison—making for some unusual and colorful shots. Because the action only occurs briefly, scout the previous day for the best locations.

After the cranes have left the wildlife area, keep the long lenses on and switch your focus (pun intended) to photographing the many smaller birds.

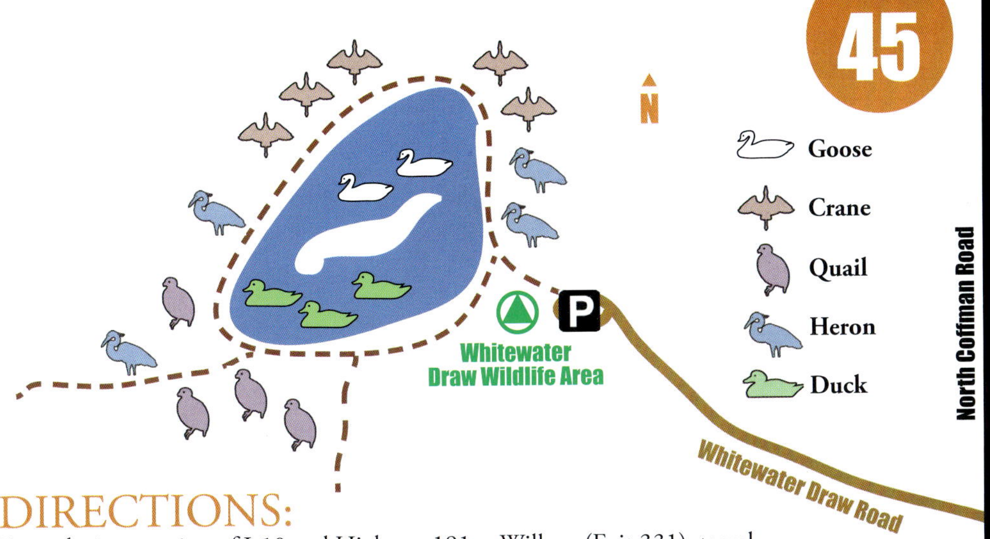

DIRECTIONS:

From the intersection of I-10 and Highway 191 at Willcox (Exit 331), travel southbound for 48.5 miles (78.1 km) to Davis Road and turn left. Drive 3 miles (4.8 km) to Coffman Road. Turn left and proceed another 2 miles (3.2 km) to the parking area.

The AZGFD closes some areas from October 15 to May 15 to protect roost sites. For more information, visit **azgfdportal.az.gov/wildlife/viewing/wheretogo/whitewater**.

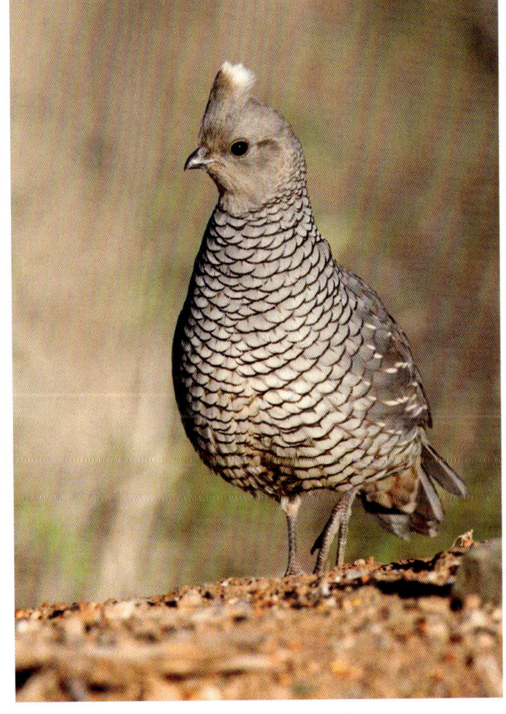

Scaled quail. Canon 50D, 500mm, f/4, ISO 200. f/8 @ 1/800 sec.

Use a pair of binoculars to locate concentrations of ducks and geese and approach quietly and slowly. Search for the incredibly colored vermilion flycatcher and small sparrows as they search the vegetation for food. Watch the water's edge for marsh wrens, avocets, and other water birds. During the winter months, the light is optimal until about 10 a.m., so there is plenty of time for images of the smaller beings before the cranes return.

Between 10 a.m. and 2 p.m., the sandhill cranes revisit the wildlife area to bathe and drink. Because their feeding areas can be as far as 25 miles (40.2 km) away, cranes tend to fly high and drop into the wildlife area in groups of hundreds. As they spiral towards the water, take images of them spreading their giant wings to slow their decent.

Many of the birds that left the wildlife area at daybreak do not return to roost until about one or two hours before sunset. Although most of these cranes drop in from hundreds of feet above, many others fly low over the refuge and offer opportunities for close-up flying images. As they approach the ponds, cranes spread their wings and glide for seemingly hundreds of feet before extending their long legs to the water.

Cochise Lake

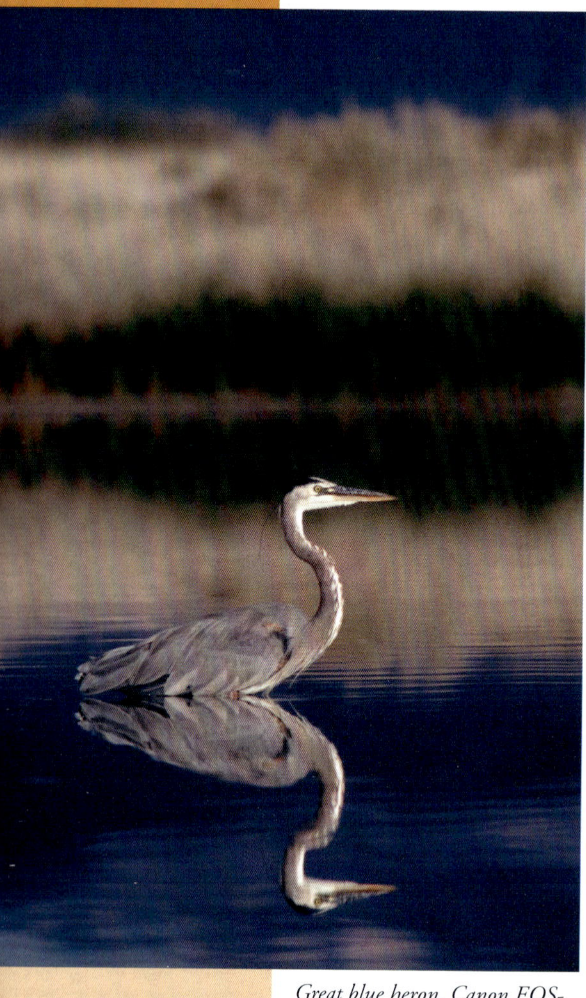

Great blue heron. Canon EOS-3, Fuji Velvia film, 500mm, f/4, shutter speed unavailable.

VIEW TIME
Year-round

IDEAL TIME OF DAY
Early morning and late afternoon

VEHICLE
Any; 4WD high-clearance when wet

HIKE
Easy

In terms of sheer numbers of water birds, Cochise Lake rivals any other place in Arizona. Cochise Lake offers a stopping off point for most migrant ducks and shorebirds as well as a breeding area for avocets and black-necked stilts. It serves as the best place in the Grand Canyon State to create breathtaking reflection images of birds with blue skies and white clouds. So what's the catch? The bird-filled lake is a sewage pond and can be odiferous while harboring thousands of pesky flies (food for many of the smaller birds).

Dress in camouflage and bring a **long telephoto lens (400mm or longer)** on your **cropped frame camera** (which can attain effective focal lengths from 600-640mm). Pairing a 500 or 600mm lens on a cropped frame camera with a 1.4x teleconverter yields a whopping 1050 to 1344mm effective focal length!

Approach the lake with the sun to your back to render the best reflections. Keep the lens axis as low to the water as possible. If you are not decked out in camo, set up a blind. The birds will likely fly away at your approach, but if you remain quiet and hidden, they will return within a few minutes. Water birds tend to act nervously and keep moving all of the time, so use your continuous autofocus setting. Set the camera mode to manual or aperture priority and keep an eye on the histogram to avoid overexposing the sky. When shooting dark waters, medium toned birds, and white skies, the camera's light meter can incorrectly set the exposure. Use a relatively high shutter speed (i.e., 1/500 second or greater) to stop the actions of the moving birds, and shoot in continuous bursts of five or more to give you a better chance of taking a tack-sharp image.

During the monsoon season, check out the ephemeral ponds surrounding the permanent lakes. Small crustaceans, such as fairy and tadpole shrimp, survive decades in the dry soil and seemingly come to life during the short monsoon season. Purchase a small glass cooking pan and fill it with a shallow bit of water (from the ephemeral ponds) and some local sand. Capture the shrimp using a small aquarium net and transfer a few to the cooking pan. Use a **macro lens** and **off-camera flash** to avoid reflections.

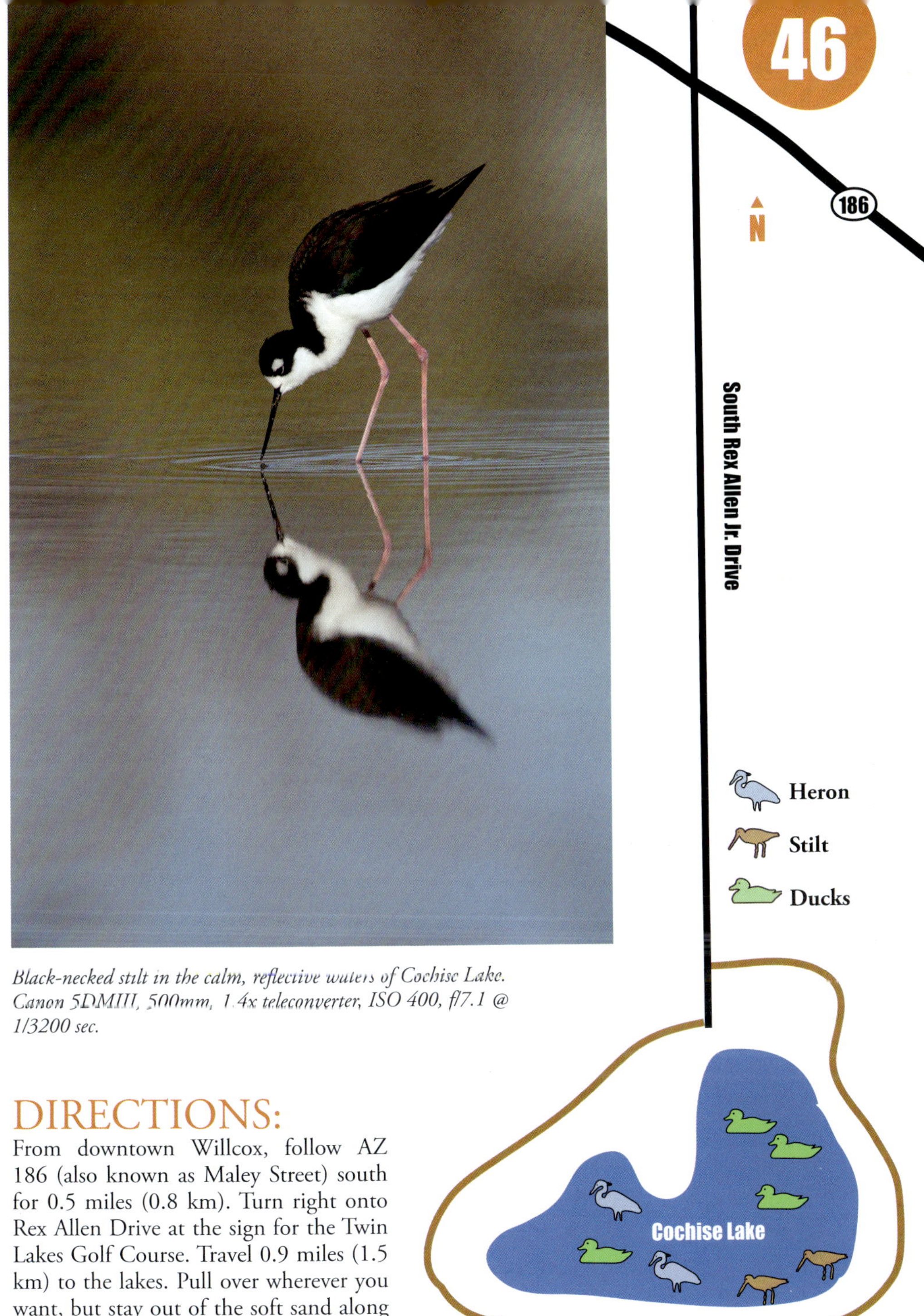

Black-necked stilt in the calm, reflective waters of Cochise Lake. Canon 5DMIII, 500mm, 1.4x teleconverter, ISO 400, f/7.1 @ 1/3200 sec.

DIRECTIONS:

From downtown Willcox, follow AZ 186 (also known as Maley Street) south for 0.5 miles (0.8 km). Turn right onto Rex Allen Drive at the sign for the Twin Lakes Golf Course. Travel 0.9 miles (1.5 km) to the lakes. Pull over wherever you want, but stay out of the soft sand along the edges.

SOUTHERN ARIZONA

SULPHUR SPRINGS VALLEY LOOP/ Willcox Playa

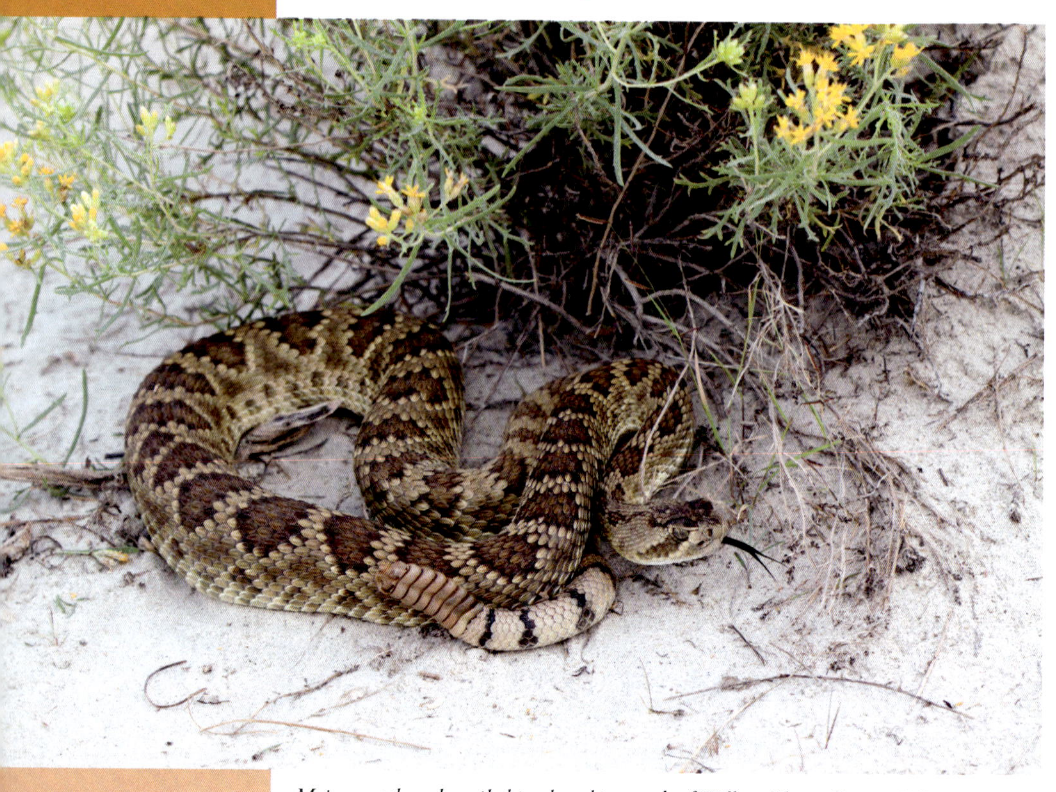

Mojave rattlesnake coiled in the white sands of Willcox Playa. Canon 50D, 100mm macro, ISO 400, f/8 @ 1/250 sec., on-camera flash @ +2/3 FEC.

VIEW TIME
June to August

IDEAL TIME OF DAY
All day; night

VEHICLE
Any

HIKE
Easy

Approximately 15,000 years ago, Cochise Lake (the precursor of Willcox Playa) filled with 46 feet (14 m) of water and covered 146 square miles (378 square km). This ancient lake resulted from one of the few North American interior-draining basins produced by tectonic pressures similar to those that created Africa's Great Rift Valley. Today, the basin sees an unusual assemblage of vegetation and substrate that support living remnants of years gone by.

During most of the year, many of the more photogenic and unique species in the valley and playa are secure in their underground dens waiting for life-giving water. Then, in late June and early July, moist air from the Sea of Cortez and Gulf of Mexico collides with the desert heat and quickly rises to thousands of feet in the air, producing great thunderheads and massive monsoon storms. Toads, frogs, estivating insects, burrowing snakes, tiger beetles, and other remnants of wetter times crawl to the surface for a brief time to enjoy the moist climate in which they evolved.

Western green toads, Great Plains toads, spadefoot toads, Sonoran desert toads, yellow mud turtles, and Sonoran tiger salamanders burst to the surface. In a matter of a few weeks, turtles and amphibians need to feed, grow, reproduce, and get ready for another nine months or more beneath the dry, desert earth. An animal frenzy ensues, and a prepared photographer can capture incredible images during this time.

To get ready, watch the weather channels and online weather prediction models for Willcox. When the first rains strike, wait a couple of days, and hope for more storms. The first rains may not bring enough moisture to fill the low-lying areas, but the next few certainly will. Get to the valley, and start driving the loop well before dark to find the ephemeral ponds where the action/breeding will take place. Note the areas with standing water near desert habitats, and return to them about 30 minutes after sunset. Listen for the ancient calls of several species of toads.

After locating the breeding ponds, gear up to get wet. Place a **macro lens** on your camera with a **macro twin or ring flash** or a flash mounted on a **macro bracket**. Before you wade into the water, set your ISO speed to 100 and your aperture to f/16. Keep your shutter speed equal to your flash's maximum sync speed. Use a small flashlight to locate and focus the camera on one of the many singing amphibians. Kneel low in the water, and get as close as possible to your subjects.

After a few short hours of sleep, get up early and drive the same route looking for Western hog-nosed snakes and Texas horned lizards crossing the road. Before moving them off the road to take pictures, remember that you will need an Arizona Game and Fish Department hunting or fishing license to legally capture them, even for the short time it takes to photograph them. Both species move relatively slowly and pose for their portraits. Get low to the ground to look them in the eye for the best images.

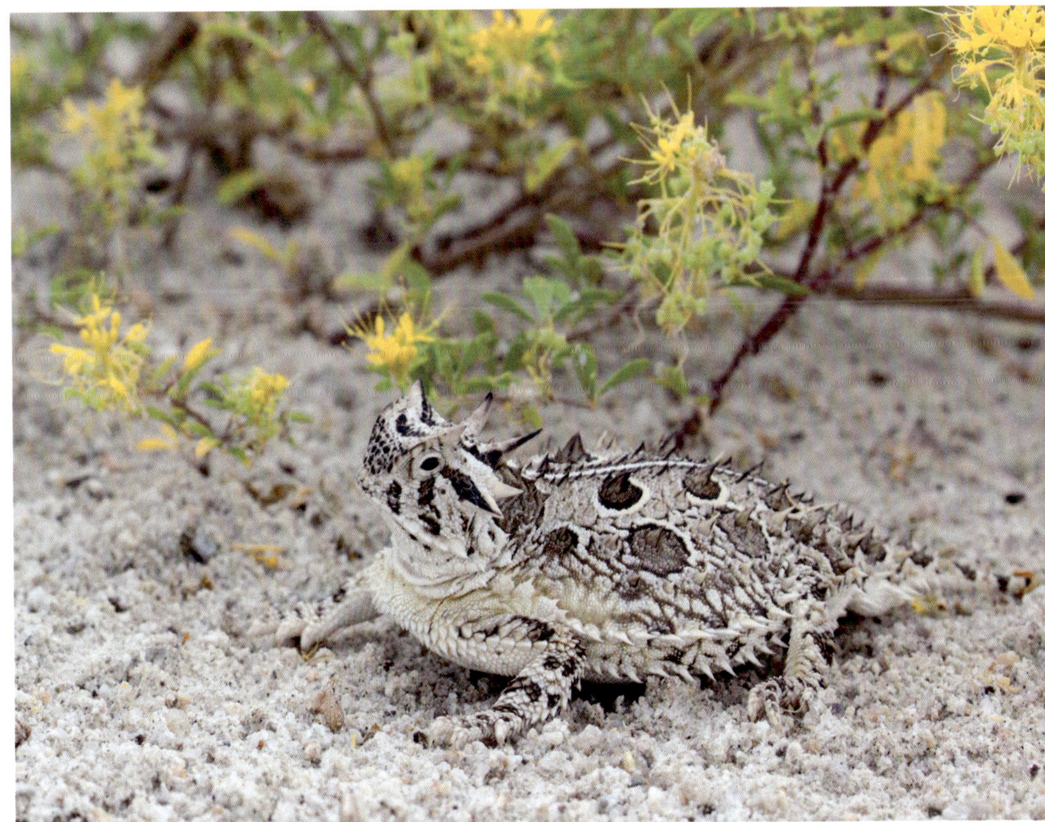

Texas horned lizard. Canon 50D, 100mm macro, ISO 400, f/10 @ 1/250 sec., on-camera flash @ +2/3 FEC.

SULPHUR SPRINGS VALLEY LOOP/ Willcox Playa

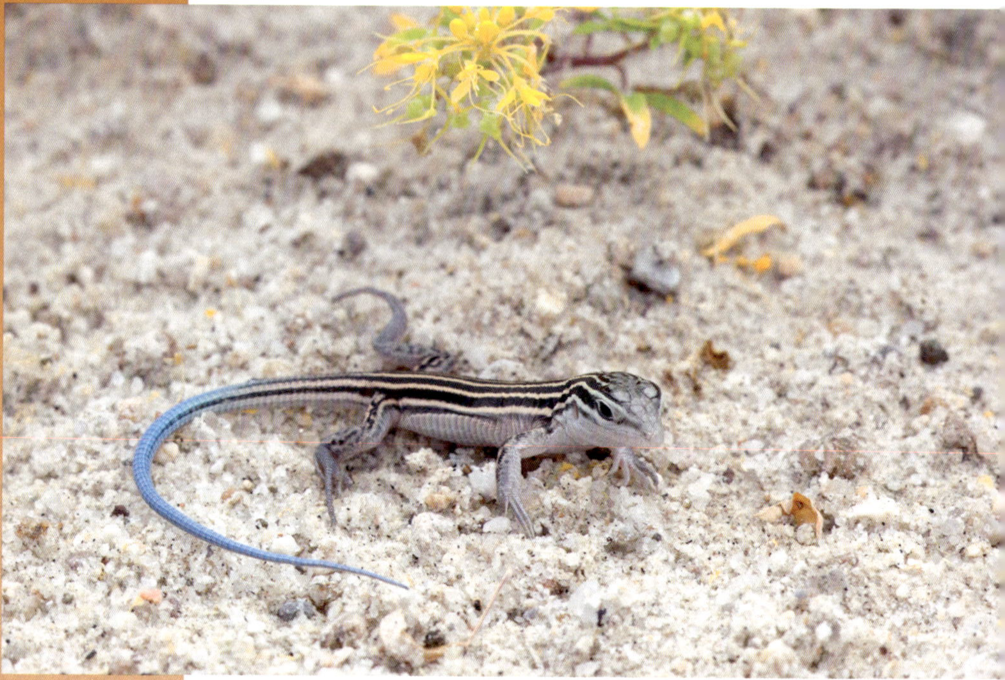

Arizona whiptail lizard shows off its grand blue tail. Canon 50D, 100mm macro, ISO 400, f/10 @ 1/250 sec., on-camera flash @ +2/3 FEC.

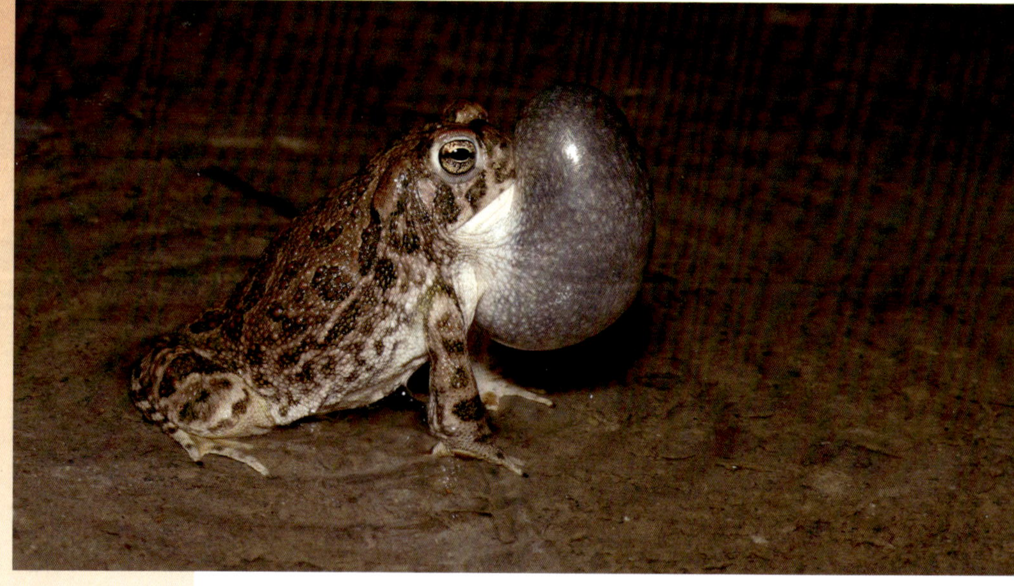

A Great Plains toad male fills its buccal pouch with air in order to call for a mate. Canon 5DMII, 100mm macro, ISO 800, f/16 @ 1/200 sec., flash mounted on macro bracket.

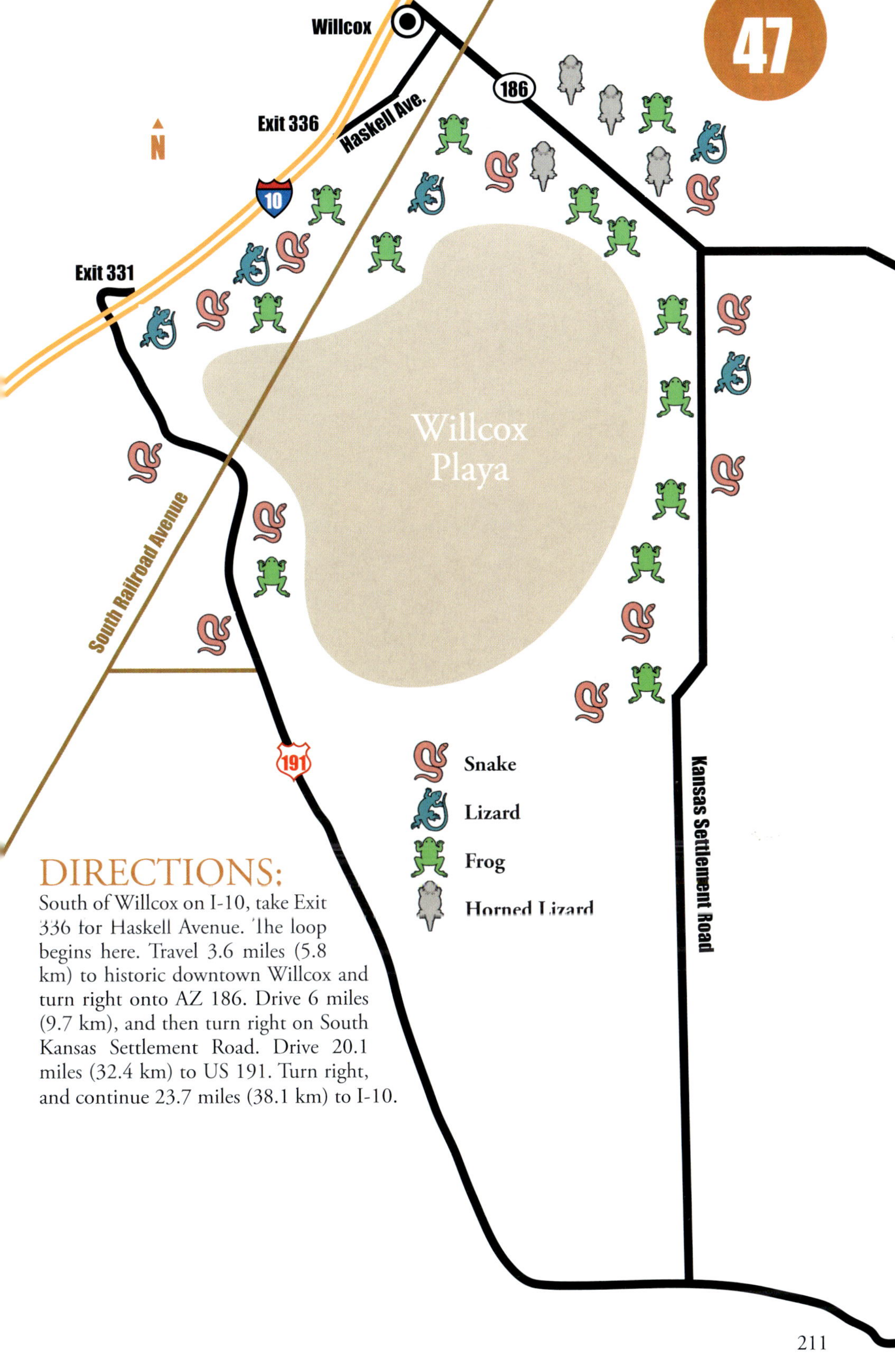

DIRECTIONS:

South of Willcox on I-10, take Exit 336 for Haskell Avenue. The loop begins here. Travel 3.6 miles (5.8 km) to historic downtown Willcox and turn right onto AZ 186. Drive 6 miles (9.7 km), and then turn right on South Kansas Settlement Road. Drive 20.1 miles (32.4 km) to US 191. Turn right, and continue 23.7 miles (38.1 km) to I-10.

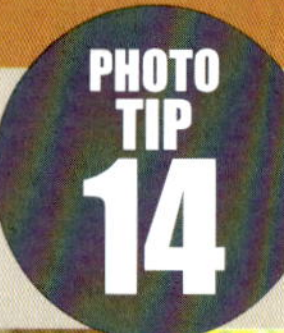

Using a Tent Diffuser

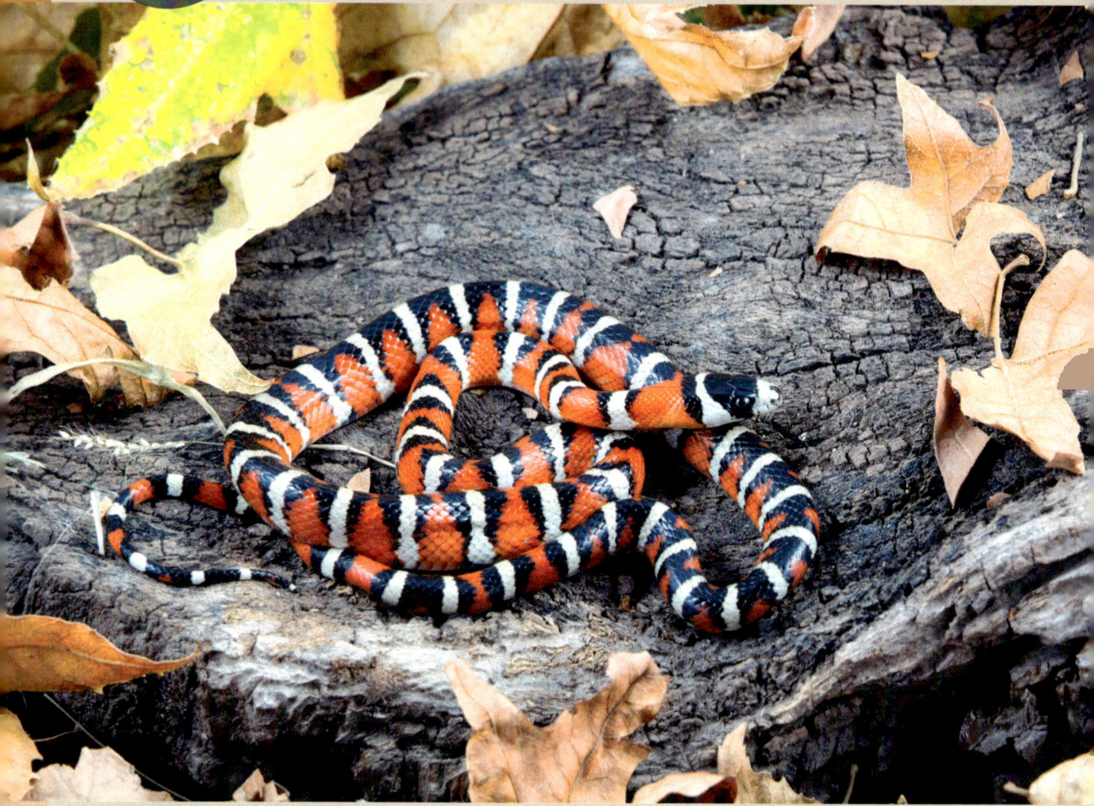

A tent diffuser containing leaves and wood creates a natural-looking setting for this Sonoran mountain kingsnake. Canon 5DMIII, 70-200mm at 140mm, 2x teleconverter, ISO 400, f/18 @ 1/200 sec., off-camera flash.

A tent diffuser, or studio soft box light tent, offers the wildlife photographer complete control over the quality of light whether you photograph a captive subject indoors or outdoors. Natural light diffused by the tent produces soft, even lighting conditions. If needed, you can also set up one or more flash units outside the tent to strengthen the diffused light source. Add some direction to the lighting by covering up part of the tent or by placing flashes in various positions around the tent.

Having control over the lighting is only half the benefit. Tent diffusers also allow you to control a small, moving subject. Try photographing a spider on a plant in the wild. If it jumps to the cluttered ground, good luck finding it for more photographs! The tent diffuser easily contains the small creatures, making recapture for another try a breeze.

To use a tent diffuser, put the expanded tent on a card table or some other device that holds it off the ground (lying on your stomach to get "eye level" with a jumping spider is not desirable, especially if the shoot takes an hour or so to finish). If you rest the tent on a table, you can sit in a nice comfortable chair and not ruin your back or neck.

Use two clamps to secure the tent to the table so it does not move (and force you to reposition a camera that took you 20 minutes to get in just the right position). Attach an appropriate artificial background (a blurry landscape photo or maybe fake plants) to the inside back of the tent using a spring clamp. Create a more natural look by placing your subject on a rock, piece of wood, or similar item close to the front of the tent. If necessary, you can attach a clamp to a GorillaPod (a flexible camera tripod made by Joby) and use it to precisely position the item that will hold the subject. Place the GorillaPod close to the front of the tent so you can easily position the subject for photography.

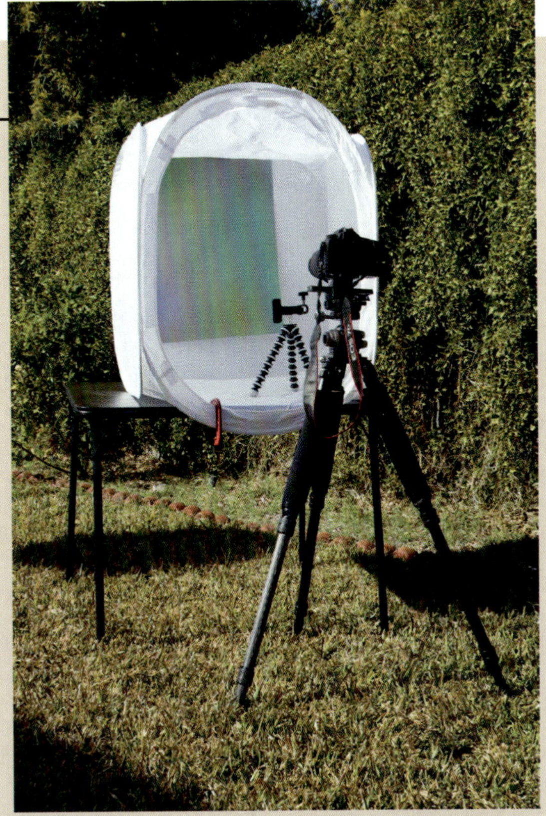

A tent diffuser helps to eliminate harsh light and contain specimens.

Set up your tripod, ball head, and camera with an appropriate modified lens or macro lens. If you are using natural light, make sure the sun is to the side or back of the tent. If you are using flashes, set them up using a remote or slave system. Some subjects, such as small snakes or terrestrial lizards, require set up on a stage rather than sitting on a piece of leaf or some other aerial platform. To make a stage, find a food tray that fits into the tent (you can place your artificial habitat on the bottom of your tent, but it may damage the tent—the food tray solves this problem). Place an appropriate substrate in the tray. Make the staged material look as much like nature as possible by using materials found in the same habitat in which your subject lives. Gather wood, small rocks, dirt or sand, grasses or other vegetation. Make sure the transition from the stage to the artificial backdrop is as natural as possible.

If you are using natural light, use aperture priority and set the aperture appropriately. Keep the subject in focus and the backdrop either out of focus or visually separate from the subject.

If you need fill flash, place the camera on sync speed or the flashes on high-speed sync. To use your flashes as the primary light source (and not just for fill), place the camera on manual mode and the flashes on TTL. Adjust the aperture to give an appropriate look to the image. Note any distracting shadows on the backdrop and reposition the flashes as needed to eliminate the shadows.

Making the Photo 11

Monsoon Madness

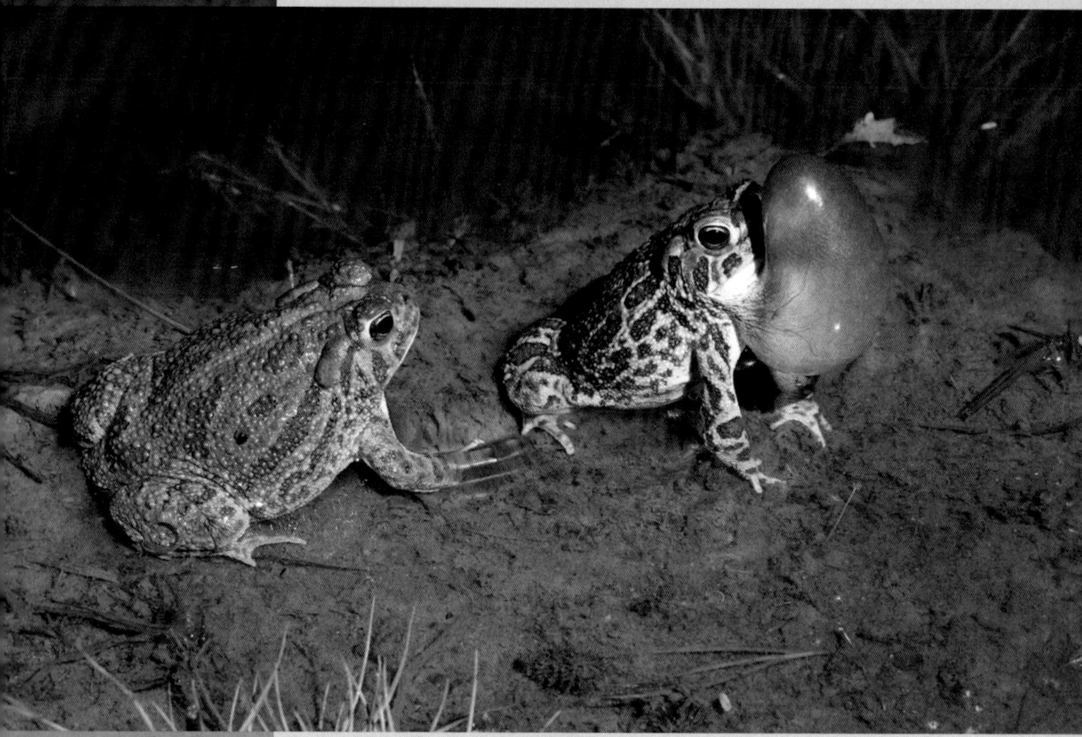

A male Great Plains toad calls to an attentive female. Canon 70D, 100mm macro, ISO 100, f/13 @ 1/200 sec.

In late June, the desert floor bakes in excess of 100 degrees F (38 degrees C), and the winds take on a southerly direction. Almost imperceptibly, the humidity begins to rise, and the upward traveling thermals take moisture with them to form welcomed clouds.

In late June and early July the White Mountains and southeast Arizona begin to experience light showers as the temperatures continue to rise. I can almost sense the ground moving from the pressure of plants and animals waiting for the life-giving monsoon storms to free them from their summer tombs. Soon, huge thunderclouds dot the landscape, and rain falls on the hard-pan ground, as desert washes begin to flow with life-giving water. The earth becomes soft enough for frogs and toads to dig their way to the surface and search for standing water in which to find mates and breed. Desert tortoises, snakes, and hordes of small mammals begin to spend more time above ground as the humidity rises. Insects feed on the new vegetation, and moths hatch from their subterranean cocoons and spread their wings for the first time. The monsoons are in full swing. Endless thrilling photographic possibilities await.

To determine where the wettest places are, I look at Intellicast (**www.intellicast.com**, click on Current, then Precipitation, and then Weekly). Here, I find a record of the amount of precipitation for the past week for the entire state. I know from experience it takes a few days of heavy rain to develop the perfect conditions for monsoon madness. I scan the maps each day looking for an area that has received sufficient rain to make a trip worthwhile.

As I watch the weather, I begin putting together my photography equipment. I favor macro photography during the monsoon, so I dust off my macro lenses, flashes, low tripod, diffusers, and reflectors. I break out the black lights and make sure the batteries are fresh. I find my rubber boots, snake sticks, containers for collecting critters, tongs for handling scorpions, lights to see my way, and a good book to read during the nonproductive heat of the day.

And then it happens! The area south of Willcox sees four days of intense rainfall. It's time to travel.

I arrive midday and begin driving the roads looking for standing water. I spend time talking to the locals to determine the extent of the storms and start to develop a game plan for that evening. After a good helping of Mexican food at the local diner, I load the vehicle with my gear and head out. My strategy is always the same—drive to the first water, stop the vehicle, and listen for calling toads (shrill, high-pitched calls). The first water only has a few active toads, so I decide to look for bigger puddles. Then I find it—the mother lode.

The water-filled depression sits about 100 yards from the edge of the road and, from the noise, I sense it is a doozy. I grab my gear, carefully cross the barbed-wire fence, and walk towards the noise. On the way, I spot several toads and a slithering desert kingsnake.

When I arrive at the pond, hundreds of calling western Great Plains toads and 40 or so western green toads croak along the edges. The noise feels so intense that I can feel the tympanum in my ear vibrating back and forth—a little unnerving. To add to the drama, lightning strikes near Dos Cabezas only a few miles away to the east.

I ignore the desire to start photographing and sit down to watch an activity that has been taking place for thousands of years. The male toads fill up their buccal (throat) pouch with air and rattle a strange, high-pitched call that fills the air and vibrates the ground and water. I can see small waves emanating from the frogs' bodies as the males call. Males crowd the pool as the females make their way to select their beau for the night. Several fights ensue as males unknowingly mistake other males for a female and attempt to mate with them. Soon the female makes her choice and the happy couple enjoys some toad love (known as "amplexing") before swimming off to the deeper pools where the female will lay her eggs while the male fertilizes them. An hour or so past dark I tire of watching and start photographing the nocturnal frenzy.

About six hours and 500 images later, I need some sleep and head for the hotel. My plans for the next night are to set out black lights to attract insects, or maybe ride the roads looking for snakes, or maybe sweep a net for tadpole shrimp, or … well, my options are many and I am too tired to think. So goes monsoon madness.

Chiricahua National Monument

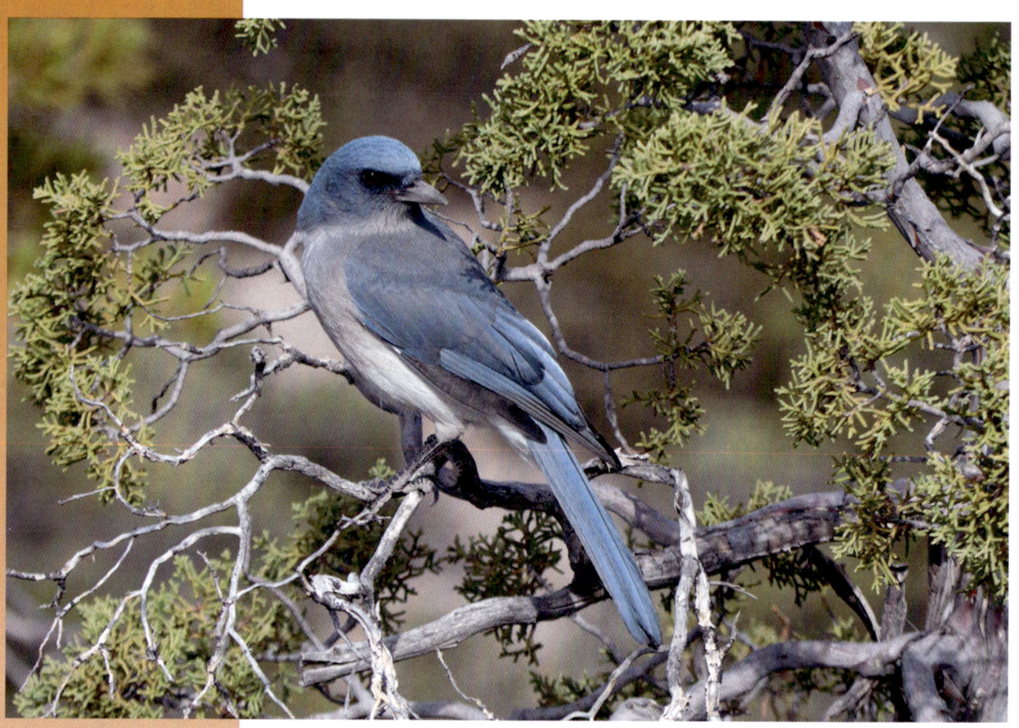

Mexican jay. Canon 50D, 100-400mm at 150mm, ISO 320, f/7.1 @ 1/640 sec.

VIEW TIME
November to February

IDEAL TIME OF DAY
Early morning and late afternoon

VEHICLE
Any

HIKE
Moderate

The Apache tribe calls the rhyolite formations (caused by the Turkey Creek Volcano some 27 million years ago) of the Chiricahua National Monument, "The Land of Standing-Up Rocks." As a sky island close to the border with Mexico, the Chiricahua Mountains offer wildlife species in unique habitats not found anywhere else. With a little effort and a lot of patience, shutterbugs can photograph many of the 170 species of birds, 8 species of amphibians, 46 reptiles, and 71 mammals. However, the park's Coues white-tailed deer and the Mexican jay steal the show.

Standing 32 inches (0.8 meters) at the shoulder, the Coues (pronounced "cows") white-tailed deer is the smallest species of deer in North America. During the rut (breeding season for deer) from December to mid-February, this elusive deer comes out in the open, making it relatively easy to photograph.

As you enter the monument, a large grassy plain on the left follows Bonita Creek upstream for a couple of miles. Drive slowly along the road or pull into the Bonita Creek and Faraway picnic areas, or into one of the other trailhead parking areas, and scan for deer with your binoculars. You will likely notice a doe deer first, but during the rut, the bucks are not far away. Two additional locations to scout for deer are the short road to the administrative site (not the visitor center) and the campground.

Once you spot any deer, relax and watch. Use a **telephoto lens** and bean bag slung over your window (to stabilize the camera and lens) and photograph the deer from the comfort of your vehicle. The deer are not

DIRECTIONS:

From downtown Willcox, drive southeast on AZ 186 for 31.3 miles (50.4 km) to Bonita Canyon Road (also referred to as AZ 181). Turn left, and follow Bonita Canyon Road for 5.1 miles (8.2 km) to the monument entrance.

The park charges an entrance fee unless you present an "America the Beautiful" annual or other pass issued by the National Park Service. For more information about the monument, visit **www.nps.gov/chir**.

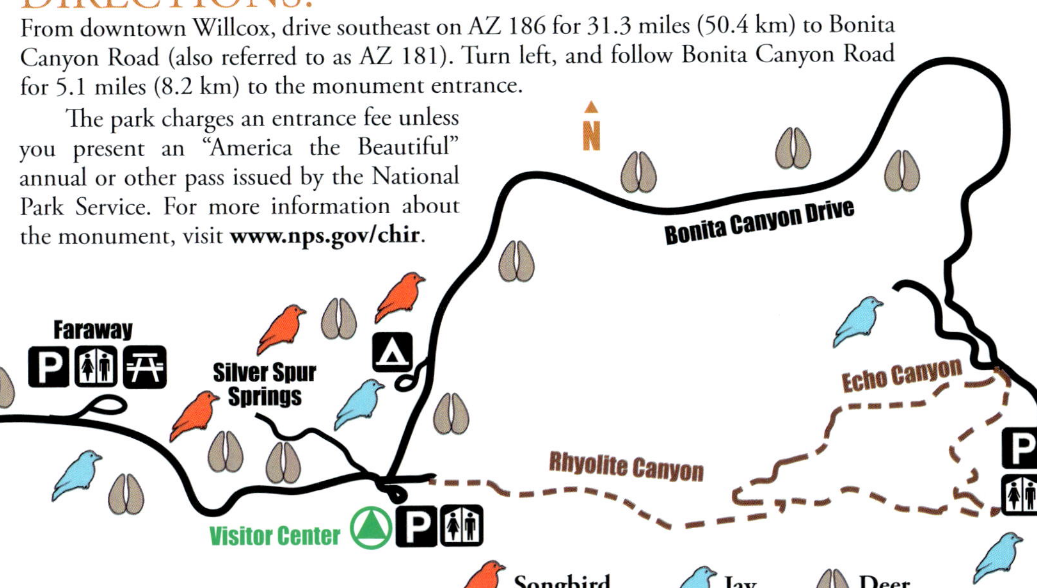

afraid of your car but will spook once you exit the vehicle. If you must exit your vehicle for a closer peek, be quiet and move as little as possible. Keep the sun to your back with your shadow pointing towards the deer. Attempt to compose an image of the deer with a pleasing background of yucca plants or oak trees. When in the tall grass habitat, watch for "false focus," which occurs when your camera focuses on the grass instead of the deer. To avoid this mishap, use manual focus to gain focus on the deer.

Save time to photograph the Mexican jays, whose distribution in the United States is limited to the southeast corner of Arizona, the southwest corner of New Mexico, and a small piece of Texas. While eating your lunch at one of the picnic tables, these curious and vocal birds will land in the surrounding trees (often in groups of three or four) and actively swoop for a closer investigation. Hand-hold a moderate telephoto lens with image stabilization (or vibration reduction) enabled.

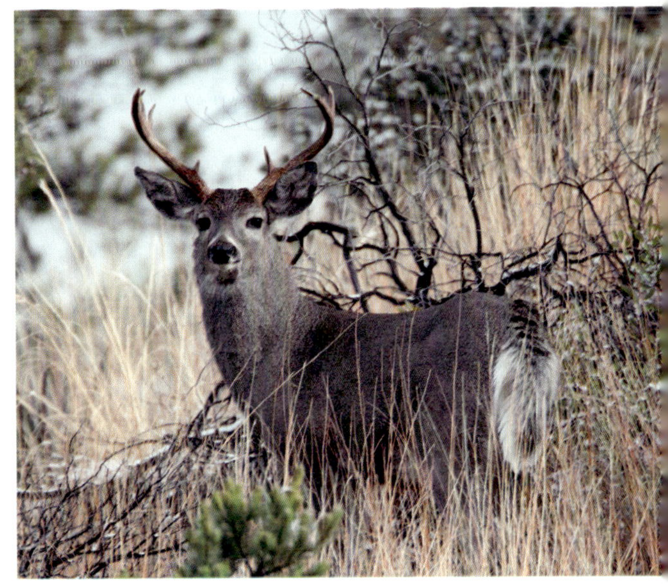

Buck Coues white-tailed deer among the late fall vegetation at Chiricahua National Monument. Canon 50D, 500mm, 1.4x teleconverter, ISO 250, f/5.6 @ 1/320 sec.

Making the Photo
12

Catch as Catch Can

A friend of mine once told me a long time ago, "Never, ever, be caught without a camera at arm's length!" I wish I had given his direction more attention. I cannot begin to describe the number of times I have wanted my camera at arm's length—but did not have it—so that I could have taken advantage of an unexpected opportunity.

Whenever I am in the field, I typically have my camera and 100-400mm telephoto lens (or something similar to this) on the passenger seat of my vehicle (or close to me if I have a passenger) along with a bean bag. When I tear down my equipment at night, I always keep at least one camera and lens at the ready just in case. At night, I have a flash filled with fresh batteries attached to a camera and lens.

Why go through all this trouble? A few years ago, I was photographing hummingbirds and bats at a friend's house in the Chiricahua Mountains. It was late, and I had removed all of the equipment so that my mess would not be visible to my friend the following morning when we came down for breakfast.

I was on my way to bed when I heard a commotion at the hummingbird feeders. I turned on my headlamp. To my surprise, a pair of ringtails were drinking sugar water from the hummingbird feeders. They seemed unafraid of me, and I really, really, really, wanted to take some images of them. Luckily, my camera at that time (a Canon 50D) with the 100-400mm lens was at arm's length. I also had two flashes on flash stands. I just needed to put my flash remote trigger on the camera, set the appropriate settings, turn the flashes on to slave, and quickly set the stands on either side of the hummingbird feeders. For the next five minutes, I had unfettered access to a scene I had longed dreamed about. The ringtails disappeared as fast as they appeared. Had I not been prepared, the cover for this book would have featured some other critter.

Having the camera at the ready has allowed me to take images of pronghorn (running and standing), several raptor species, Rocky Mountain bighorn sheep, Uinta chipmunks, Kaibab squirrels, lizards on rocks, many species of snakes, and almost uncountable numbers of great weather events.

You are never fast enough to stop the vehicle, get the right camera gear out, get back in the vehicle, and take an image. The animals hardly ever stay still long enough. If you are sitting around a campfire at dusk and an elk just happens to present itself, you have seconds—not minutes—to take the image before it disappears.

So now I pass this advice to you: "Never, ever, be caught without a camera at arm's length!"

RIGHT: While photographing bats at a nearby hummingbird feeder, this ringtail came down to feed at this same feeder. I quickly grabbed two flashes (already mounted on flash stands), placed one flash stand on each side of the critter, and set them to ETTL mode. Then, I placed the flashes' remote device on my camera and managed to record a few images before the ringtail left for the night. Canon 5DMIII, 70-200mm at 70mm, 2x teleconverter, ISO 800, f/16 @ 120 sec., multiple off-camera flashes.

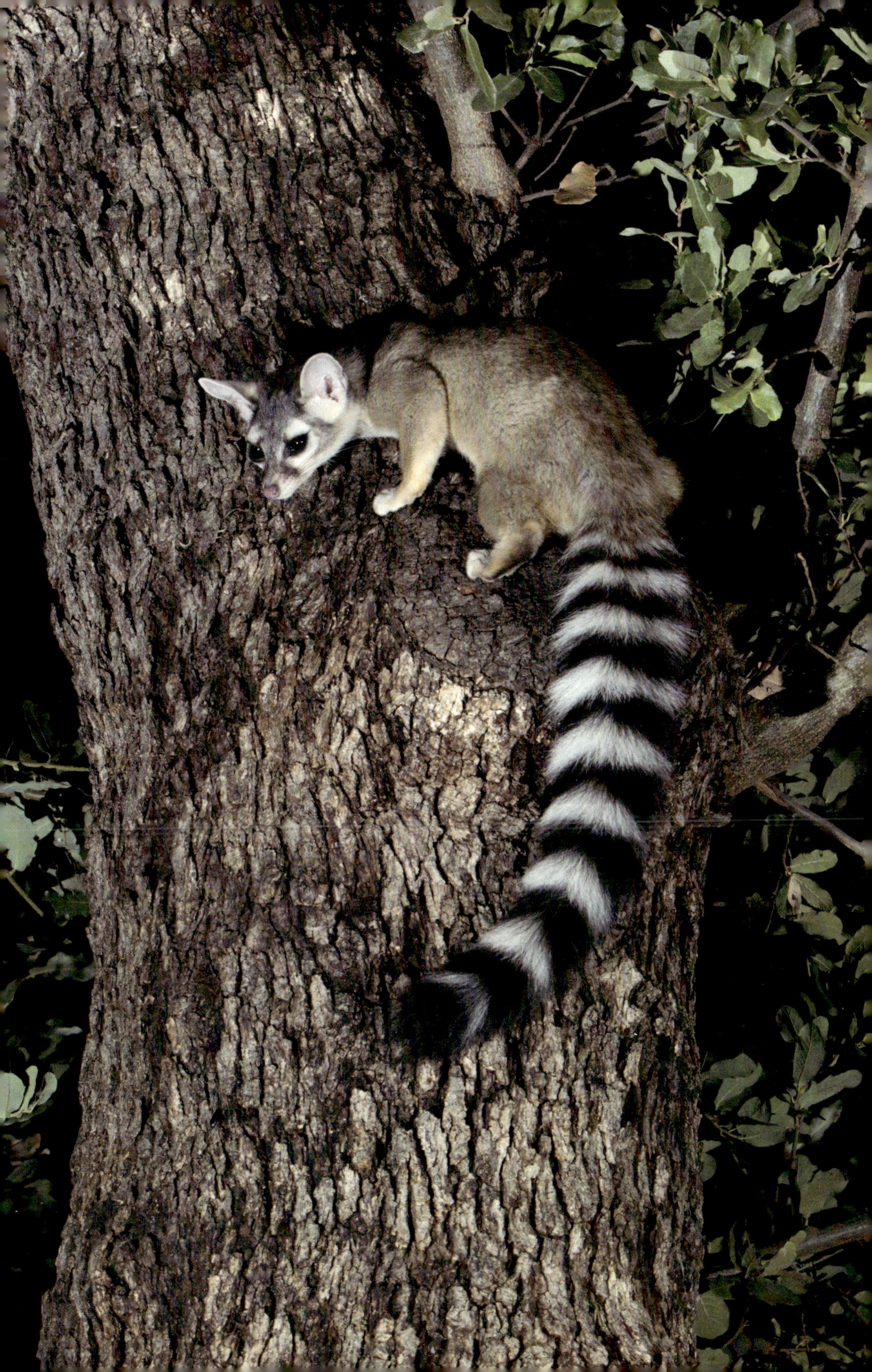

Cave Creek Ranch

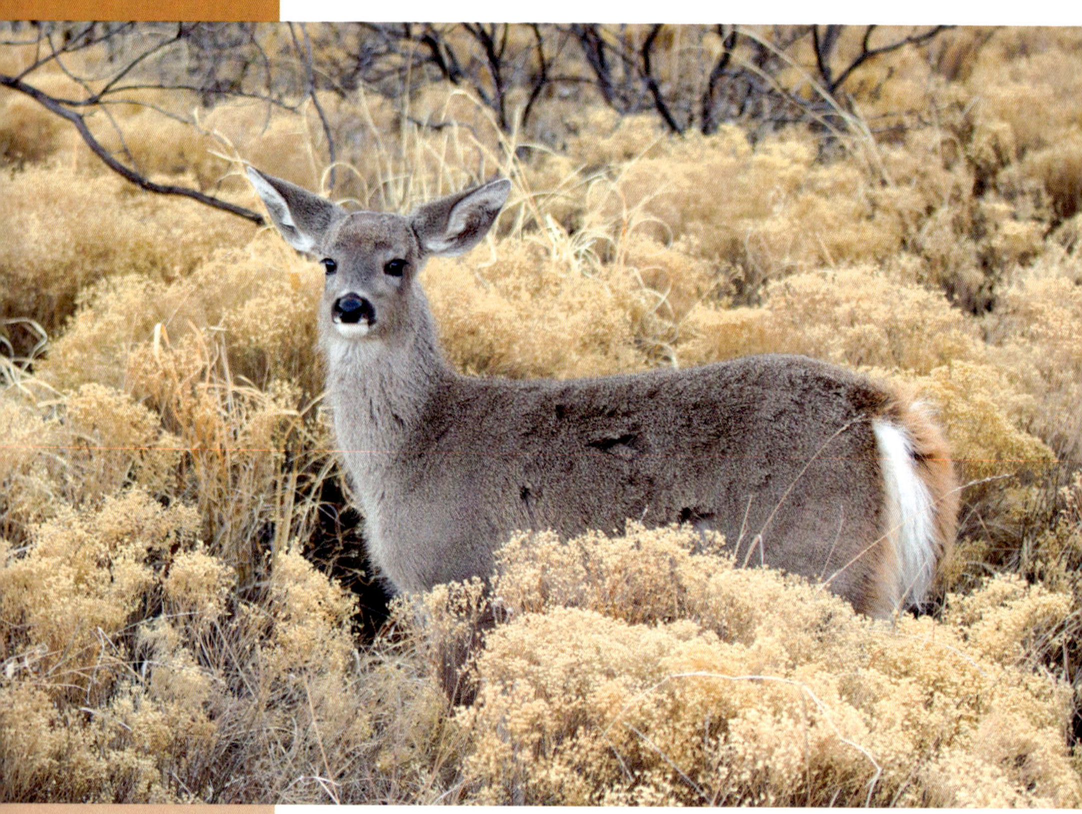

Coues white-tailed deer doe at Cave Creek Ranch. Canon 50D, 70-200mm at 70mm, ISO 400, f/4 @ 1/200 sec.

VIEW TIME
Year-round

IDEAL TIME OF DAY
Sunrise to sunset

VEHICLE
Any

HIKE
Easy

Among the sycamore and oak trees on the eastern side of the Chiricahua Mountains, Cave Creek Ranch presents unmatched opportunities to photograph many of Arizona's unique species including javelina, Coues white-tailed deer, coatimundi, magnificent hummingbirds, blue-throated hummingbirds, Arizona woodpeckers, and acorn woodpeckers. The ranch's owner, Reed Peters, labored for years to attract birds, mammals, butterflies, and insects to his property solely for the enjoyment of wildlife watchers and photographers.

The ranch rents several restored cabins, each with a kitchen and private bathroom (the ranch closes to non-guests from 4 p.m. to 10 a.m.). Because it is 25 miles (40.2 km) to the nearest gas station, arrive with a full tank of gas. There is a small store/café that has great food but check when they are open and closed during different seasons.

Cave Creek Ranch offers a year-round wildlife photography adventure with the most action from April to September. Reed puts out a variety of bird seed, cracked corn, peanut butter, suet, and hummingbird feeders in front of the ranch office. A small pond with running water also attracts abundant wildlife.

Because the concentrated feeding area covers approximately 150 feet (45.7 m) by 75 feet (22.8 m), the photographic strategy differs here from other sites. You never know what critter will come into the feeders, so be ready for everything and anything.

If you enjoy photographing the larger mammals (coatimundi, white-tailed deer, or javelina), bring a **telephoto zoom lens (ranging from around 100mm to 400mm)**. If the birds at the feeders are in your sights, then use the largest telephoto lens you have. Grab your armada of telephoto lenses, a sturdy **tripod** with **ball or gimbal head**, and **fill flash**. Wait patiently while sitting in a comfortable chair. If you are fortunate enough to own two cameras, put a long telephoto lens on one to shoot distant birds and a shorter telephoto zoom lens on the other to photograph a herd of javelina (when they come close to feed). Changing lenses when the action hits will frustrate you and you might miss a chance to record a great image.

Zoom in tightly on your subject to keep the cottages and dirt road out of your photograph. If you follow javelina or the white-tailed deer into the surrounding woods after they leave the feeding area, you can make full body images in a natural setting.

Coatimundi come to the feeding area to lick the grape jelly and eat bird suet. You can get very close to them, but get ready for a moving, active subject before you approach them. Keep your camera on continuous autofocus and high-speed continuous shooting mode to give yourself the best chance for making a usable image. As always, keep your distance from animals that can bite or kick you.

In April, birds begin the migration from their winter homes in Mexico, and other areas south, to their northern summer homes to feed, breed, and grow. Grand species such as painted redstart, Townsend's warbler, sulphur-bellied flycatcher, broad-billed hummingbird, yellow warbler, western tanager, summer tanager, Scott's oriole, and many others stop at Cave Creek

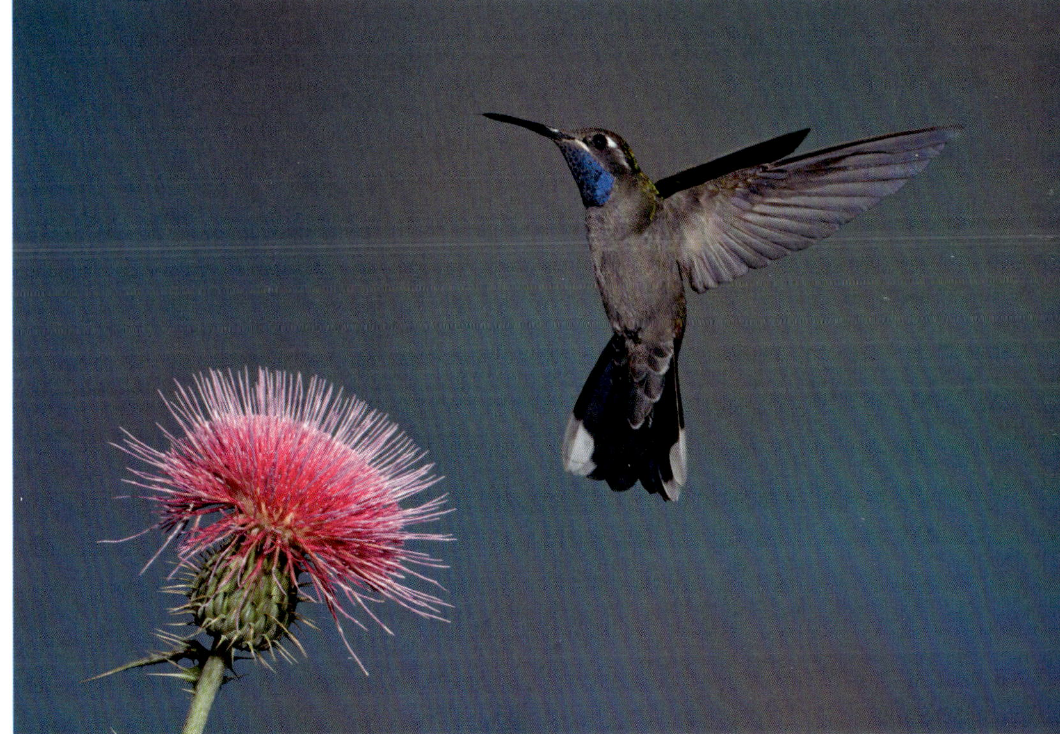

A blue-throated hummingbird feeds at a thistle flower. Canon 50D, 100-400mm at 180mm, ISO 200, f/18 @ 1/250, high-speed hummingbird set-up.

Cave Creek Ranch

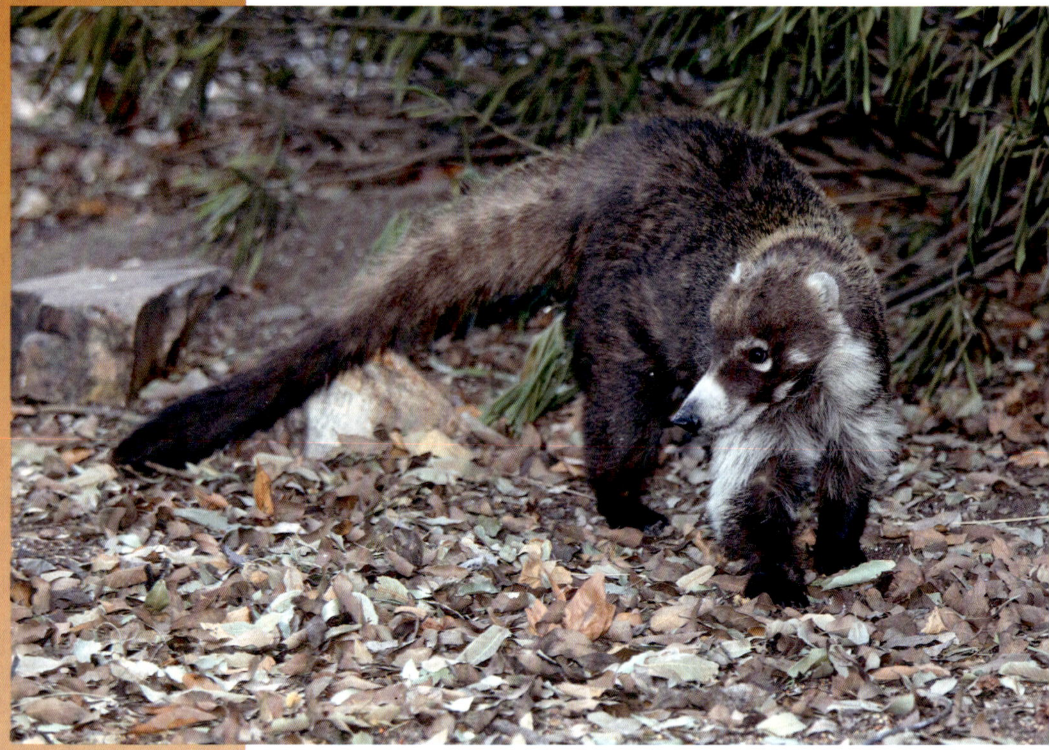

A coatimundi searches for grape jam and suet at Cave Creek Ranch. Canon 5DMII, 70-200mm at 110mm, 1.4x teleconverter, ISO 1600, f/5.6 @ 1/160 sec., on-camera flash @ -1/3 FEC.

Ranch for food and water. Some breed in the surrounds, but many stop by only for a few days to refuel before continuing their travels to more northern climates.

By late May, many of the migrant birds have gone north, but Arizona serves as the northernmost summer home for several species of birds. Painted redstarts, ruby-crowned kinglets, summer tanagers, band-tailed pigeon, white-winged dove, several species of hummingbirds, multiple species of vireos, black-headed grosbeak, and other summer residents begin the process of breeding and raising their young. From time to time, these birds visit the rich feeders at Cave Creek Ranch.

To reduce any friction between photographers and bird-watchers, the ranch limits high-speed hummingbird photography to the front of a rented cabin. This restriction is of very minor consequence since blue-throated and magnificent hummingbirds are year-round residents. In addition, the largest numbers of other species occurs from July through September during the southward migration when there are plenty of cottages available (see "Hummingbird Photography" on page 154).

Since the birds frequently fly in shaded areas, add a little **fill flash** to your scene. Many birds perch on the large pyracantha bush in the feeding area. During winter, the pyracantha fruit appears bright red, making a unique opportunity to record photos of birds eating fruit. If luck is on your side, an elegant trogon or gray fox will make one of their infrequent visits to the pyracantha bush to eat its ripe berries.

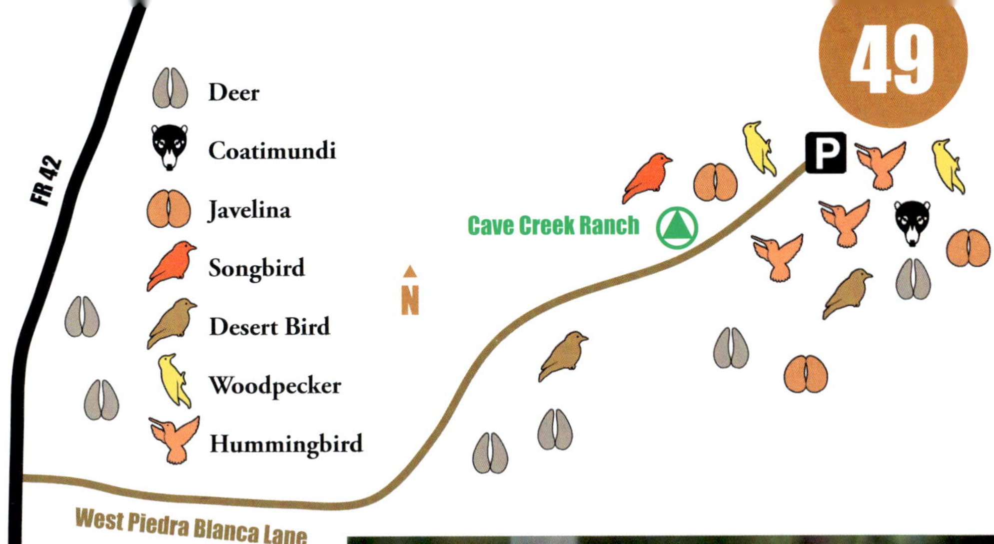

DIRECTIONS:

Follow I-10 over the Arizona-New Mexico border (about 56 miles (90.1 km) east of Willcox) and take Exit 5 for NM80. Travel 27.5 miles (44.3 km) south on US80 and turn right on Portal Road (also referred to NM533). Go 7.7 miles (12.4 km) west on Portal Road (where you will cross back into Arizona) to Forest Road 42 and turn left. After 1 mile (1.6 km), turn left onto West Piedra Blanca Lane at the sign for Cave Creek Ranch. Travel about 300 feet (91.4 m) to the ranch.

For more information about the Cave Creek Ranch, visit **www.cavecreekranch.com**.

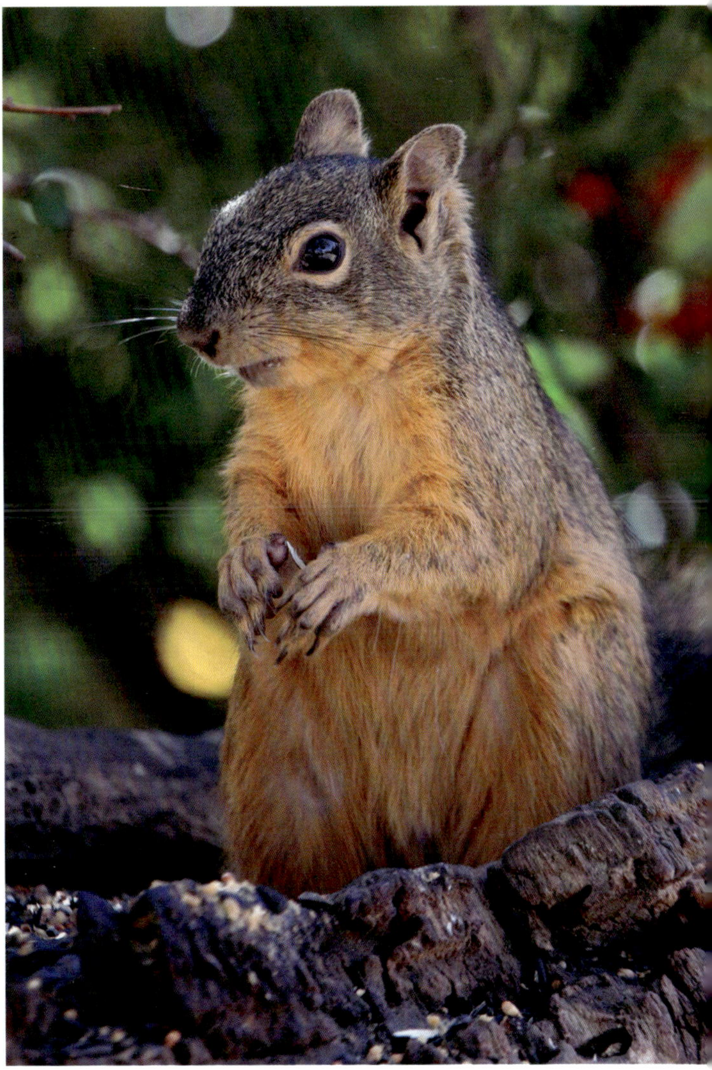

A rare look at a Chiricahua fox squirrel at Cave Creek Ranch. Canon 40D, 100-400mm at mm, ISO 400, f/7.1 @ 1/100 sec.

Cave Creek Canyon

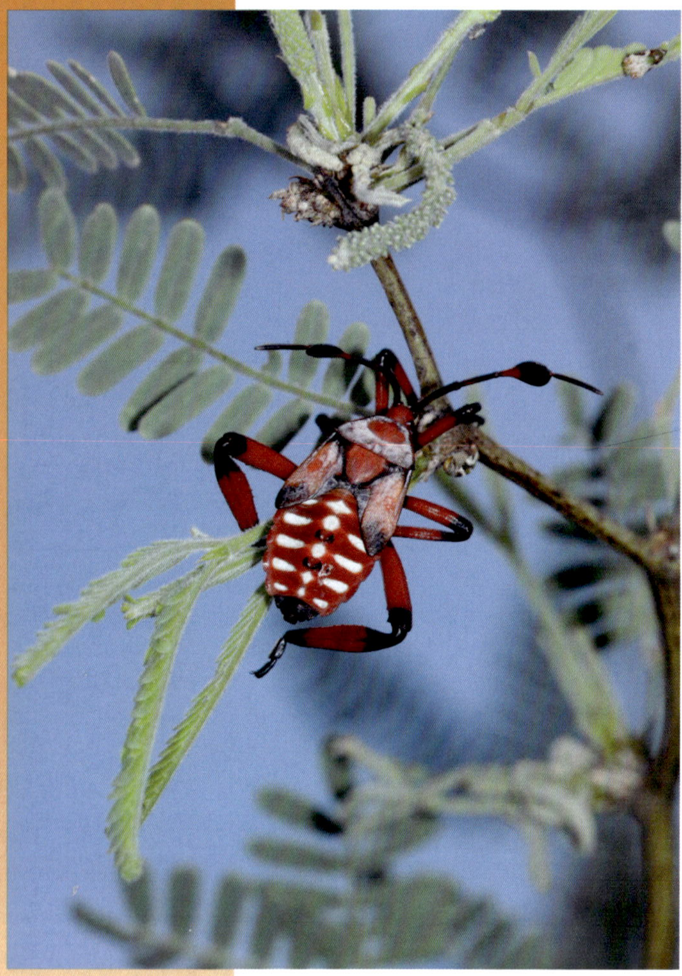

Mesquite beetle. Canon 50D, 100mm macro, ISO 400, f/9 @ 1/160 sec., on-camera flash @ -1 FEC.

Protected by a national forest, Cave Creek Canyon offers flowing water and abundant stands of oak, sycamore, pinyon, juniper and pine trees. The canyon forged its way through rock walls hundreds of feet high after much volcanic action 10,000 years ago. This diverse habitat makes the canyon an internationally-renowned jewel of Arizona for scientists, wildlife watchers, and photographers.

The most sought after species in this spot (and in the southwestern United States) is the elegant trogon which is only found in three of Arizona's mountain ranges (Chiricahua, Huachuca, and Santa Rita mountains). Several pair breed in Cave Creek Canyon with one or two regularly nesting in the South Fork of Cave Creek. As cavity-nesters, once the trogon occupies a nest, the birds appear unafraid of humans and photographers. Their nests usually reside 30 feet (9.1 m) or more from the forest floor and are often obstructed by thick vegetation. Use a **telephoto lens** and sturdy **tripod** to hone in on males attempting to attract females to breed with or when the pair feed their young.

Other notable birds found in Cave Creek Canyon are the sulphur-bellied flycatcher, Arizona woodpecker, painted redstart, hepatic tanager, band-tailed pigeon, a variety of owls, several hummingbird species, and orioles, to name just a few. To spot these flying creatures, walk the trails with a telephoto lens and tripod.

For the most part, using a blind or sitting and waiting for the bird to get close enough to use even the longest telephoto lens will not work. Unless you hike in South Fork (where bird calls are not permitted), you will need to locate your species of choice and then use its artificial bird call to tempt your targeted subject close enough to take an image.

VIEW TIME
April to July

IDEAL TIME OF DAY
Early morning and late afternoon

VEHICLE
Any

HIKE
Easy to moderate

DIRECTIONS:

Follow I-10 over the Arizona-New Mexico border (about 56 miles (90.1 km) east of Willcox) and take Exit 5 for NM80. Travel 27.5 miles (44.3 km) south on NM80 and turn right on Portal Road (also referred to NM533). Go 7.7 miles (12.4 km) west on Portal Road (where you will cross back into Arizona) to Forest Road 42 and turn left. Travel 2.7 miles (4.4 km) to the intersection of FR 42 and South Fork Road. Stay on FR 42 and travel 2 miles (3.2 km) to the Southwest Research Station.

Return to South Fork Road and travel 1.3 miles (2.1 km) to the Cave Creek Trailhead parking lot. Walk for up to 2 miles (3.2 km) south on the South Fork Cave Creek Trail.

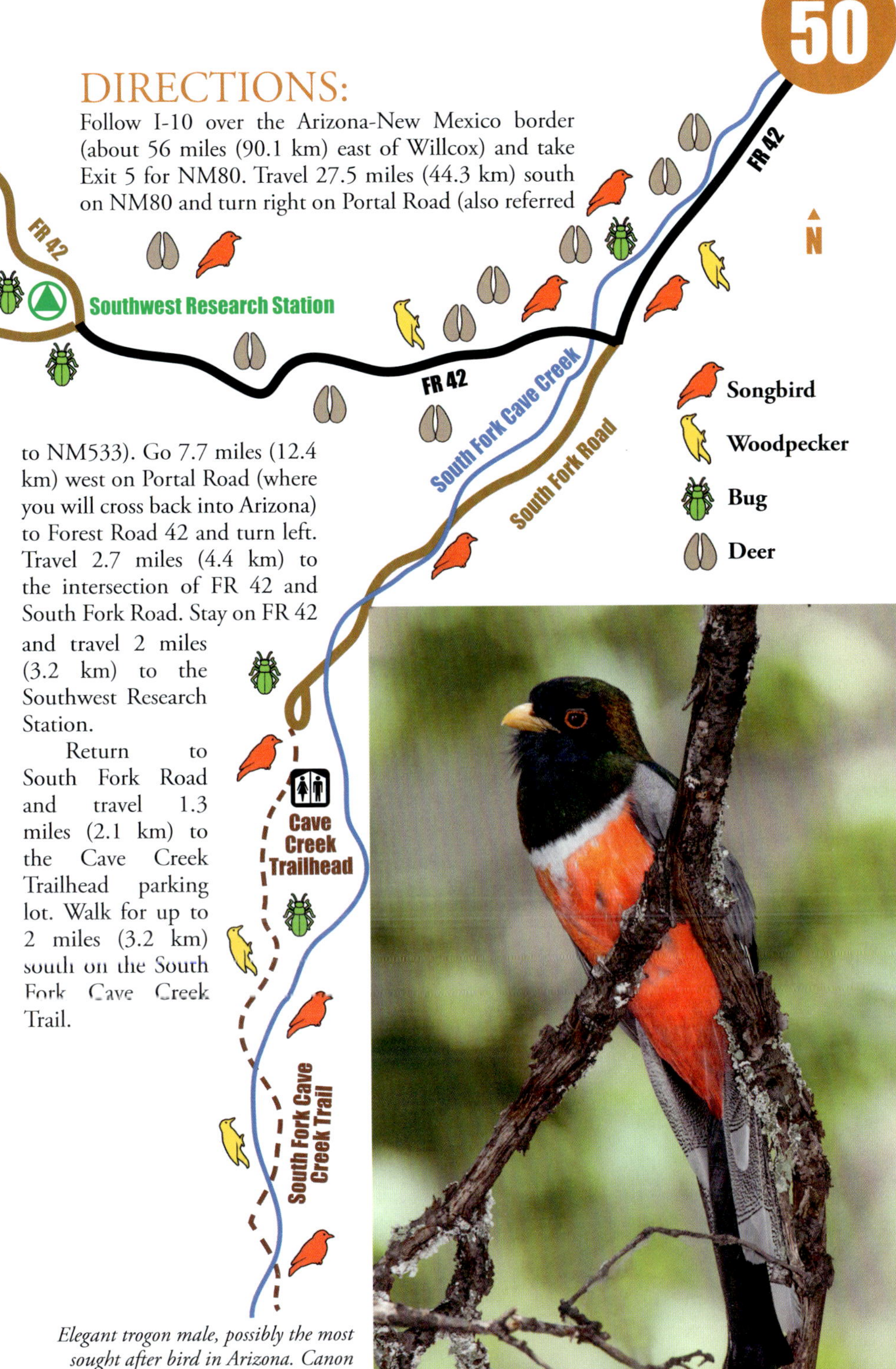

Elegant trogon male, possibly the most sought after bird in Arizona. Canon 50D, 500mm, 1.4x teleconverter, ISO 400, f/10 @ 1/250 sec., on-camera flash @ -2/3 FEC.

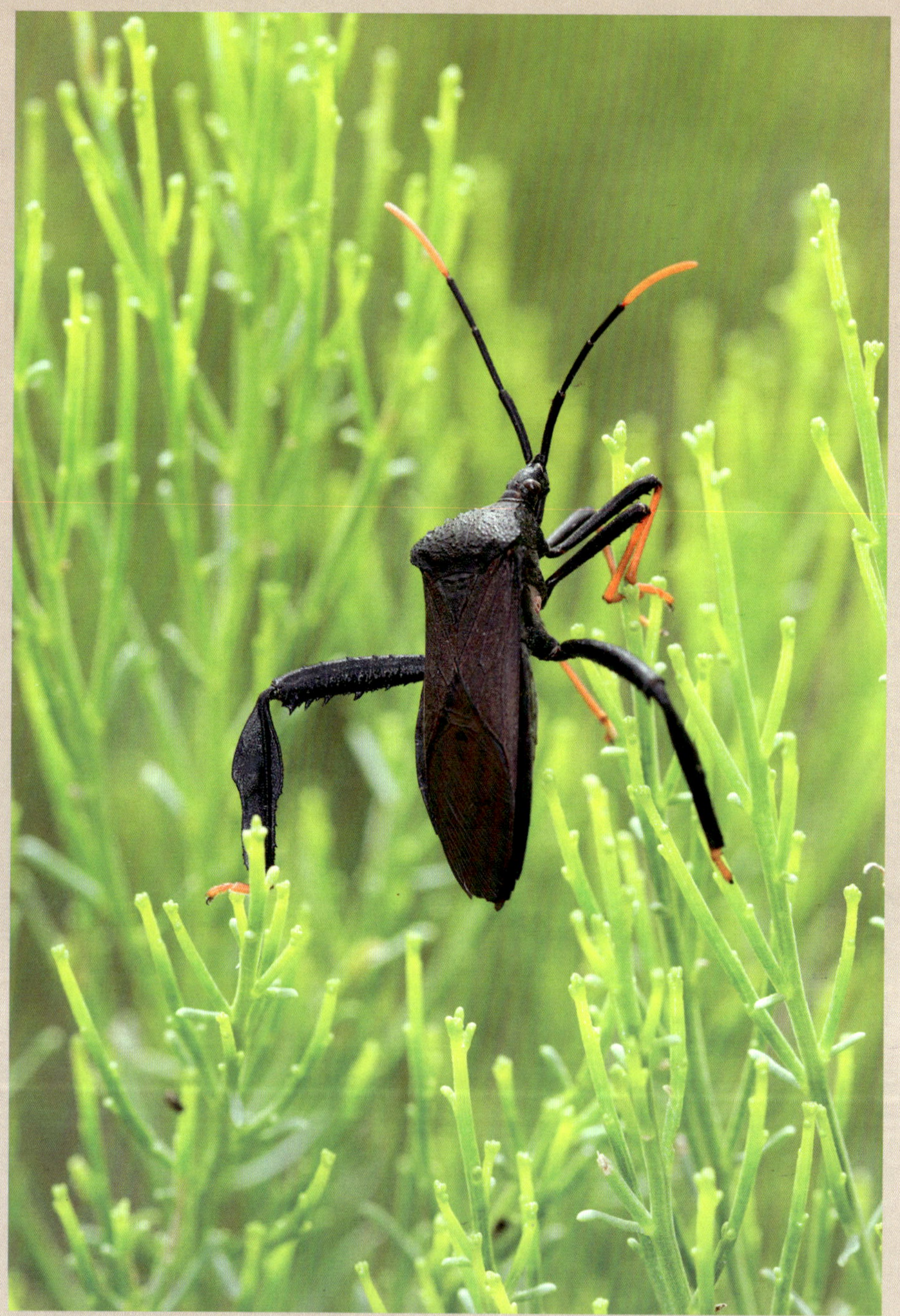

Acanthocephala (Hemiptera). Canon 1DMIV, 100mm macro, ISO 400, f/8 @ 1/300 sec., on-camera flash.

APPENDIX

Additional Resources

Adams, Sharen and Mallman, Sharon, eds., *Arizona Wildlife Viewing Guide* (Cambridge, MN: Adventure Publications, 2007).

Brennan, Thomas C. and Holycross, Andrew T., *A Field Guide to Amphibians and Reptiles in Arizona* (Arizona Game and Fish Department, 2006).

Corman, Troy E., and Wise-Gervais, Cathryn, eds., *Arizona Breeding Bird Atlas* (Albuquerque, NM: University of New Mexico Press, 2005).

Hoffmeister, Donald F., *Mammals of Arizona* (Tucson, AZ: University of Arizona Press, 1986).

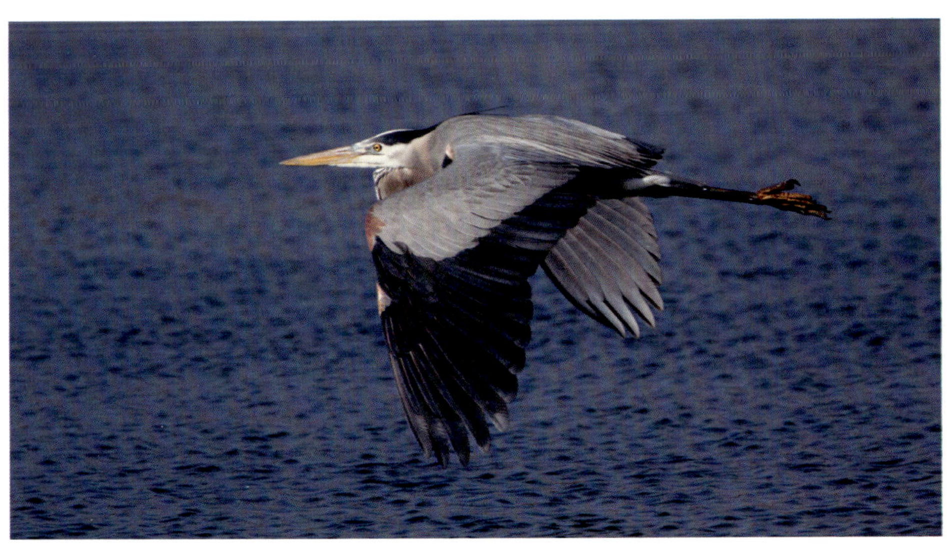

Great blue heron. Canon 50D, 500mm, 1.4x teleconverter, ISO 200, f/5.6 @ 1/1600 sec.

Shoot Summary

#	Location	Pg.	Jan	Feb	Mar	Apr	May	Jun
1	Mount Trumbull	40				X	X	X
2	Grand Canyon Highway	42					X	X
3	South Rim-Grand Canyon National Park	46	X	X	X	X	X	X
4	Flagstaff Nordic Center	50					X	X
5	Page Springs and Bubbling Ponds Hatcheries	52	X	X	X	X	X	X
6	Willow Beach	58					X	X
7	Fain and Williamson Valley Roads	62	X	X	X	X	X	X
8	Watson Lake	64	X	X	X	X	X	
9	Hassayampa River Preserve	68	X	X	X	X	X	X
10	Barry M. Goldwater Range	70	X	X	X	X	X	
11	Cibola National Wildlife Refuge	72	X	X	X	X	X	
12	Bartlett Dam Road	78			X	X	X	
13	South Verado Way	82	X	X	X	X	X	X
14	South Mountain Park and Preserve	86		X	X	X	X	X
15	Desert Botanical Garden	90	X	X	X	X	X	X
16	Ponds at Papago Park	94	X	X				
17	Scottsdale Lakes	100	X	X	X			
18	Butterfly Wonderland	104	X	X	X	X	X	X
19	Veterans Oasis Park	110	X	X	X			
20	Riparian Preserve at Water Ranch	114	X	X	X	X	X	X
21	Lower Salt River Recreation Area	120	X	X	X	X	X	X
22	Canyon Lake	122					X	X
23	Boyce Thompson Arboretum	126	X	X	X	X	X	X
24	Green Valley Park	132	X	X	X	X		
25	Rim Road	134						X

Jul	Aug	Sep	Oct	Nov	Dec	Ideal Time of Day	Vehicle	Hike
X						Sunrise to sunset	4WD HC	E/M
X	X	X	X			Sunrise and sunset	Any	E/M
X	X	X	X	X	X	Sunrise to sunset	Any	E/M
X	X	X	X			Late morning to sunset	Any	E
X	X	X	X	X	X	Early morning to late afternoon	Any	E/M
X	X					Early morning	Any	Boat
X	X	X	X	X	X	Early morning	Any	E
			X	X	X	Early afternoon	Any	E
				X	X	Early morning	Any	E/M
					X	Early morning, late afternoon, and night	4WD HC	E/M
					X	Sunrise and sunset	Any	E
X						Night	Any	M/S
X	X	X	X	X	X	Early morning; sunset	Any	E
						Sunrise to late morning	Any	E/M/S
X	X	X	X	X	X	Early to late morning	Any	E/M
			X	X	X	Sunrise; late afternoon to sunset	Any	E
			X	X	X	Sunrise; late afternoon to sunset	Any	E
X	X	X	X	X	X	Early morning	Any	E
			X	X	X	Early morning and late afternoon	Any	E/M
X		X	X	X	X	Sunrise to early morning; late afternoon	Any	M
X	X	X	X	X	X	Early morning and late afternoon	Any	M
X						Late morning to early afternoon	Any	Boat
X	X	X	X	X	X	Early morning and late afternoon	Any	M
			X	X	X	Sunrise to early morning	Any	E
X						Night	Any	E

Shoot Summary

(Continued)

#	Location	Pg.	Jan	Feb	Mar	Apr	May	Jun
26	Pintail Lake	142					X	X
27	Jacques Marsh Wildlife Area	144				X	X	X
28	Woodland Lake Park	146				X	X	X
29	Becker Lake Wildlife Area	150					X	X
30	Sipe White Mountain Wildlife Area	152	X	X	X	X	X	X
31	Coronado Trail	156						
32	Sonoita	162	X	X	X	X	X	X
33	San Pedro River Road	166				X	X	
34	Sweetwater Wetlands Park	170	X	X	X	X	X	
35	Tucson Botanical Gardens	174	X	X	X	X	X	
36	Mount Lemmon	176				X	X	X
37	Saguaro National Monument–West	178					X	X
38	Arizona-Sonora Desert Museum	182	X	X	X	X	X	X
39	The Pond at Elephant Head	188					X	X
40	Santa Rita Lodge	190			X	X	X	X
41	Madera Canyon	194				X	X	X
42	Fort Huachuca	198				X	X	X
43	Beatty's Miller Canyon Guest Ranch	200				X	X	X
44	Battiste's Bed, Breakfast, and Birds	202			X	X	X	X
45	Whitewater Draw Wildlife Area	204	X	X	X			
46	Cochise Lake	206	X	X	X	X	X	X
47	Sulphur Springs Valley Loop/Willcox Playa	208						X
48	Chiricahua National Monument	216	X	X				
49	Cave Creek Ranch	220	X	X	X	X	X	X
50	Cave Creek Canyon	224				X	X	X

Jul	Aug	Sep	Oct	Nov	Dec	Ideal Time of Day	Vehicle	Hike
X	X	X				Early morning and late afternoon	Any	E/M
X						Sunrise and sunset	Any	M
X	X	X	X			Early morning and late afternoon	Any	M
X	X	X				Early morning	Any	E
X	X	X	X	X	X	Sunrise; late afternoon	2WD HC	E/M
X	X	X				Early morning and late afternoon	Any	E
X	X	X	X	X	X	Early morning and late afternoon	Any	E/M
X						Night	2WD HC	E/M
			X	X	X	Early morning and late afternoon	Any	E/M
			X	X	X	Early morning to late afternoon	Any	E
						Early morning or late afternoon	Any	E/M
X						Sunrise to late morning	Any	E/M
X	X	X	X	X	X	Early morning and late afternoon	Any	E/M
X	X	X	X			Night	Any	E
X	X	X				Sunrise to late morning; early afternoon to sunset	Any	E
X	X	X	X			Sunrise to sunset; night	Any	E/M
X	X	X	X			Early morning and late afternoon	2WD HC	E/M
X	X					Early morning	Any	E
X	X					Early morning, sunset, and night	Any	E
			X	X	X	Sunrise to sunset	Any	E/M
X	X	X	X	X	X	Early morning and late afternoon	Any, 4WD if wet	E
X						Sunrise to sunset; night	Any	E
				X	X	Early morning and late afternoon	Any	M
X	X	X	X	X	X	Sunrise to sunset	Any	E
X						Early morning and late afternoon	Any	E/M

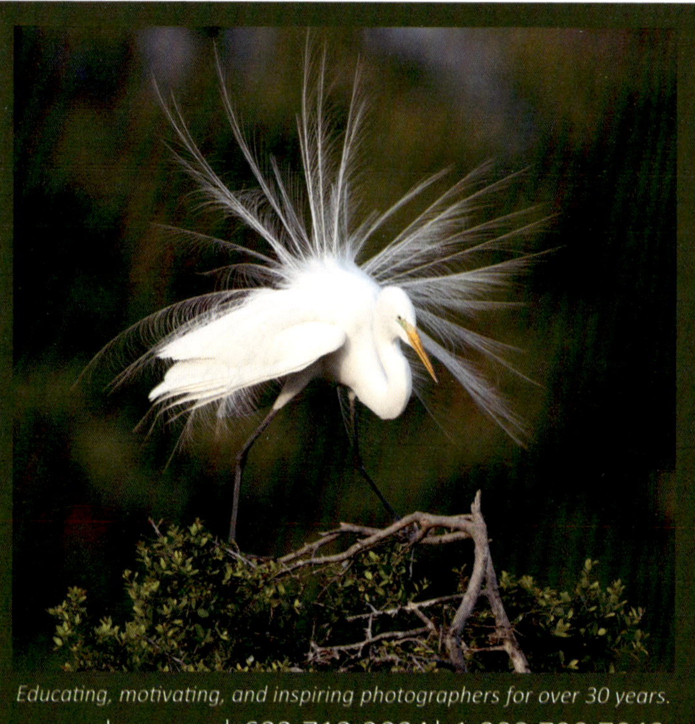

ARIZONA
HIGHWAYS
Photo Workshops
EDUCATE MOTIVATE INSPIRE

CHALLENGE your creative potential, **IMPROVE** your photographic skills and **CAPTURE** stunning images--that's our promise to you.

Our learning based curriculum focuses in the creative, technical and professional practices of photography. Whether you like landscape, wildlife, macro, portrait or post-processing, we have a workshop just right for you.

Educating, motivating, and inspiring photographers for over 30 years.

www.ahpw.org | 602.712.2004| 1.888.790.7042

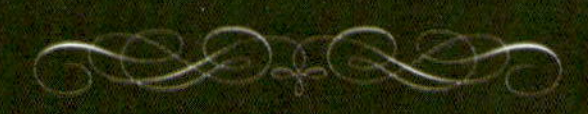

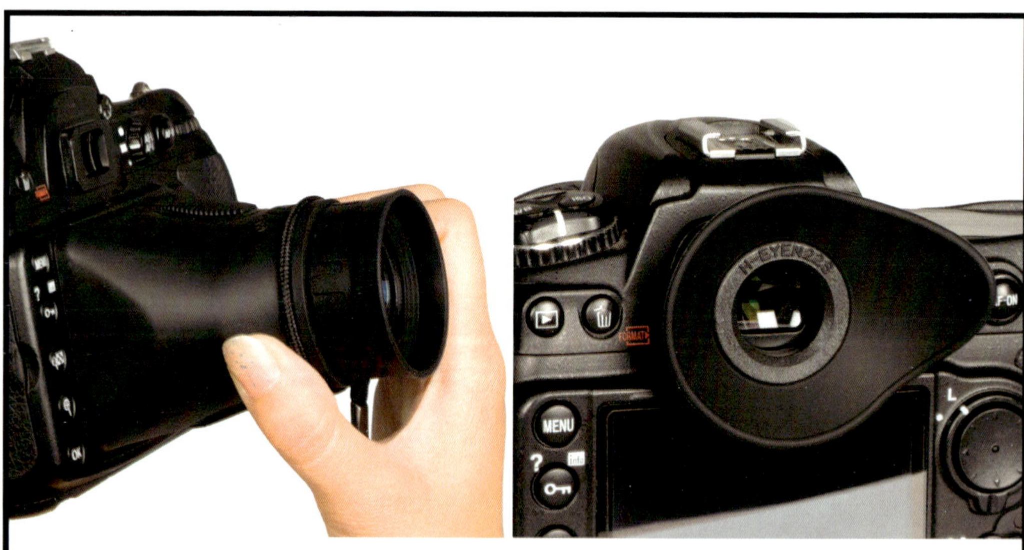

Wildlife Viewing Duo

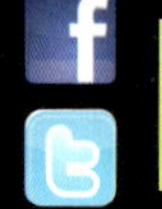

Inspired Imaging Toolstm

www.HoodmanUSA.com 800.818.3946

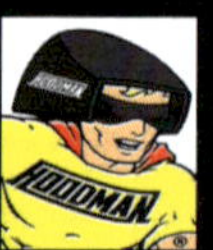

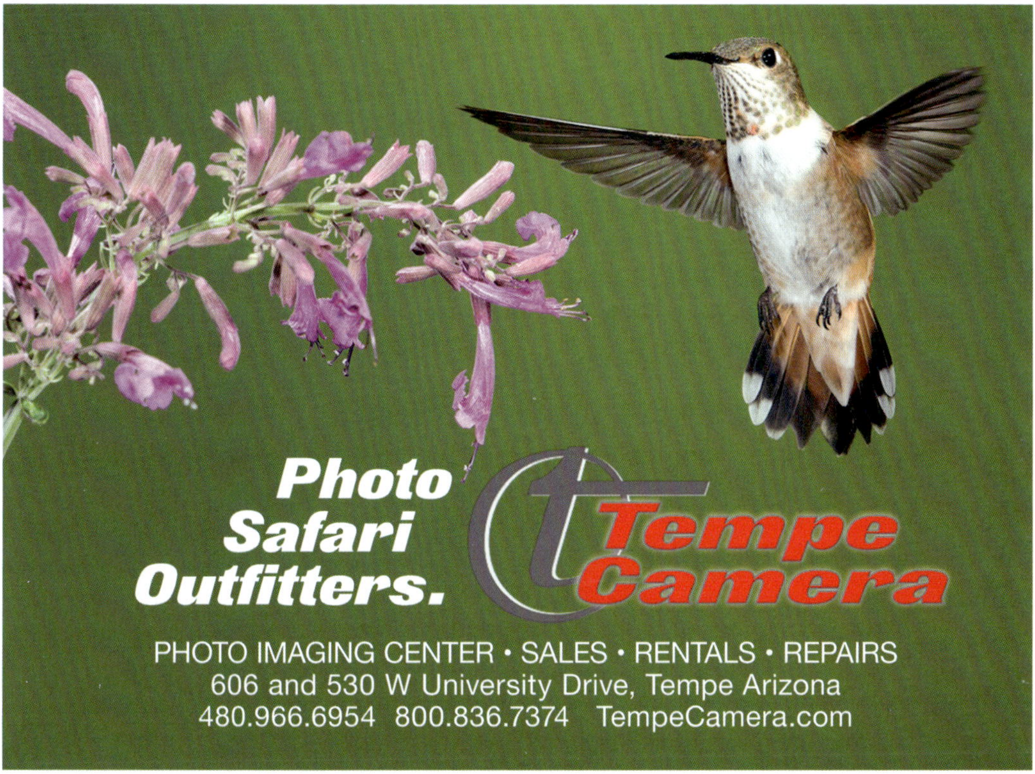

Photo
Safari
Outfitters.

Tempe
Camera

PHOTO IMAGING CENTER • SALES • RENTALS • REPAIRS
606 and 530 W University Drive, Tempe Arizona
480.966.6954 800.836.7374 TempeCamera.com

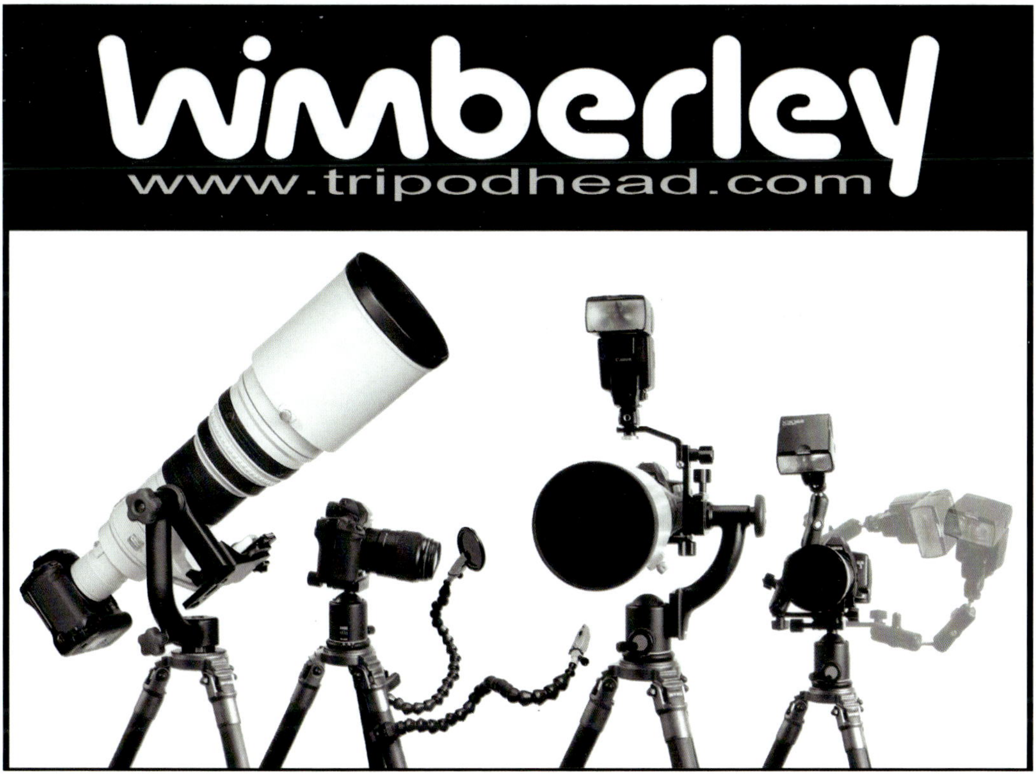

wimberley
www.tripodhead.com

Boyce Thompson Arboretum

arboretum.ag.arizona.edu
37615 E. US Highway 60
Superior, AZ 85173
(520) 689-2723
BTAinfo@ag.arizona.edu

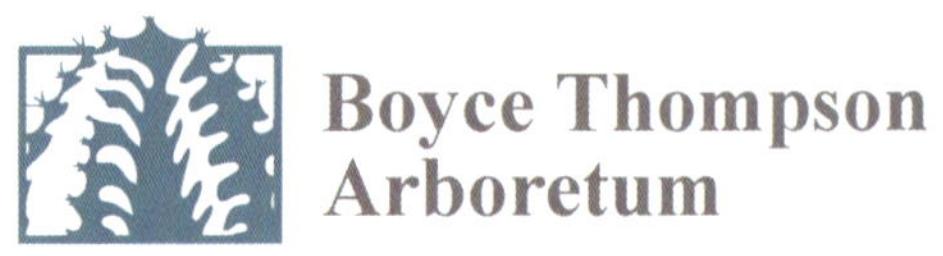

Boyce Thompson Arboretum, founded in 1924, is considered the first institution of its kind to study plants of the world's arid regions. Its mission is to instill in people an appreciation of plants through the fostering of educational, recreational, research, and conservation opportunities associated with the world's arid land plants.

More than 270 birds and 72 terrestrial species have been seen at the Arboretum and adjacent areas. Weekend walks and tours led by experts interpret the plants, geology, butterflies, birds, reptiles, dragonflies, and other wildlife found on the Arboretum grounds.

Phototrap

www.phototrap.com
(520) 444-6649
phototrap@aol.com

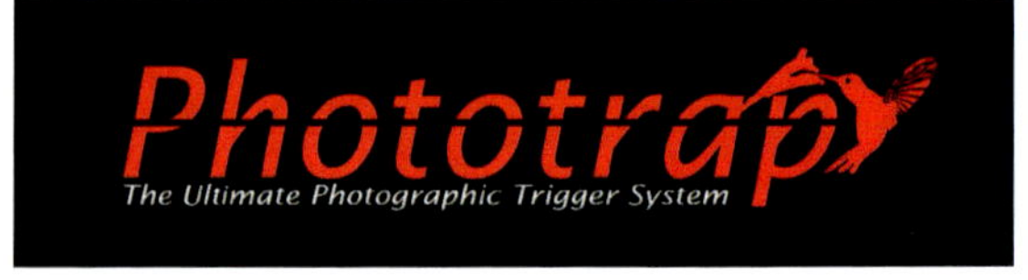

The Phototrap is a high-speed camera trigger system. The standard infrared sensors work in any lighting situation and allow you to capture images of fast-moving wildlife, from insects to elephants—even people finishing races or obstacles courses at the finish line! It also works as a lightning trigger.

The easy-to-use Phototrap system is made in the USA and offers durable construction in a compact transport case.

If you visit the Pond at Elephant Head to photograph the hummingbirds, mammals, and bats, be sure to request a demo of the Phototrap (reservations required prior to your arrival). The unit works with any camera that can be triggered by a remote cord.

Cave Creek Ranch

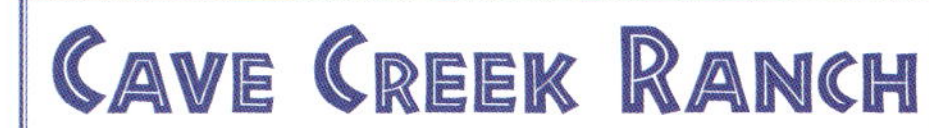

www.cavecreekranch.com
P.O. Box 16554
1396 W. Piedra Blanca Lane
Portal, AZ 85632
(520) 558-2334
info@cavecreekranch.com

Located in majestic Cave Creek Canyon in southeastern Arizona, Cave Creek Ranch is a seven-acre secluded oasis of peaceful quiet, birds, animals, trees, and famous Cave Creek. Our hummingbird feeders, seed feeders, water, shrubs, and trees attract many Arizona specialty birds, including the elegant trogon, Montezuma quail and 12 species of hummingbirds. Nearly half of the species of birds in North America can be seen at or near Cave Creek Ranch.

We're at 5,000 feet, and open all year. Come stay in our cottages or apartments, all with private bathrooms and fully equipped kitchens.

Santa Rita Lodge

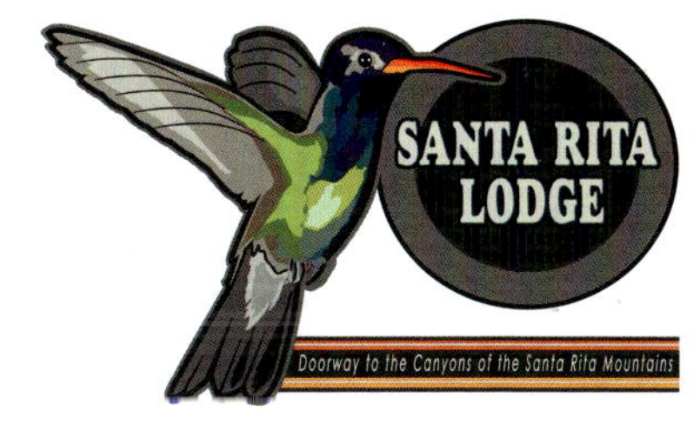

www.santaritalodge.com
1218 S. Madera Canyon Rd.
Madera Canyon, AZ 85614
(520) 625-8746
birds@santaritalodge.com

As the city of Tucson grew, Madera Canyon, formerly Whitehouse Canyon, became an escape from the summer heat in the valley. By the 1900s, summer cottages and year-round residents called the canyon home. Today, Madera Canyon continues to be a unique natural resource for birders, hikers, picnickers, artists, photographers, researchers and nature lovers alike. In 1922, Mr. C.R. Dusenberry built the Santa Rita Trails Resort, now known as the Santa Rita Lodge. The Holt family purchased the Santa Rita Lodge in 2006. Since that time, the casitas and cabins were refurbished and a new gift shop and visitor's office was built in 2012.

The Santa Rita Lodge is located at the heart of Madera Canyon, in the Coronado National Forest. Just 13 miles southeast of Green Valley at 5,000 foot elevation, the Lodge is 10+ degrees cooler on any given day. Come visit us for world-renowned bird watching, hiking or a quiet retreat.

Discover and Photograph

ARIZONA'S WILDFLOWERS

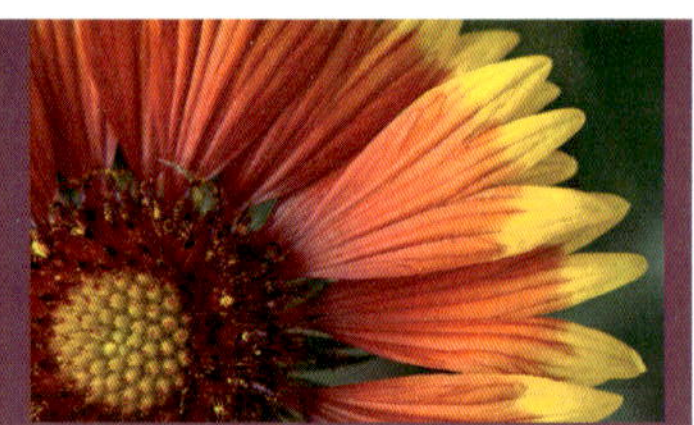

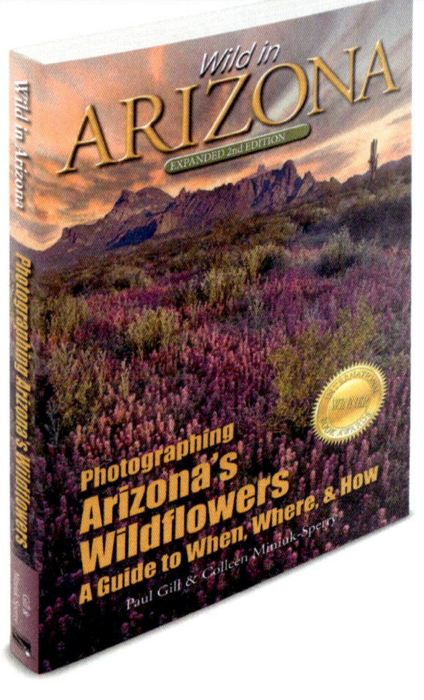

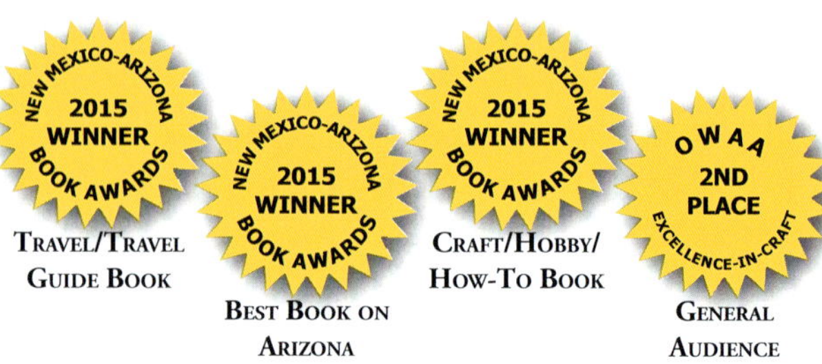

"...Wild in Arizona: Photographing Arizona's Wildflowers *is packed with vital information and tips on the best places to find and photograph Arizona's bountiful blooms. It belongs on every naturalist's reference shelf and in every photographer's camera bag*"

~Peter Ensenberger,
former *Arizona Highways* Director of Photography

Find, identify, and make your own stunning photographs of Arizona's magnificent wildflowers with the expanded 2nd edition of the award-winning book, ***Wild in Arizona: Photographing Arizona's Wildflowers: A Guide to When, Where, & How*** as your guide.

Tapping into co-authors/photographers **Paul Gill** and **Colleen Miniuk-Sperry's** 45+ years of collective experience and expertise, this book features:

- 248 pages with over 280 color photographs
- 60 locations with maps & driving directions
- 60 featured flowers
- 17 instructional photo tips
- 10 "Making the Photo" stories
- A "Bloom Calendar"

Makes a great gift for photographers, hikers, painters, gardeners, & visitors to Arizona!

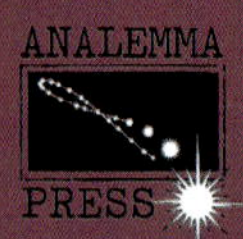

Order the book/eBook today at

WWW.WILDINARIZONA.COM

Discover the most photogenic places in

ACADIA NATIONAL PARK!

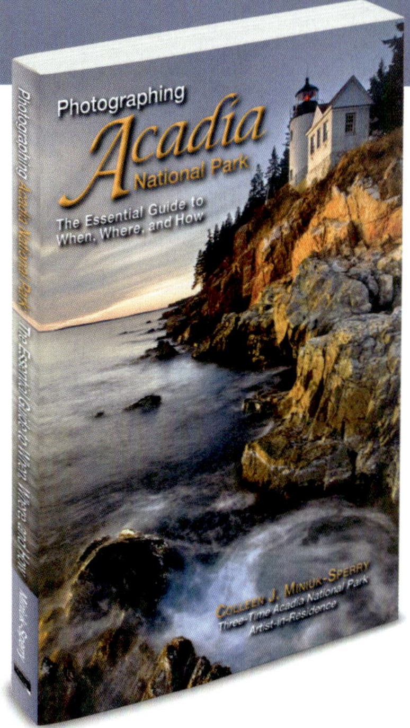

"If you tote a camera with you to Acadia National Park, this guide can only enhance both your experience and your photos."

~National Parks Traveler

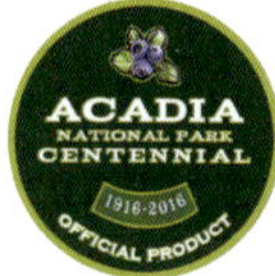

Create stunning photographs of Acadia no matter your camera or photographic skill level with the award-winning guidebook, ***Photographing Acadia National Park: The Essential Guide to When, Where, and How!***

Written by three-time Acadia Artist-in-Residence and pro photographer Colleen Miniuk-Sperry to help you **get in the right place at the right time,** this jam-packed guide features:

- 50 can't-miss classic and lesser-known locations on Mount Desert Island, the Schoodic Peninsula, and Isle au Haut
- Location-specific tips on ideal time of day and year, as well as recommended photo gear
- 18-page instructional Photography Basics section
- 12 insightful "Making the Photo" stories
- A Shoot Calendar to help plan your visit
- 224 pages with over 180 inspiring color photographs

New Mexico-Arizona 2014 Winner Book Awards

Travel/Travel Guide Book

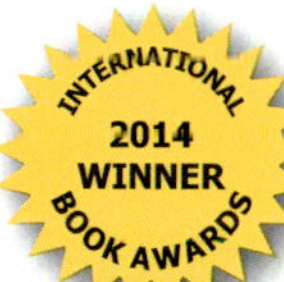

Best Travel Guide/Essay

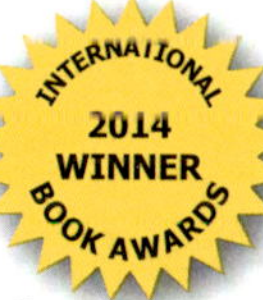

Best Interior Design

10% of this book's profits is donated to the ***SCHOODIC EDUCATION ADVENTURE*** *program in Acadia*

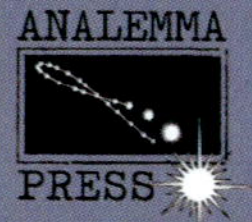

Order the book/eBook today at

WWW.PHOTOACADIA.COM

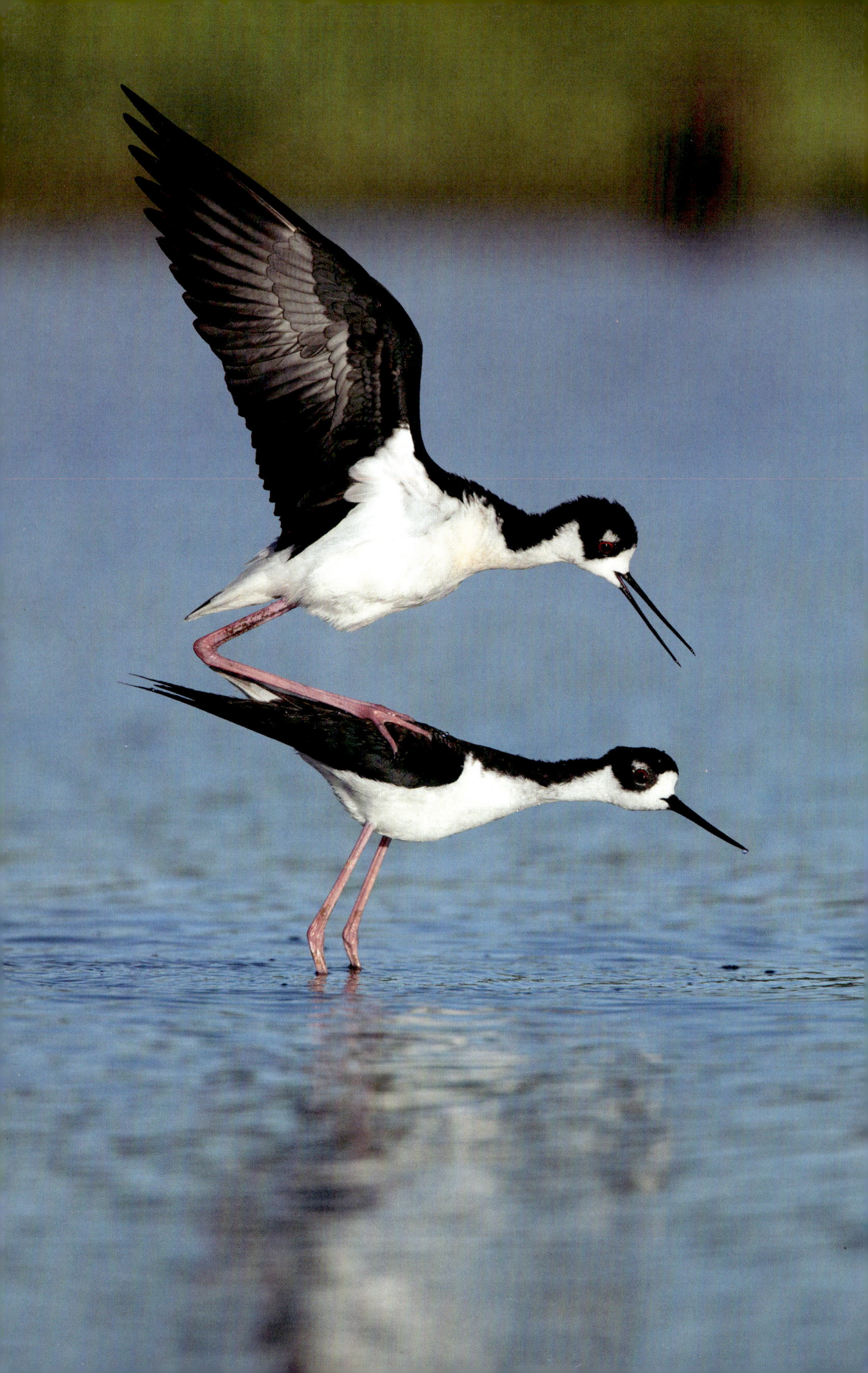

Valued Individual Contributors

Anonymous
Ambika Balasubramaniyan
Stan and Shirley Bormann
Stu Glenn
Ted Fleming
Shawn Ger
David Goodell
David Halgrimson
Christina Heinle
Scot and Yvonne Jamison
Jeff Maltzman
Muriel S. McClellan
Robert and Jacqueline Miniuk
Matt Nickell
Bill Stuart
Sun Lakes Camera Club
Carol and Joe Webster
Vern and Barb West
Dan and Denise Womack

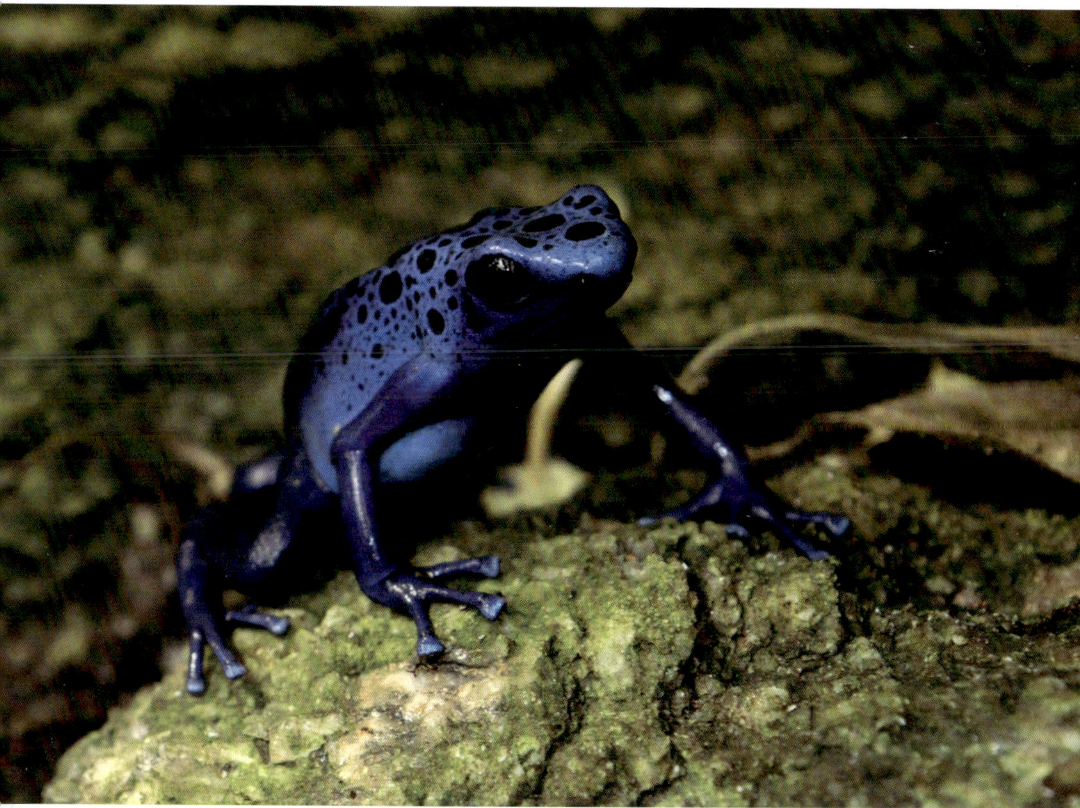

ABOVE: Poison dart frog. Canon 7DMII, 70-200mm at 140mm, 2x teleconverter, ISO 800, f/7.1 @ 1/50 sec., on-camera flash @ -2/3 FEC.
LEFT: A pair of adult black-necked stilts mate at the Riparian Preserve at Water Ranch. Canon 5DMIII, 500mm,1.4 teleconverter, ISO 400, f/7.1 @ 1/2000 sec.

Index

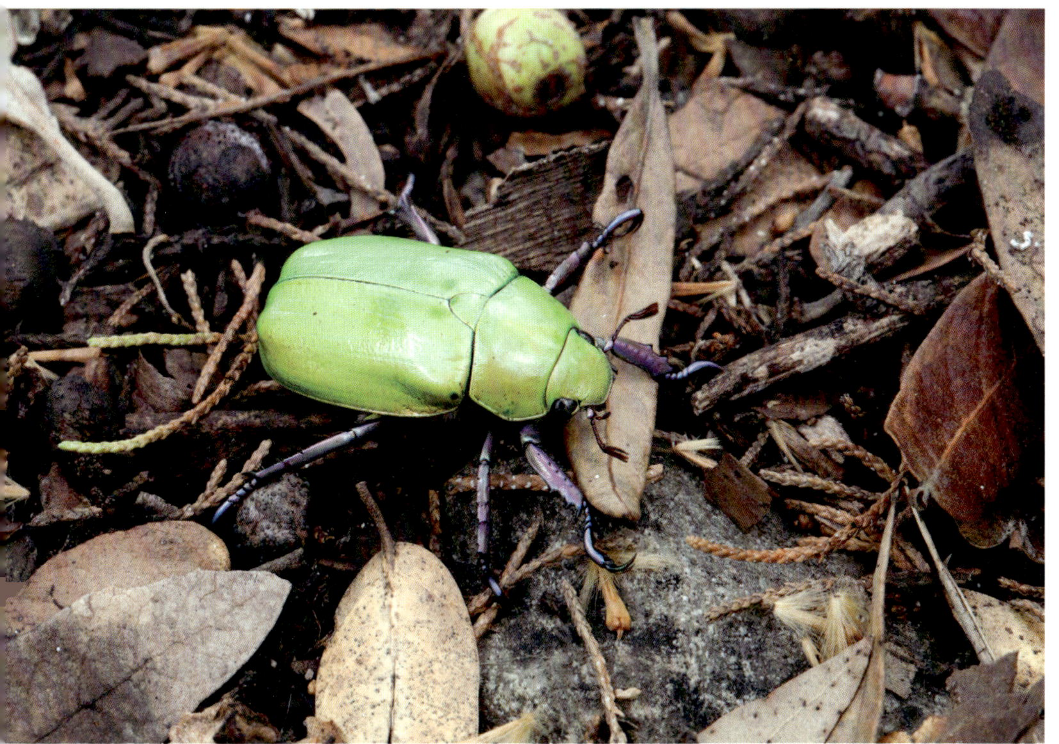

A purple and bright green Chrysina beyeri crawls in the oak leaves in Madera Canyon. Canon 5DMII, 100mm macro, ISO 500, f/18 @ 1/4 sec.

Index (continued)

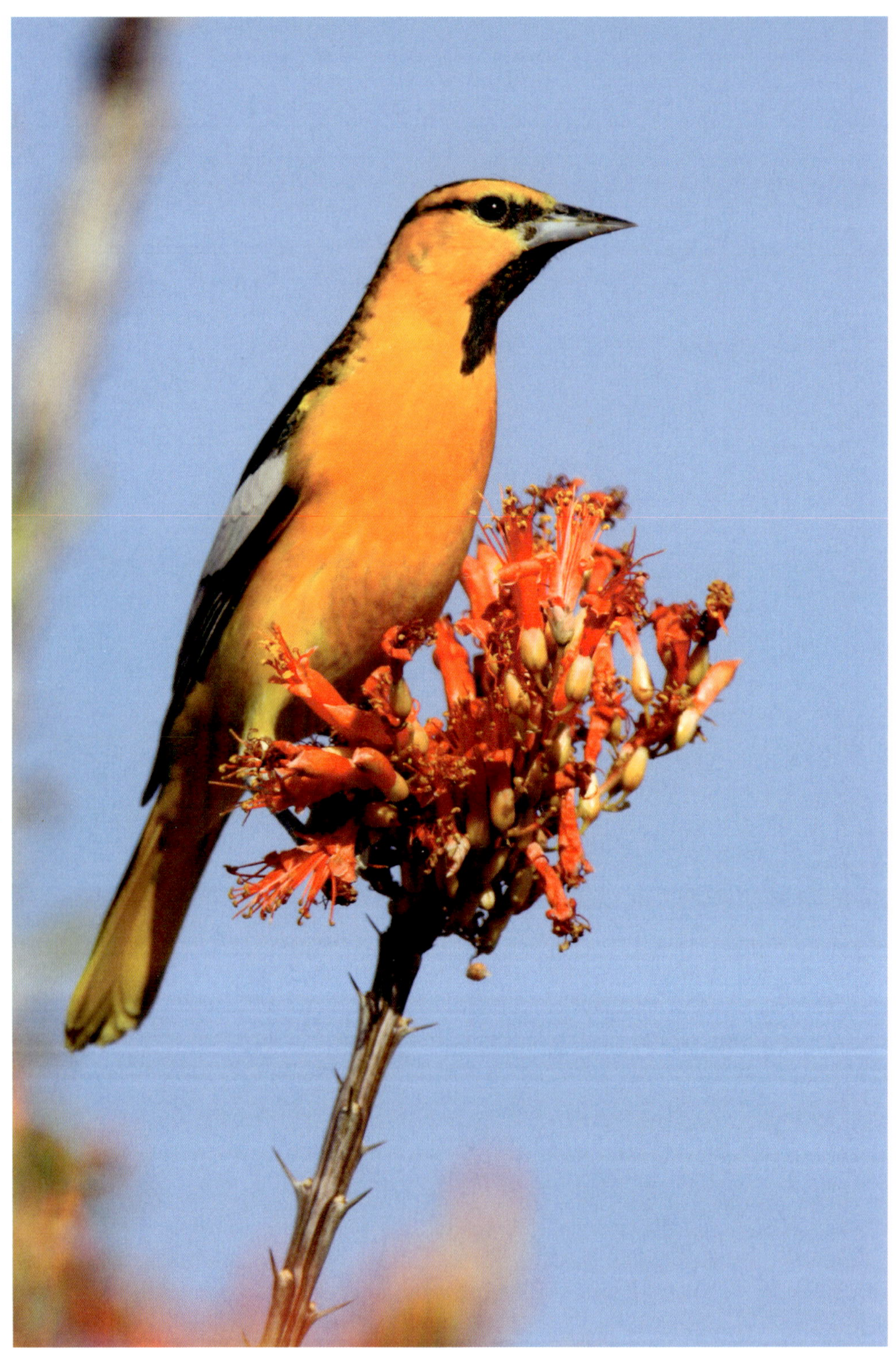

Male Bullock's oriole feeds from ocotillo flowers. Canon Digital Rebel XTi, 500mm, 1.4x teleconverter, ISO 200, f/8 @ 1/500 sec.

Index (continued)

Index (continued)

Riparian Preserve at Water Ranch

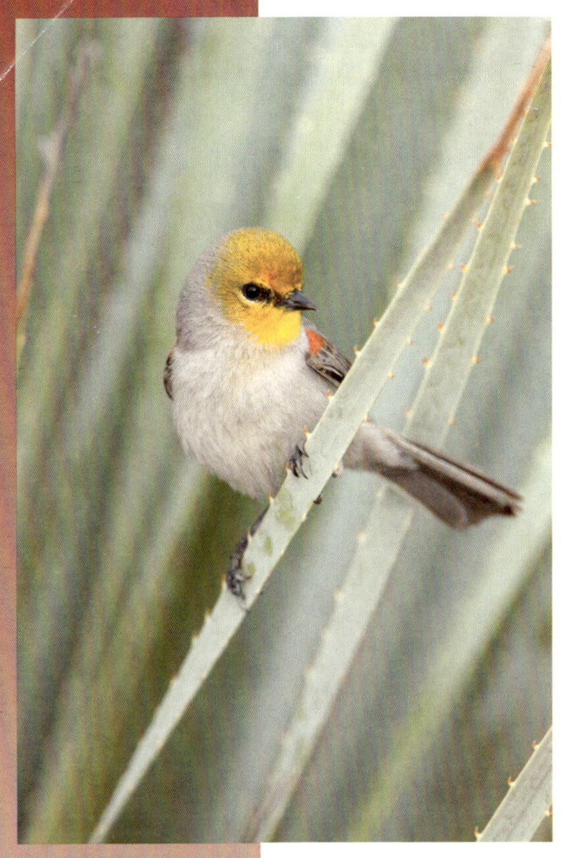

quiet. Within minutes, the birds will forget your presence and go about their normal activities. Many of the early migrant birds are quite small and require the magnification power of a **telephoto lens** (400mm or greater, especially when combined with a **cropped frame camera**) to record frame-filling images. When you are using telephoto lenses, mount your camera and lens on a sturdy **tripod** with a stout **ball or gimbal head**. Keep your perspective as low to the water as possible. Finally, watch your backgrounds. At Water Ranch, beautiful green or orange reflections frequently surround your subject.

Also in mid-September, the many resident birds that have left the urban oasis for cooler summer climates begin to come home. Great blue herons, snowy egrets, great egrets, little green herons, neotropic cormorants, black-crowned night herons, osprey, and others make appearances and create a photographer's bonanza. In addition, in late October and November, species such as white pelicans, Wilson's snipes, peregrine falcons, American kestrels, and several warbler species take center stage.

Photograph them feeding, fighting, squabbling, flying, and basically enjoying life as these larger birds go about their daily activities seemingly oblivious to people. Use a short telephoto lens (e.g., 300mm) to obtain images of groups of birds flying or defending their territories. Longer lenses can help create frame-filling portraits. Use the continuous focus mode on your camera to focus, as it is almost impossible to take sharp images of moving birds using single-shot autofocus.

ABOVE: Verdin. Canon 50D, 500mm, 1.4x teleconverter, ISO 400, f/7.1 @ 1/250 sec. RIGHT: Pair of American avocets just prior to breeding. Canon 1DMIV, 500mm, 1.4x teleconverter, ISO 500, f/9 @ 1/1600 sec.

As the temperatures cool in the Canadian provinces and northern United States, migrant ducks and additional shorebirds arrive by the hundreds. October through February presents excellent opportunities to photograph wigeon, northern pintail, green-winged teal, cinnamon teal, northern shoveler, ring-necked duck, mallard duck, and others. Along with the ducks, greater numbers of American avocets and black-necked stilts arrive.

In mid to late February, migrants begin to travel to their northern homes, and the number of birds drops significantly. Fortunately, the remaining birds (egrets, herons, avocets, stilts, etc.) are very photogenic. The egrets show off their nuptial plumage and offer even better images than during the winter. American avocets molt into their beautiful tan breeding plumage, and both avocets and stilts begin to breed and lay eggs. From March through May, photographers can easily attain great images of both species breeding, nesting, and raising their young.

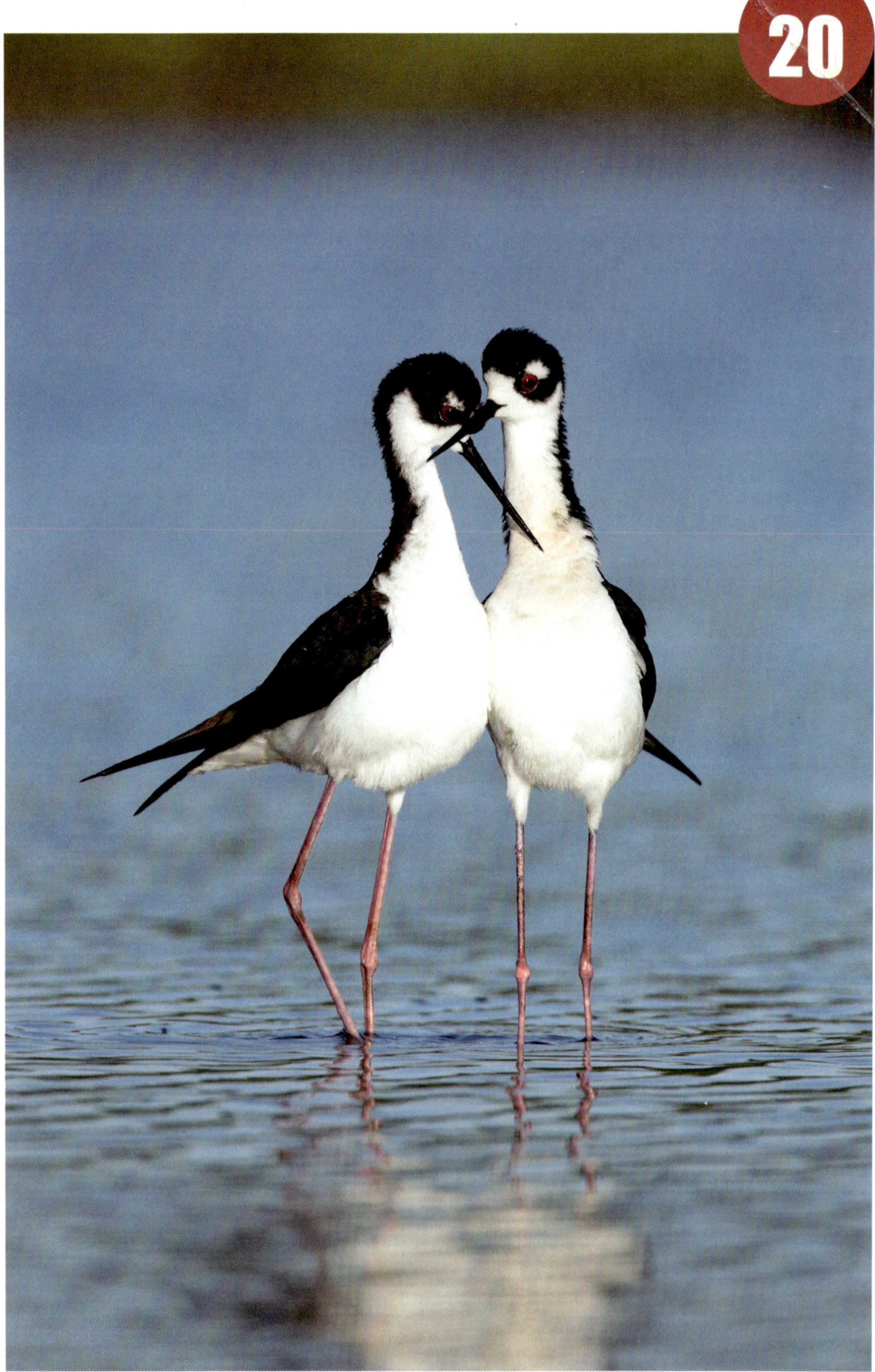

Pair of black-necked stilts courting each other. Canon 5DMIII, 500mm, 1.4x teleconverter, ISO 400, f/7.1 @ 1/2000 sec.

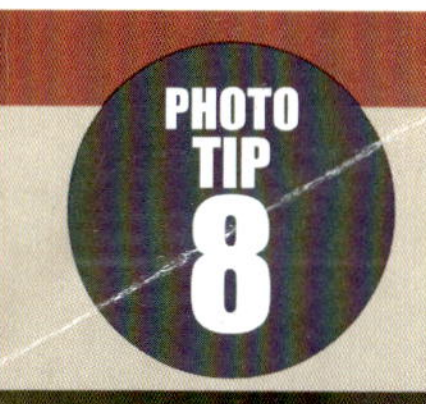

Photographing Flying Birds

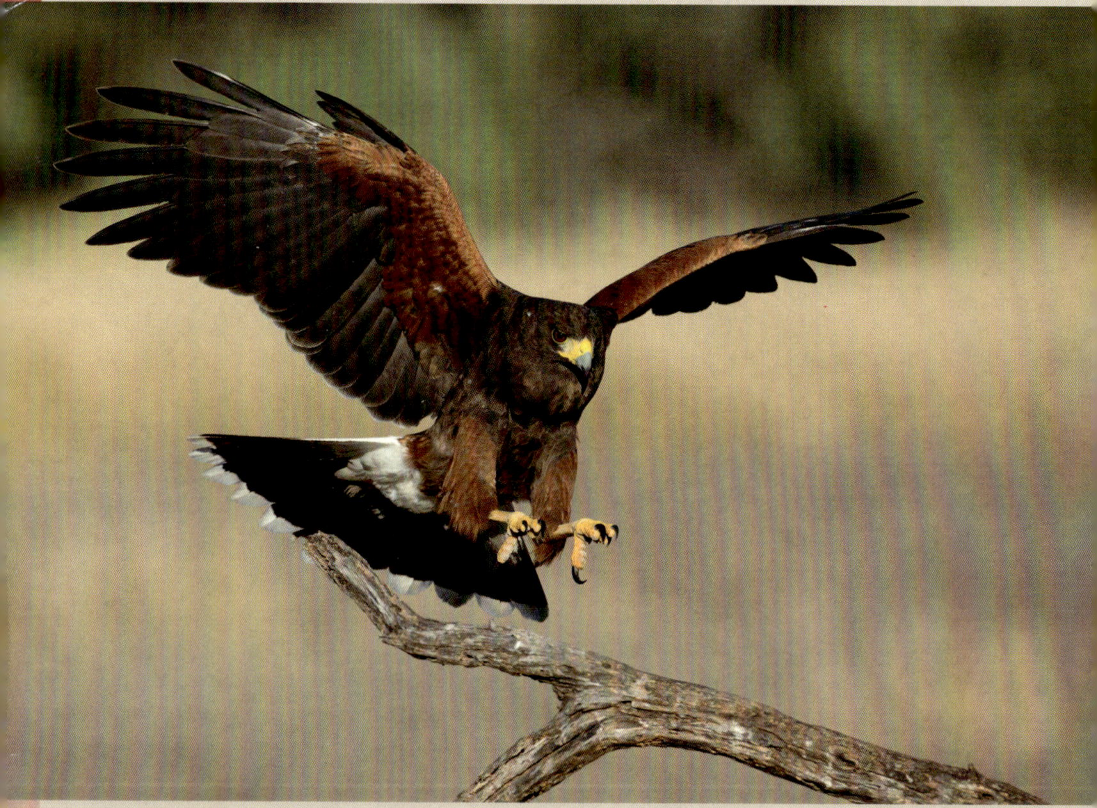

Harris's hawk. Canon 5DMIII, 500mm, ISO 500, f/7.1 @ 1/2500 sec.

Photographing flying birds requires a combination of skill, practice, and patience. Getting the bird in the frame of a telephoto lens takes lots of practice and determining the correct combination of camera settings for each situation requires some experience. Prepare to take thousands of images to become efficient and confident in your approach with flying birds.

To shoot birds in action, set your long telephoto lens (preferably 300mm or longer) to continuous autofocus mode (also referred to as AI Servo or follow focus). Continuous autofocus allows your camera to keep focus on your subject as it moves.

If you plan on hand-holding your gear to photograph flying birds, use the equivalent to a 300mm or 400mm lens. For small subjects or animals appearing at a distance, choose a 500mm lens and shoot with the aid of a tripod and gimbal head.

In addition, place the camera on high-speed continuous shooting mode which enables you to not only take many frames in a rapid-fire, machine-gun style, but it also helps you take sharper images.

All cameras have several focusing points (usually ranging from 5-63) that the camera uses

DIRECTIONS:

From downtown Chandler, drive 4 miles (6.4 km) east on Chandler Boulevard (which turns into Williams Field Road). Turn right onto Lindsay Road, and drive an additional 5 miles (8.1 km). The park is on the northeast corner of the intersection of South Lindsey Road and East Chandler Heights Road.

For more information, visit **www.chandleraz.gov/default.aspx?pageid=682**.

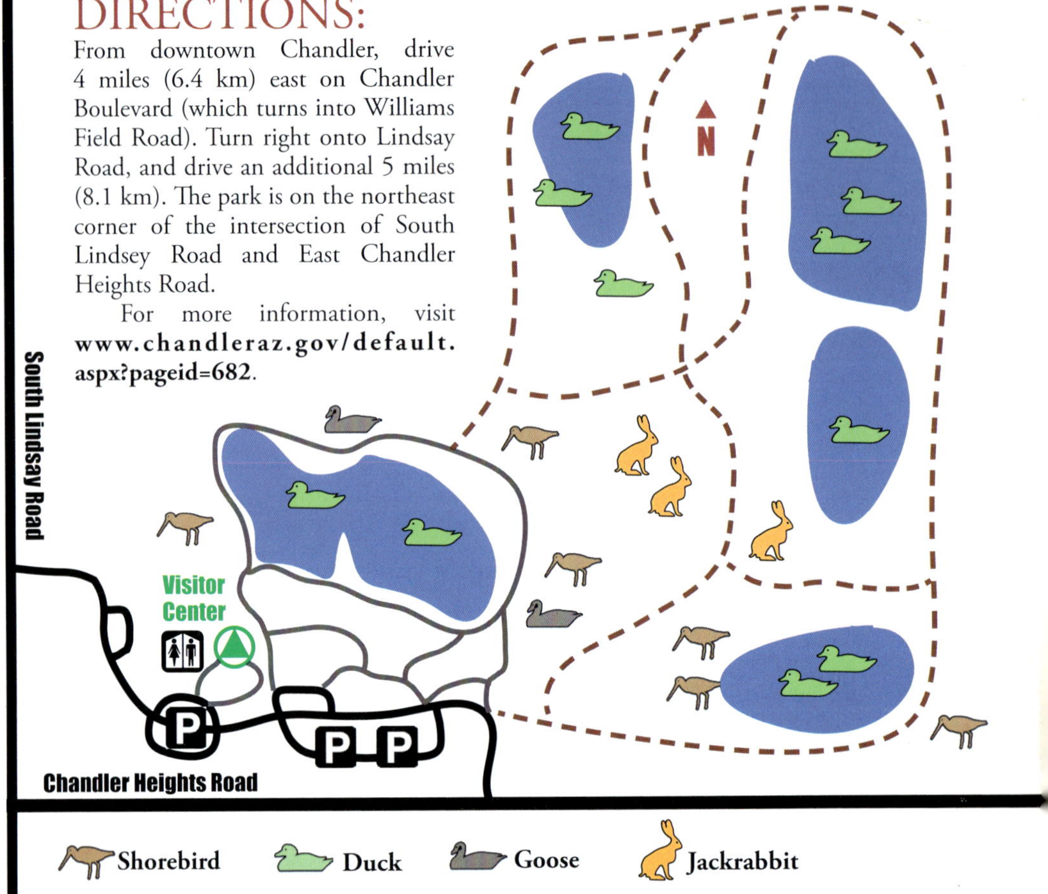

South Lindsay Road

N

Visitor Center

P

P P

Chandler Heights Road

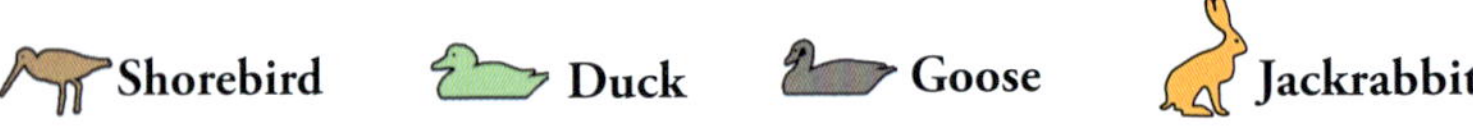

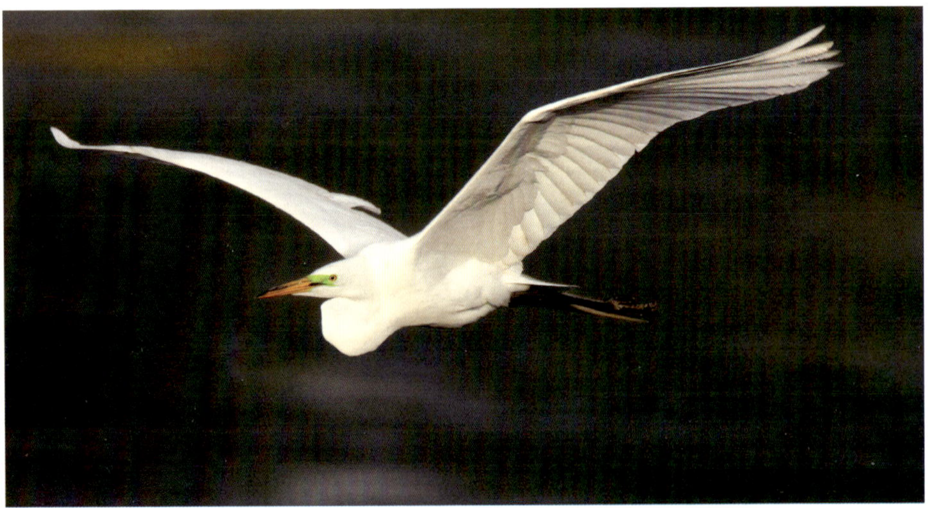

Great egret. Canon 50D, 500mm, 1.4x teleconverter, ISO 400, f/7.1 @ 1/1250 sec.

CENTRAL ARIZONA

Riparian Preserve at Water Ranch

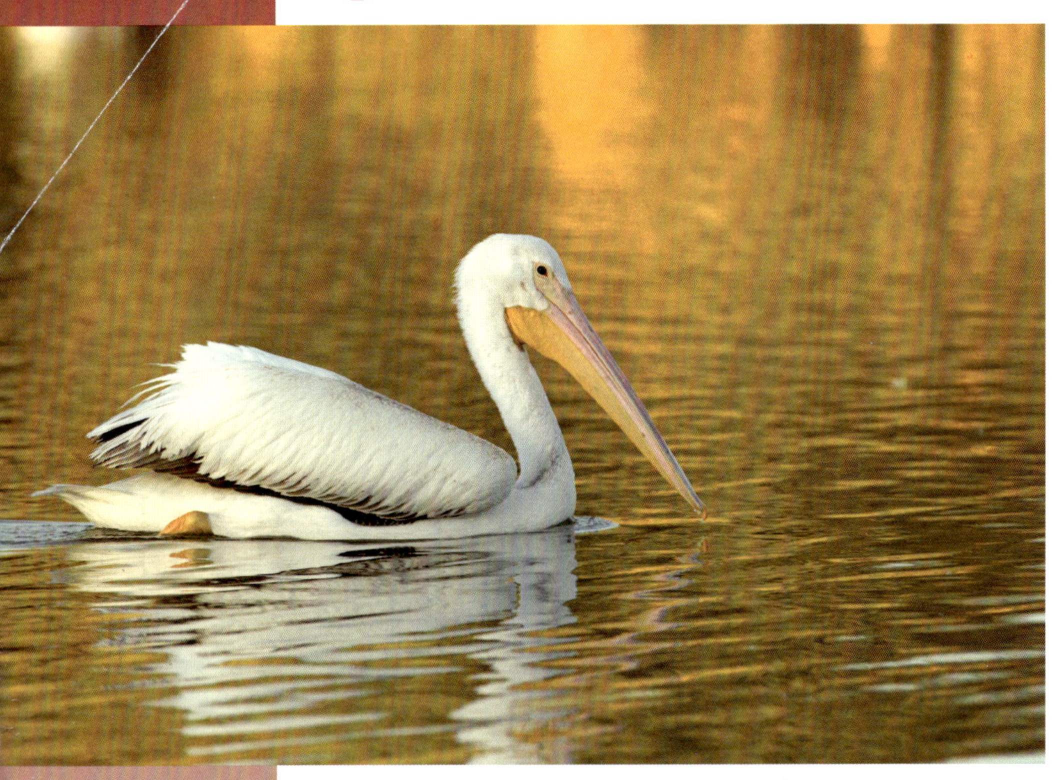

White pelican in the reflective late afternoon waters. Canon 1DMIV, 70-200mm at 200mm, 2x teleconverter at 400mm, ISO 400, f/7.1 @ 1/400 sec.

VIEW TIME
September to July

IDEAL TIME OF DAY
Sunrise to early morning; late afternoon

VEHICLE
Any

HIKE
Moderate

Riparian Preserve at Water Ranch offers some of the best aquatic bird photography in Arizona. Eight reclaimed water recharge ponds draw over 200 species of migrating and resident birds. Because the birds at Water Ranch are habituated to the many hikers, dog walkers, anglers, and equestrian traffic that share this recreational park, photographers have no problem approaching them close enough to take amazing images. If you can only go to one place in Arizona to photograph birds, head to the Water Ranch.

As with most bird photography, the best light occurs in the early morning and late afternoon. Because the ponds primarily operate to benefit water recharge and delivery systems, the photographic game always changes. The best pond today may appear empty in a week and full the following month. If you have the opportunity, arrive a day early to scout bird locations. Before you set out, check **www.desertriversaudubon.org** for times and dates of birdwatching tours.

All months offer outstanding bird photography with the exception of July and August (during the desert's scorching hot weather). Beginning in mid-September, migrant birds such as least sandpiper, long-billed dowitcher, greater yellowleg, gadwall, and others arrive from the freezing north.

Put the sun at your back to obtain front lighting on your subjects (your shadow pointing towards your subjects), find a comfortable place next to the water's edge, sit down (on a comfortable waterproof mat), and stay to determine where to focus on the subject within the frame. Unless the photographer limits the position or number of focus points, the camera uses all of them. Sometimes, this causes the camera to inadvertently focus on the closest element to your camera and not your subject. For example, while photographing a bird on a perch with all of the focusing points active, your camera will likely focus on the closet part of the scene. If a portion of the perch sits closer to the camera than the bird, the camera will focus on the perch and not on the bird. To avoid this, select either the center focus point or a small group of focus points in the center area for photographing flying birds.

Finally, dial in your exposure using the camera's manual mode. This enables you to record consistent exposures as you photograph birds moving from one location to the next. In order to stop most, if not all, wing movement, start with a shutter speed of 1/1500 of a second or faster. If you are photographing in strong natural light, try an ISO speed around ISO 400 and an aperture of about f/8. If the light is not favorable, modify the ISO speed and aperture to obtain the fastest shutter speed you can.

After preparing your camera to shoot, observe and learn the flight pattern of your subject. For example, if you are photographing a group of sandhill cranes leaving their roost, you will know the exact direction from which they are flying. As soon as you see the flying bird, focus on it, even if it looks too far and small in the frame for an image. As the bird moves closer, keep your finger on the shutter button and keep it depressed halfway. In the half-depressed position, the camera initiates the autofocus and follow focus. As soon as you feel the bird has approached close enough, push the shutter button to take bursts of at least five images. Keep firing away, five or so frames at a time, until the bird is gone.

Now, what do you do with the 10,000 images you bring home from three days of photographing flying birds? Do not delete images from the card when it is in the camera. Give yourself a couple of days to relax your "mind's eye" (that place in your brain that predetermines what you will get before you go on a trip or have a chance to look at your final images).

Editing software like BreezeBrowser, Adobe Lightroom, or Adobe Bridge provides the fastest way of going through hundreds or thousands of images in a short time frame. These applications create small JPEG copies of each image that allow me to quickly review and perform a technical and creative critique of each to determine if my photo is out of focus, underexposed, etc.

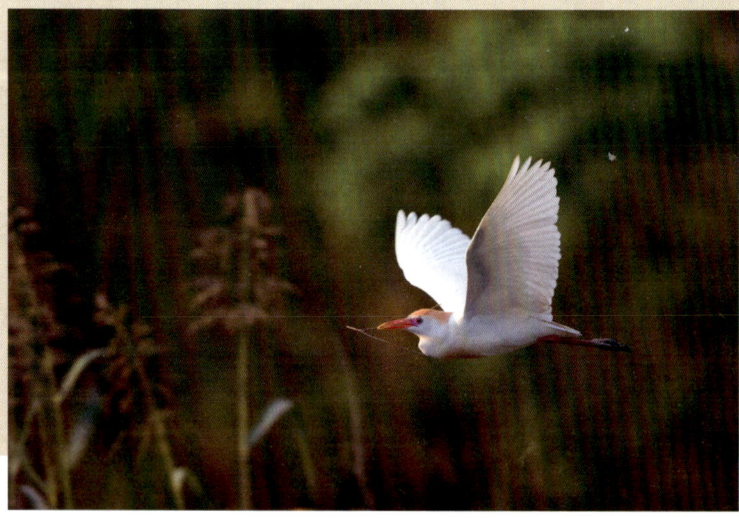

Cattle egret carries nesting material in the early morning sun. Canon 20D, 500mm, ISO 400, f/4.5 @ 1/800 sec.